Lecture Notes in Artificial Intelligence 7842

Subseries of Lecture Notes in Computer Science

LNAI Series Editors

Randy Goebel
University of Alberta, Edmonton, Canada
Yuzuru Tanaka
Hokkaido University, Sapporo, Japan
Wolfgang Wahlster
DFKI and Saarland University, Saarbrücken, Germany

LNAI Founding Series Editor

Joerg Siekmann
DFKI and Saarland University, Saarbrücken, Germany

T0242766

Fabrizio Riguzzi   Filip Železný (Eds.)

Inductive
Logic Programming

22nd International Conference, ILP 2012
Dubrovnik, Croatia, September 17-19, 2012
Revised Selected Papers

 Springer

Volume Editors

Fabrizio Riguzzi
University of Ferrara
Department of Mathematics and Computer Science
Via Saragat 1, 44122 Ferrara, Italy
E-mail: fabrizio.riguzzi@unife.it

Filip Železný
Czech Technical University in Prague
Department of Computer Science and Engineering
Faculty of Electrical Engineering
Karlovo namesti 13
12135 Prague 2, Czech Republic
E-mail: zelezny@fel.cvut.cz

ISSN 0302-9743   e-ISSN 1611-3349
ISBN 978-3-642-38811-8   e-ISBN 978-3-642-38812-5
DOI 10.1007/978-3-642-38812-5
Springer Heidelberg Dordrecht London New York

Library of Congress Control Number: 2013939736

CR Subject Classification (1998): F.4, I.2, D.1.6, F.3, F.1, F.2

LNCS Sublibrary: SL 7 – Artificial Intelligence

Typesetting: Camera-ready by author, data conversion by Scientific Publishing Services, Chennai, India

Printed on acid-free paper

Springer is part of Springer Science+Business Media (www.springer.com)

Preface

This volume contains the proceedings of ILP 2012: the 22nd International Conference on Inductive Logic Programming held during September 17–19, 2012, in Dubrovnik. The ILP conference series, started in 1991, is the premier international forum on learning from structured data. Originally focusing on the induction of logic programs, it broadened its scope and attracted a lot of attention and interest in recent years. The conference now focuses on all aspects of learning in logic, multi-relational learning and data mining, statistical relational learning, graph and tree mining, relational reinforcement learning, and other forms of learning from structured data.

This edition of the conference solicited three types of submissions:

1. Long papers (12 pages) describing original mature work containing appropriate experimental evaluation and/or representing a self-contained theoretical contribution
2. Short papers (6 pages) describing original work in progress, brief accounts of original ideas without conclusive experimental evaluation, and other relevant work of potentially high scientific interest but not yet qualifying for the above category
3. Papers relevant to the conference topics and recently published or accepted for publication by a first-class conference such as ECML/PKDD, ICML, KDD, ICDM etc. or a journal such as MLJ, DMKD, JMLR etc.

We received 20 long and 21 short submissions, and one previously published paper. Each submission was reviewed by at least three Program Commitee members. The short papers were evaluated on the basis of both the submitted manuscript and the presentation at the conference, and the authors of a subset of papers were invited to submit an extended version. Eventually, 18 papers were accepted for the present proceedings, nine papers were accepted for the Late-Breaking Papers volume published in the CEUR workshop proceedings series, and eight papers were invited for the Machine Learning journal special issue dedicated to ILP 2012, and subjected to further reviewing. The paper entitled "Plane-Based Object Categorization Using Relational Learning" by Reza Farid and Claude Sammut won the Machine Learning journal best student paper award and appears in the mentioned special issue. The prize is kindly sponsored by the Machine Learning journal (Springer).

The papers in this volume represent well the current breadth of ILP research topics such as propositionalization, logical foundations, implementations, probabilistic ILP, applications in robotics and biology, grammatical inference, spatial learning, and graph-based learning.

The conference program included three invited talks. In the lecture entitled "Declarative Modeling for Machine Learning," Luc De Raedt proposed applying the constraint programming methodology to machine learning and data mining and specifying machine learning and data mining problems as constraint satisfaction and optimization problems. In this way it is possible to develop applications and software that incorporates machine learning or data mining techniques by specifying declaratively what the machine learning or data mining problem is rather than having to outline how the solution needs to be computed.

Ben Taskar's talk "Geometry of Diversity and Determinantal Point Processes: Representation, Inference and Learning" discussed approaches to inference and learning in graphical models using determinantal point processes (DPPs) that offer tractable algorithms for exact inference, including computing marginals, computing certain conditional probabilities, and sampling. He presented recent work on a novel factorization and dual representation of DPPs that enables efficient inference for exponentially sized structured sets.

Geraint A. Wiggins spoke about "Learning and Creativity in the Global Workspace" and presented a model based on Baars Global Workspace account of consciousness, that attempts to provide a general, uniform mechanism for information regulation. The key ideas involved are: information content and entropy, expectation, learning multi-dimensional, multi-level representations and data, and data-driven segmentation. The model was originally based on music, but can be generalized to language. Most importantly, it can account for not only perception and action, but also for creativity, possibly serving as a model for original linguistic thought.

The conference was kindly sponsored by the Office of Naval Research Global, the *Artificial Intelligence* journal and the *Machine Learning* journal. We would like to thank EasyChair.org for supporting submission handling. Our deep thanks also go to Nada Lavrač, Tina Anžič, and Dragan Gamberger for the local organization of the conference and Radomír Černoch for setting up and maintaining the conference website.

March 2013
Fabrizio Riguzzi
Filip Železný

Organization

Program Committee

Erick Alphonse	LIPN - UMR CNRS 7030, France
Dalal Alrajeh	Imperial College London, UK
Annalisa Appice	University Aldo Moro of Bari, Italy
Ivan Bratko	University of Ljubljana, Slovenia
Rui Camacho	LIACC/FEUP University of Porto, Portugal
James Cussens	University of York, UK
Saso Dzeroski	Jozef Stefan Institute, Slovenia
Floriana Esposito	Università di Bari, Italy
Nicola Fanizzi	Università di Bari, Italy
Daan Fierens	Katholieke Universiteit Leuven, Belgium
Nuno Fonseca	CRACS-INESC Porto LA, Portugal
Tamas Horvath	University of Bonn and Fraunhofer IAIS, Germany
Katsumi Inoue	National Institute of Informatics, Japan
Nobuhiro Inuzuka	Nagoya Institute of Technology, Japan
Andreas Karwath	University of Freiburg, Germany
Kristian Kersting	Fraunhofer IAIS and University of Bonn, Germany
Ross King	University of Wales, Aberystwyth, UK
Ekaterina Komendantskaya	University of Dundee, UK
Stefan Kramer	TU München, Germany
Nada Lavrač	Jozef Stefan Institute, Slovenia
Francesca Alessandra Lisi	Università degli Studi di Bari, Italy
Donato Malerba	Universita' di Bari, Italy
Stephen Muggleton	Imperial College London, UK
Ramon Otero	University of Corunna, Spain
Aline Paes	COPPE/PESC/UFRJ, Brasil
David Page	University of Wisconsin, USA
Bernhard Pfahringer	University of Waikato, New Zealand
Ganesh Ramakrishnan	IIT Bombay, India
Jan Ramon	K.U. Leuven, Belgium
Oliver Ray	University of Bristol, UK
Fabrizio Riguzzi	University of Ferrara, Italy
Celine Rouveirol	LIPN, Université Paris 13, France
Chiaki Sakama	Wakayama University, France
Jose Santos	Imperial College London, UK
Vítor Santos Costa	Universidade do Porto, Portugal
Michèle Sebag	Université Paris-Sud, CNRS, France

Jude Shavlik	University of Wisconsin – Madison, USA
Takayoshi Shoudai	Kyushu University, Japan
Ashwin Srinivasan	Indraprastha Institute of Information Technology, India
Prasad Tadepalli	Oregon State University, USA
Alireza Tamaddoni-Nezhad	Imperial College, London, UK
Tomoyuki Uchida	Hiroshima City University, UK
Christel Vrain	LIFO - University of Orléans, France
Stefan Wrobel	Fraunhofer IAIS and University of Bonn, Germany
Akihiro Yamamoto	Kyoto University, Japan
Gerson Zaverucha	PESC/COPPE - UFRJ, Brasil
Filip Zelezny	Czech Technical University in Prague, Czech Republic

Additional Reviewers

Di Mauro, Nicola	Stein, Sebastian
Duboc, Ana Luisa	Trajanov, Aneta
Ferilli, Stefano	

Table of Contents

A Relational Approach to Tool-Use Learning in Robots

Solly Brown and Claude Sammut

School of Computer Science and Engineering
The University of New South Wales
Sydney, Australia 2052
`claude@cse.unsw.edu.au`

Abstract. We present a robot agent that learns to exploit objects in its environment as tools, allowing it to solve problems that would otherwise be impossible to achieve. The agent learns by watching a single demonstration of tool use by a teacher and then experiments in the world with a variety of available tools. A variation of explanation-based learning (EBL) first identifies the most important sub-goals the teacher achieved using the tool. The action model constructed from this explanation is then refined by trial-and-error learning with a novel Inductive Logic Programming (ILP) algorithm that generates informative experiments while containing the search space to a practical number of experiments. Relational learning generalises across objects and tasks to learn the spatial and structural constraints that describe useful tools and how they should be employed. The system is evaluated in a simulated robot environment.

Keywords: robot learning, planning, constrain solving, explanation-based learning, version space.

1 Introduction

A tool is any object that is deliberately employed by an agent to help it achieve a goal that would otherwise be more difficult or impossible to accomplish. We would like a robot to be able to learn to use a new tool in a manner similar to the way a human would learn. That is, after the learner observes the tool being used correctly by another agent, the learner forms an initial hypothesis about the properties and use of the tool. The hypothesis is refined by active experimentation, selecting different types of objects as the tool and manipulating them in different ways. By analysing the teacher's demonstration and then experimenting, the robot learns a new tool action that will allow the agent to perform variations of the task. We limit the robot to starting from a single observation since we are most interested in investigating active learning in an incremental manner.

We define a *tool action* to be one that changes the properties of one or more objects in a way that enables the preconditions of one or more other actions in the agent's plan library. The changed properties of the objects that result from the tool action are called the sub-goals of the tool action.

F. Riguzzi and F. Železný (Eds.): ILP 2012, LNAI 7842, pp. 1–15, 2013.

There are four things that the robot must learn so that it consistently solves a tool-use task: (1) what useful effect the tool action achieves, (2) how to identify a good tool (i.e. what object properties are necessary), (3) the correct position in which the tool should be placed, (4) how the tool should be manipulated after being placed. In this paper, we focus on learning the first three of these. For manipulation, the learnt action is translated into an executable behaviour by a spatial constraint solver and a motion planner.

The general setting for learning is as follows:

1. Observe trainer using a tool to achieve a goal
2. Form an initial hypothesis explaining the tool use
3. Construct an experiment to test the hypothesis
4. Update the hypothesis according to the success of failure of the experiment
5. repeat until no further successful hypotheses can be constructed

One of the challenges in active learning is how to choose an experiment that will yield useful information. In a batch learning setting, where many training examples are available, we typically aim to find the simplest concept description that has high accuracy. A general-to-specific search of the hypothesis space is often employed to achieve this. In active learning by a robot, concept formation is done incrementally, with few training example available because they are usually costly to obtain since the robot must expend time and resources to perform an experiment. As we are learning the description of an action, an experiment is an instance of the hypothesis. Were we to employ a purely general-to-specific search, we might generate an instance that is close to the boundary of the hypothesis. The problem with this approach is that we are likely to generate many more negative examples than positive. More importantly, when an experiment fails, we want to know why it failed. The usual scientific method perform experiments that change just one variable at a time so that we know what was the cause of the experiment's outcome. This is more easily done if we employ a specific-to-general search in which we perform a least general generalisation [1]. However, a specific-to-general search to tends yield a complex concept description. Thus, we have developed a bi-directional search, related to Mitchell's [2] version space, where the concept description is derived from the most general hypothesis consistent with the data, but experiment generation is based by the most specific hypothesis.

In the following section, we give an example of tool use, which we use to illustrate the learning method described in the following sections. We then describe our experimental method in section 6, followed by a discussion of related work and our conclusions. The methods described here are applicable to real robots but the experiments were conducted in simulation to avoid the time and cost of dealing with robot hardware. The simulation models a real robot and is sufficiently accurate that the results should translate to the real world.

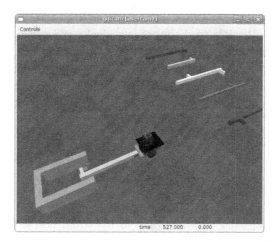

Fig. 1. Using a tool to reach an object in a closed "tube"

2   An Example of Learning to Use a Tool

A robot is set the goal of obtaining an object, in this case a small box, that is placed in a tube lying on the ground. The tube is open at one end and closed at the other, as shown in Figure 1. The robot is unable to reach directly into the tube to pick up the box because the tube is too narrow. The robot must use a hooked stick tool to pull the box out of the tube before it can be picked up.

As shown in Figure 1, the robot is provided with a selection of different objects that can potentially be used as a tool for accomplishing the task. Some of these objects are clearly inappropriate since they cannot be inserted into the tube, lack a suitable "hook" affordance or are not long enough. However, the robot does not know ahead, in advance, which objects make good tools. The only information the robot has is provided by its sensors, i.e. the object's dimensions and the relative orientation of its surfaces.

The robot is also provided with a set of behaviours that it can use to effect changes in the world. In this example, the robot is given **goto**, **pickup-object** and **drop-object** behaviours. We assume that the agent does not already possess any sort of **pull-with-tool** behaviour. Learning this behaviour and using it to obtain the box is the objective of the problem.

The behaviours are represented by action models that describe how executing each behaviour affects the world. The action models are written as STRIPS-like operators [3]. These are used by the robot's planner to create sequences of behaviours that achieve the robot's goals. A robot already competent in solving the tube problem might construct the following plan:

goto(stick-tool), pickup(stick-tool), goto(tube),
pull-with-tool(stick-tool, box),
drop(stick-tool), pickup(box).

The difficulty for our robot is that not only does it lack the appropriate tool-using behaviour, **pull-with-tool**, but it also lacks the action model for this behaviour. This means that the agent is initially unable to form a plan of how to achieve the goal of obtaining the box. It must learn this missing behaviour. We simplify learning by providing the agent with an observation trace of a "teacher" performing the same task. In the case of the tube, our teacher demonstrates by simply picking up an appropriate stick-hook and using it to pull the box from the tube. The robot uses this example to create an initial hypothesis for the tool action, which is the starting point for the robot's experimentation.

In the tube problem the necessary properties of the tool include having a right-angled hook at the end of the handle and that the hook is on the correct side of the handle (if the box is sitting in the left side of the tube, the agent needs a left-sided hook stick). Learning the correct spatial and structural relations for tool use requires experimenting with a variety of tools and poses. We will describe this process after giving details about the action representation.

3   Action Representation

As in the STRIPS representation, the robot's action model includes the preconditions and effects of the action. A **pickup** action is specified as:

pickup(Obj)

PRE	forall(Tube:tube, ¬in(Obj,Tube)),
	empty-gripper,
	forall(x:obj, ¬obstructing(x,Obj))
ADD	gripping(Obj)
DEL	empty-gripper,
PRIMTV	fwd, back, rotleft, rotright
MOVING	robot

The robot must approach an object so that the gripper surrounds the object and can close on it. The precondition states that the object must not be in any tube, the robot's gripper must be empty and there must not be any other object obstructing the target object. The additional lists of primitive motor actions and objects moved by this action are needed to turn the action into motor commands to the robot.

Action models do not specify behaviours in sufficient detail that they can be immediately executed by the robot. In the example above, the action model says nothing about the precise position that the robot must be in after the action is executed, nor does it specify the path that the robot must take to get itself into the goal position. However, actions can be made operational by using a constraint solver to create a range of solutions that satisfy the goal (e.g. any position such that the gripper surrounds the target object) then a motion planner can output a set of primitive motor actions to the robot (in this example, manoeuvring the robot to a position within the bounds set by the constraint solver). This process is illustrated in Figure 2.

We use Hoffmann and Nebel's FF [4] to generate plans and the constraint solving facilities in *ECLiPSe* [5] to generate the constraints. A Rapid Random Tree (RRT) path planner [6] produces the primitive actions for the robot. RRT requires a set of motion primitives, a list of the objects that are manipulated by the planner, and a specification of their relative spatial pose during the manipulation. This is the reason why we extend the STRIPS representation to include a list of the objects involved in the action and a set of robot motion primitives. The motion planner outputs a path for the objects that leads to the desired goal state. A simple controller outputs the motor commands required to follow the path. Brown [7] gives a detailed discussion of the implementation.

In the architecture depicted in Figure 2, nondeterminism in the world is handled by the control layer. The controller implements a feedback loop that monitors the progress of the robot and adjusts the motor actions to minimise errors. This hides much of the nondeterminism of the world from the higher, deliberative layers. Thus, at the abstract level of action models, we can use a deterministic planner and learn deterministic action models. We do not address the problem of nondeterminism at the planning level. For the class of problems described here, this is unnecessary. Future research may investigate learning for nondeterministic planning, possibly incorporating probabilistic ILP.

4   Learning by Explanation

The aim of learning by explanation is to identify novel tool actions in the teacher's demonstration and to construct an action model that describes it. This includes identifying the sub-goal that the tool achieves and some of the necessary preconditions for using the tool. It involves the following steps: (1) watch a teacher using a tool to complete a task; (2) identify actions in the teacher's demonstration by matching them against actions stored in the robots plan library; (3) any sections of the demonstration that cannot be matched to known actions are labelled as components of a novel action; (4) a STRIPS model for the novel action is constructed by identifying the subset of literals in the action's start and end states that are relevant to explaining how the teacher achieved the goal.

To recognise novel behaviours, the robot begins by trying to explain what the teacher is doing during its demonstration. The robot constructs an explanation by trying to match its own set of action models to the execution trace of the teacher's demonstration. Gaps in the explanation, where the agent is unable to match an existing behaviour to what the teacher is doing, are designated novel behaviours, which the robot can try to learn.

Fig. 2. Behaviour generation

The first difficulty faced by the learner is in labelling the parts of the demonstration that it recognises. The demonstration trace is provided to the agent as a discrete time series $w_1, w_2, \ldots, w_n$ of snapshots of the low-level world state taken every tenth of a second. This world state consists of the positions and orientations of each object in the world at a particular point in time. We do not provide the learner with a demonstration trace that is already segmented. Rather, the trace is a sampling of a continuous time series of object poses and the learner must arrive at a segmentation of the trace by itself.

The trace is segmented into discrete actions by applying an heuristic, that action boundaries occur at points at which objects start or stop moving. Thus, when the agent grips a stick-tool and starts moving it towards the tube, an action boundary is recognised. When the tool makes contact with the box and causes it to start moving also, another segment boundary is recognised. In this way, the robot constructs a very general movement-based description of the actions of the teacher. In the case of the tube problem, the segments correspond to the robot moving to pick up the stick, placing it against the box, pulling the box from the tube, placing the stick down and finally, moving to pick up and carry away the box.

Once the learner has segmented the trainer's trace, it attempts to match segments to its existing set of action models. Each of the robot's action models incorporates a STRIPS model along with a list of the objects that are moved during the action. A model is matched to a segment by checking that three conditions are met: the moving objects in the demonstration match the model; the preconditions of the model are true at the start of the action segment; the effects of the model have been achieved by the end of the action segment.

In the unusual case where more than one abstract action matches a segment, the segment is labelled with the action with the most specific set of preconditions. If the two sets cannot be distinguished, one is chosen at random. This may affect the learning time if a poor choice is made but will ultimately be corrected if an experiment fails. Segments that cannot be matched to any existing action model are labelled as components of a novel action. In the tube example, this produces the following labelling of the teacher's demonstration:

> goto(stick), pickup(stick),
> **novel-action(stick, box)**,
> drop(stick), goto(box), pickup(box).

where **novel-action(stick,box)** is actually a compound action composed of two unrecognised actions. The first is a positioning action in which the stick is moved by the teacher. The second is the actual tool action that pulls the box from the tube, i.e., the stick and box are moved together. The learner must now try to explain how these two consecutive actions were used to achieve the goal of acquiring the box. It does so by constructing two action models, for positioning and tool use, that are consistent with the other actions in its explanation of the demonstration.

We use an heuristic that actions occurring before a novel action should enable the novel action's preconditions. Similarly, the effects of the novel action should help enable the preconditions of actions occurring later in the demonstration. This heuristic is based upon the assumption that the teacher is acting rationally and optimally, so that each action in the sequence is executed in order to achieve a necessary sub-goal. The program constructs a STRIPS model by examining the start and end states of the novel action and identifies relevant literals in the actions executed prior to and after the novel action. The effects of the novel action are defined as any state changes that occur during the action that support an action precondition later in the demonstration.

In the case of the tube problem, ¬**in(box,tube)** becomes true during the novel action segment. This effect enables the preconditions of the **pickup(box)** action that occurs later in the demonstration. In general, there may be many irrelevant effects. However, this explanation-based reasoning allows us to eliminate all but the effects that were important for achieving the goal.

The preconditions of the novel tool action are constructed by a similar argument. The learner examines the world state at the start of the novel action and identifies the subset of state literals that were produced by the effects of earlier actions. The effect **gripping(stick)** occurs before the novel action and is still true at the start of the action. Since it is a known effect of the **pickup(stick)** action, it is considered a relevant literal to include in the novel action preconditions.

A common pattern for many tool-use problems is that there is a positioning step, followed by the application of the tool. We use this as a template for constructing two action models for the tube problem. The preconditions and effects identified in the demonstration are converted to literals with parameters by simply substituting each instance of an object with a unique variable.

position-tool(Tool, Box)
PRE   in-gripper(Tool), gripping,
ADD   **tool-pose(Tool, Box)**, obstructing(Tool, Box)
DEL   –

The first action represents the robot getting itself into the correct position so that the tool can be applied. Note that we have not yet determined what that position is. The predicate **tool-pose(Tool,Box)**, which expresses this position, will be learned in the experimentation stage. A side-effect of this action is that the tool is obstructing the object. Later, when the robot tries to pick up the object, this side-effect will have to be undone. The tool action is:

pull-from-tube(Tool, Box, Tube)
PRE   **tool-pose(Tool, Box)**,
         gripping(Tool), in-tube(Box,Tube)
ADD: –
DEL: in-tube(Box, Tube)

The **tool-pose(Tool,Box)** effect of the position-tool action becomes a precondition of the tool action. There is only one subgoal of the tool action in this case, corresponding to an object being removed from the tube. This model becomes the starting point for learning the **tool-pose(Tool,Box)** predicate, which we describe next.

5   Learning by Experimentation

The robot must learn the *tool pose state*, i.e. the state in which the correct object has been selected as the tool and the pose in which it can be applied successfully. It does so by testing a variety of tools and tool poses in a series of learning tasks. Thus, the trial-and-error learning has two components: generating new learning tasks and updating the robot's hypothesis for the tool-pose concept depending on the outcome of the experiment. The process for learning the concept describing the tool pose state is as follows:

1. Select a tool that satisfies the current hypothesis and place it in a pose defined by the spatial constraints in the hypothesis. Generate a motion plan that solves the task from this state and execute it, observing whether the action sub-goal is achieved.
2. If the action achieved the desired sub-goal, label the initial state as a positive example. If the action failed to achieve the desired sub-goal, label it as a negative example.
3. If the example was positive, generalise the hypothesis. If the example was negative specialise the hypothesis.
4. If the current task was successfully solved, a new learning task is generated and presented to the robot. If the robot failed then the current task is reset.
5. The robot continues its experimentation until the agent is able to solve a pre-defined number of consecutive tasks without failure.

The robot's world contains a selection of tool objects available for solving the problem. The dimensions and shapes of the tools are chosen randomly according to a type definition specified by the user. Figure 1 shows some tools that are available for the tube task. Only the hook tool is suitable for solving the task.

We use a version-space representation for the hypothesis [2], where we maintain both a most-specific and most-general clause representing the concept.

$$h_S \leftarrow \text{saturation}(e_0).$$
$$h_G \leftarrow \text{true}.$$

e_0 is the initial example derived from the teacher's demonstration. The saturation of e_0 is the set of all literals in e_0 plus all literals that can be derived from e_0 and the robot's background knowledge. The most-general clause, initially, covers every possible instance of the hypothesis space.

Since our learner must generate experiments to find positive examples, we generalise conservatively by trying to test examples that are close to the most-specific boundary, where positive examples are more likely to be found. However, the initial most-specific hypothesis contains many irrelevant literals. We need a way of creating a specific example that avoids these irrelevant literals as much as possible. We do this by starting from with most-general clause and then specialise it by adding literals from the most-specific clause. This is similar to Progol, except that we take advantage some domain knowledge.

This process is divided into two. The tool pose hypothesis incorporates both structural literals (describing the physical composition, structure, and shape of the tool) and spatial literals (describing how the tool should be placed relative to other objects). We first select a tool that best matches h_S and then try to select a tool pose that matches as closely as possible that specified in h_S. To select a tool, each available tool object is scored for suitability according to the number of literals in h_S it satisfies. It must also satisfy all of the structural constraints in h_G.

Pose selection, similarly, tries to maximise the number of satisfied literals in h_S. In this case, however, we are interested in satisfying the greatest number of spatial literals. This allows the agent to place its selected tool in a pose that is as similar as possible to the previous positive examples it has observed. The process for selecting a suitable subset of spatial literals requires that all spatial constraints in h_G are satisfied. This ensures that the most important constraints are applied first. We then successively apply spatial constraint literals from h_G, using the constraint solver to check, at each step, whether the cumulative constraints can be satisfied.

Once the tool and pose have been selected, a plan is generated and executed by the robot. Figure 3 shows three examples generated for the tube problem. If the plan succeeds, we have a new positive example, which is used to generalise h_S. If it fails, h_G must be specialised to exclude the new negative example.

The most-specific clause is built using only positive examples. We use a constrained form of least general generalisation [1] to find the minimal generalisation that covers the positive examples. The most-general clause is built by generalising the most-specific clause. Negative-based reduction is used to simplify the most-specific clause without covering additional negative examples. This is a process of successively eliminating literals that do not change the coverage of the clause. We use negative-based reduction to recreate h_G when a new negative example arrives.

The resulting algorithm borrows heavily from GOLEM [8]. The main difference is that GOLEM's bias is to learn the shortest, most-general hypothesis

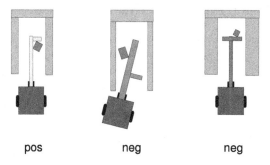

pos   neg   neg

Fig. 3. Positive and negative examples of tool use

possible. In contrast, we maintain both a most-specific and most-general boundary on the hypothesis space. The algorithm is rerun after each example is received and the current hypothesis is rebuilt.

6   Evaluation

This system has been implemented and tested on several learning problems using the Gazebo robot simulator [9]. The robot is a Pioneer 2, equipped with a simple gripper. In this section, we trace the robot's learning on the pull-from-tube-task. After the robot observes the trainer's demonstration, the first step is to segment the trace of the trainer's behaviour into discrete actions. Figure 4 shows the results of clustering object motions, as described earlier.

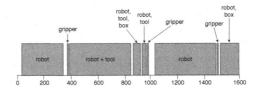

Fig. 4. Robot and object motion during the teacher's demonstration. Units are 10ths of seconds.

The *primitive state* of the world is given by the positions and orientations of all objects and is shown in Figure 5. From this, the robot builds an *abstract state* description that describes the properties and relationships between objects and agents in the world. The abstract state is expressed in predicate logic and is generated from a set of definitions provided to the agent as background knowledge, which consists of action models for known actions. These are shown in Figure 6. In addition to these abstract actions, the agent also has two manipulation recognition models, shown in Figure 7.

The robot constructs an explanation of the teacher's activities (Table 1) by matching action models to the motion segments it has identified. A model

Table 1. Explanations of teacher's demonstration

Seg	Moving objects	Explanation
1	robot	*put_in_gripper(hookstick)*
2	gripper	*grip(hookstick)*
3	robot, hookstick	*recognise_carry_obj(hookstick)*
4	robot, hookstick, box	*??*
5	robot, hookstick	*move_obstacle(hookstick,box)*
6	gripper	*ungrip(hookstick)*
7	robot	*remove_from_gripper(hookstick),put_in_gripper(box)*
8	gripper	*grip(box)*
9	robot, box	*recognise_carry_obj(box)*

```
tool_pose_S(Tool,Box,State):-

% static literals:

attached(Tool,Hook),                                    narrower(Tool,TubeLeft),
num_attachments(Tool,1),                                narrower(TubeLeft,Box),
num_attachments(Box,0),                                 narrower(Hook,TubeLeft),
longest_component(Tool),                                narrower(Tool,TubeRight),
narrower(Tool,Box),                                     narrower(TubeRight,Box),
shorter(Box,Tool),                                      narrower(Hook,TubeRight),
shape(Tool,sticklike),                                  narrower(Tool,TubeBack),
shape(Box,boxlike),                                     narrower(TubeBack,Box),
closed_tube(Tube,TubeLeft,TubeRight,TubeBack),          narrower(Hook,TubeBack),
attached_side(Tool,Hook,right),                         narrower(TubeBack,TubeLeft),
attached_side(TubeLeft,TubeBack,right),                 narrower(TubeBack,TubeRight),
attached_side(TubeRight,TubeBack,left),                 shorter(Hook,Tool),
attached_end(Tool,Hook,front),                          shorter(Tool,TubeLeft),
attached_end(TubeLeft,TubeBack,front),                  shorter(Tool,TubeRight),
attached_end(TubeRight,TubeBack,front),                 shorter(TubeBack,Tool),
attached_angle(Tool,Hook,right_angle),                  shorter(Box,Hook),
attached_angle(TubeLeft,TubeBack,right_angle),          shorter(Box,TubeLeft),
attached_angle(TubeRight,TubeBack,right_angle),         shorter(Hook,TubeLeft),
attached_type(Tool,Hook,end_to_end),                    shorter(TubeBack,TubeLeft),
attached_type(TubeLeft,TubeBack,end_to_end),            shorter(Box,TubeRight),
attached_type(TubeRight,TubeBack,end_to_end),           shorter(Hook,TubeRight),
num_attachments(Hook,0),                                shorter(TubeBack,TubeRight),
num_attachments(TubeLeft,1),                            shorter(TubeBack,TubeLeft),
num_attachments(TubeRight,1),                           shorter(Box,TubeBack),
num_attachments(TubeBack,0),                            shorter(Hook,TubeBack),
narrower(Hook,Tool),                                    shape(TubeLeft,sticklike),
narrower(Hook,Box),                                     shape(TubeRight,sticklike),
                                                        shape(TubeBack,sticklike),

% dynamic literals:

in_gripper(Tool,State),                                 at_oblique_angle(Box,Hook,State),
touching(Tool,Box,right,State),                         at_oblique_angle(Box,TubeLeft,State),
at_oblique_angle(Tool,Box,State),                       at_oblique_angle(Box,TubeRight,State),
in_tube(Box,Tube,State),                                at_oblique_angle(Box,TubeBack,State),
in_tube_end(Box,Tube,front,State),                      parallel(Tool,TubeLeft,State),
in_tube_side(Box,Tube,right,State),                     parallel(Tool,TubeRight,State),
touching(Hook,Box,back,State),                          parallel(Hook,TubeBack,State),
at_right_angles(Hook,TubeLeft,State),                   in_tube(Hook,Tube,State),
at_right_angles(Hook,TubeRight,State),                  in_tube_end(Hook,Tube,front,State),
at_right_angles(Tool,TubeBack,State),                   in_tube_side(Hook,Tube,right,State).
```

Fig. 5. The initial most-specific hypothesis clause h_S, formed from the trainer's example

grip(Obj)

PRE	¬gripping, in_gripper(Obj)
ADD	gripping
DEL	-
PRIMTV	closeGrip
MOVING	-

ungrip(Obj)

PRE	gripping, in_gripper(Obj)
ADD	-
DEL	gripping
PRIMTV	openGrip
MOVING	-

remove_from_gripper(Obj)

PRE	in_gripper(Obj), ¬gripping
ADD	empty_gripper
DEL	in_gripper(Obj)
PRIMTV	back
MOVING	robot

put_in_gripper(Obj)

PRE	forall(Tube:tube, ¬in(Obj,Tube)), empty_gripper, ¬gripping, forall(Obstacle:obj, ¬obstructing(Obstacle,Obj)
ADD	in_gripper(Obj)
DEL	empty_gripper
PRIMTV	fwd,back,rotleft,rotright
MOVING	robot

move_obstacle(ObjA,ObjB)

PRE	moveable_obj(ObjA) obstructing(ObjA,ObjB) in_gripper(ObjA), gripping
ADD	-
DEL	obstructing(ObjA,ObjB)
PRIMTV	fwd,back,rotleft,rotright
MOVING	robot, ObjA

Fig. 6. Action models provided as background knowledge for the pull-tool problem

recognise_goto

| PRE | empty_gripper |
| MOVING | robot |

recognise_carry_obj(Obj)

| PRE | in_gripper(Obj), gripping |
| MOVING | robot, Obj |

Fig. 7. Manipulation recognition models provided to the agent for the pull-tool problem

matches a segment if the action's preconditions are true at the beginning of the segment and the effects have been achieved by the end of the segment. The moving objects in the segment must also match the moving objects named in the corresponding action model.

At this point, the explanations are turned into the incomplete action models **position-tool** and **pull-from-tool**. The robot now enters its experimentation phase in which it constructs h_G and h_S for the definition of the **tool-pose** predicate. Each new example causes the learner to update the version space. The sequence of experiments is show in figure 8. After twelve experiments, the h_G is:

tool-pose$_G$(Tool,Box,State) ←
    in-tube-side(Box,Tube,Side,State),
    attached-side(Tool,Hook,Side),
    touching(Hook,Box,back,State),
    attached-angle(Tool,Hook,rightangle),
    attached-end(Tool,Hook,back).

This states that the tool must have a hook attached at a right angle and be on the same side as the target object in the tube. The hook must be touching the back of the object.

We evaluate the robot's performance by the number of experiments required before it can consistently solve a new task. Twelve experiments were required to learn **pull-from-tube**. A similar task, learning to extract an object from an open tube by pushing it through the tube, requires between 10 and 15 experiments, depending on random variations in the tools made available to the robot for its experiments. A third experiment was the classic Shakey problem of pushing a ramp up to a platform so that the robot can push a block off the platform.

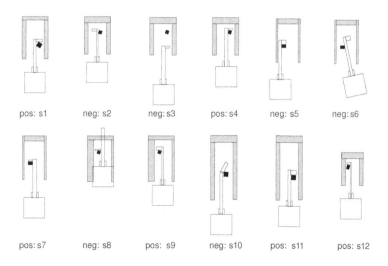

pos: s1   neg: s2   neg: s3   pos: s4   neg: s5   neg: s6

pos: s7   neg: s8   pos: s9   neg: s10   pos: s11   pos: s12

Fig. 8. Examples generated during learning by experimentation

Between 9 and 13 experiments were required to learn that the ramp can be used as a tool to mount the platform.

7   Related Work

Tool use has received relatively little attention in Artificial Intelligence and robotics research. Bicici *et al* [10] provides a survey of earlier AI research related to the reasoning and functionality of objects and tools. Perhaps the first attempt to build a robot agent specifically tailored towards learning and tool-use was report by Wood *et al.*, [11]. In this work an artificial neural network was used to learn appropriate postures for using reaching and grasping tools, on board the Sony Aibo robot platform. Stoytchev [12] has implemented an approach to learning tool-using behaviours with a robot arm. The robot investigates the effects of performing its innate primitive behaviours (including pushing, pulling, or sideways arm movements) whilst grasping different reaching tools provided by the user. The results of this exploratory learning are used to solve the task of using the tool to move an object into a desired location.

Although there is little other research that directly addresses tool use learning, work in the area of learning models of agent actions is relevant to our approach. Gil [13] used learning by experimentation to acquire planning knowledge. Benson [14] used ILP to learn action models of primitive actions and then combined them into useful behaviours. Other work has since focused on learning action models for planning in more complex environments, allowing for stochastic action [15] or partial observability [16]. Levine and DeJong [17] also used an explanation-based approach to acquiring planning operators.

8   Conclusion

This research contains several novel features. We integrate tool-use learning and problem solving in a robot that learns new tool actions to help it solve a planning problem. We have devised a novel action representation that integrates symbolic planning, constraint solving and motion planning. The system incorporates a novel method for incremental learning that uses relative least general generalisation but adopts a version-space representation. This is the basis for a new method for generating examples in an incremental setting where cautious generalisation is desirable. Our approach is based upon sampling near the most-specific hypothesis boundary in the version space. In addition, our robot is able to infer the approximate form of a new action model by observing a teacher and examining the context of the plan in which it is executed. The robot also uses trial-and-error to refine this approximate model.

There are several important issues that we do not address in this work. Some tool-use behaviours require complex non-linear control of the tool or target object. Learning these types of behaviours is a field of research in itself and we do not address it here. Instead, we have chosen tasks requiring simple manipulation that can be achieved with a generic controller. The learning tasks we attempt are restricted to those that can be easily implemented in a rigid-body simulation. Problems involving liquids or deformable bodies are not considered. Nevertheless, the general learning system is intended to be applicable to a wide range of scenarios. We do not try to handle learning by analogy. For example, using a ladder to reach a light bulb and using a stick to reach an object on a shelf are conceptually similar problems.

References

1. Plotkin, G.: A note on inductive generalization. In: Meltzer, B., Mitchie, D. (eds.) Machine Intelligence, vol. 5, pp. 153–163. Edinburgh University Press (1970)
2. Mitchell, T.: Generalization as search. Artificial Intelligence 18, 203–266 (1982)
3. Fikes, R., Nilsson, N.: STRIPS: A new approach to the application of theorem proving to problem solving. Artificial Intelligence 2(3-4), 189–208 (1971)
4. Hoffmann, J., Nebel, B.: The FF planning system: Fast plan generation through heuristic search. Journal of Artificial Intelligence Research 14 (2001)
5. Apt, K.R., Wallace, M.G.: Constraint Logic Programming using ECLiPSe. Cambridge University Press (2007)
6. Kuffner, J., LaValle, S.: RRT-Connect: An efficient approach to single-query path planning. In: International Conference on Robotics and Automation (April 2000)
7. Brown, S.: A relational approach to tool-use learning in robots. PhD thesis, School of Computer Science and Engineering, The University of New South Wales (2009)
8. Muggleton, S., Feng, C.: Efficient induction of logic programs. In: Muggleton, S. (ed.) Inductive Logic Programming, pp. 281–298. Academic Press (1992)
9. Koenig, N., Howard, A.: Design and use paradigms for Gazebo, an open-source multi-robot simulator. In: International Conference on Intelligent Robots and Systems, vol. 3, pp. 2149–2154 (September 2004)
10. Bicici, E., St Amant, R.: Reasoning about the functionality of tools and physical artifacts. Technical Report 22, NC State University (2003)
11. Wood, A., Horton, T., St Amant, R.: Effective tool use in a habile agent. In: IEEE Systems and Information Engineering Design Symposium, pp. 75–81 (April 2005)
12. Stoytchev, A.: Behaviour-grounded representation of tool affordances. In: International Conference on Robotics and Automation (April 2005)
13. Gil, Y.: Learning by experimentation: Incremental refinement of incomplete planning domains. In: International Conference on Machine Learning (1994)
14. Benson, S.: Learning action models for reactive autonomous agents. PhD thesis, Department of Computer Science, Stanford University (1996)
15. Pasula, H.M., Zettlemoyer, L.S., Kaelbling, L.P.: Learning symbolic models of stochastic domains. Journal of Artificial Intelligence Research 29, 309–352 (2007)
16. Amir, E.: Learning partially observable deterministic action models. In: International Joint Conference on Artificial Intelligence (IJCAI 2005), Edinburgh, Scotland, UK, pp. 1433–1439 (August 2005)
17. Levine, G., DeJong, G.: Explanation-based acquisition of planning operators. In: ICAPS, pp. 152–161 (2006)

A Refinement Operator for Inducing Threaded-Variable Clauses

Angelos Charalambidis[1,2] and Stasinos Konstantopoulos[1]

[1] Institute of Informatics and Telecommunications,
NCSR 'Demokritos', Ag. Paraskevi 153 10, Greece
{acharal,konstant}@iit.demokritos.gr
[2] Dept. of Informatics and Telecommunications,
University of Athens, Ilissia 157 84, Greece

Abstract. In Logic Programming, a thread of input/output variables is often used to carry state information through the body literals of the clauses that make up a logic program. When using Inductive Logic Programming (ILP) to synthesise logic programs, the standard refinement operators that define the search space cannot enforce this pattern and non-conforming clauses have to be discarded after being constructed. We present a new refinement operator that defines a search space that only includes Horn clauses that conform to this pattern of input/output variable threads, dramatically narrowing the search space and ILP run times. We further support our theoretical justification of the new operator with experimental results over a variety of datasets.

1 Introduction

A common *Logic Programming (LP)* technique is to use variables to carry state information through the body literals that make up clauses. These variables follow a very specific pattern where one literal's output variable is another literal's input, forming a *thread* that runs through the clause's body. Input/output variable threading is most prominent in *Definite Clause Grammars*, but many variations have been applied to different LP settings.

Inductive Logic Programming (ILP) is the machine learning discipline that deals with the automatic synthesis of logic programs from examples. ILP methods are very rich in providing for ways to declare syntactic and semantic constraints that valid results should respect; these are used both to define the space of valid hypotheses and to apply checks during the search through this space. Naturally, it is preferable that constraints are used to define the search space; one can easily see that constructing and then rejecting invalid hypotheses can greatly deteriorate execution efficiency compared to avoiding invalid areas of the search space altogether.

The *refinement operator* is the piece of ILP machinery that constructs the next candidate clause to be considered given the current state of the search. In this article, we introduce the *Variable Threading Operator (VT Operator)*, an

F. Riguzzi and F. Železný (Eds.): ILP 2012, LNAI 7842, pp. 16–31, 2013.

ILP refinement operator that only constructs clauses that conform to the pattern of input/output variable threading.

More specifically, after a short introduction to ILP in general and the theory and practice of refinement operators (Section 2), we proceed to formally define and justify the VT Operator (Section 3). We then present experimental results in DCG induction (Section 4) and in the induction of *Description Logics* (Section 5). Finally, we draw conclusions and outline future research (Section 6).

2 Background

In the work described here, we use the ALEPH (Srinivasan, 2004) implementation of the PROGOL algorithm (Muggleton, 1995), one of the most successful ILP systems in the literature. PROGOL operated within the *example setting* under *definite semantics*: background knowledge and constructed hypotheses are definite Horn clauses and the empirical evidence comprises fully ground positive and negative examples.

The importance of this setting is that key theoretical results and algorithms have been established within, including the reduction of the task to a search problem in a lattice of clauses organized by the (efficient to compute) θ-subsumption operator (Muggleton, 1995).

In PROGOL, background predicates provided to the system define prior world knowledge and are the building blocks for the constructed hypothesis; that is to say, they define the literals that will be used in the body of the hypothesised clauses. The search for clauses within the search lattice defined by the background predicates is driven by the refinement operator, but also by semantic prior knowledge that encodes restrictions on the types of the variables of a clause and on the types that each predicate may accept.

2.1 Threaded Variables

Threading input/output variables is a popular technique of Logic Programming. Each literal has an input argument and an output argument. We use the term input and output when we are expecting from the predicate to use the input argument as an input and yield the result of the computation as a binding to the output argument. The variable threading between body literals happens when the output argument is a variable that occurs as input argument in another literal. In other words, the literals are pipelined and the output of one literal becomes the input of the next.

Consider the following logic program that is a simple *parser* and is constructed by threading other simpler parsers:

$$sentence(S_1, S_3) \quad \leftarrow nounphrase(S_1, S_2) \land verbphrase(S_2, S_3)$$
$$nounphrase(S_1, S_3) \leftarrow det(S_1, S_2) \land noun(S_2, S_3)$$
$$verbphrase(S_1, S_3) \quad \leftarrow verb(S_1, S_2) \land nounphrase(S_2, S_3)$$

The variables S_i are lists of tokens. Each parser has two arguments: the first is the input to consume and the second is the remaining list of tokens after the successful parsing. In order to create a parser by combining simpler parsers, the input of the head literal is used as the input of the first body literal while the output of the last body literal is the final output of the parser. Moreover, this is a *single variable threading* since each output is connected to only one input and the input and the output of the first and last literal of the thread are connected to the head literal variables. Notice that by design, any clause that is not formed as a single thread of parsers is not itself a parser.

Input/output threading itself can be achieved by mode declarations using the standard refinement operator.

However, the complete requirement that the body comprises a single thread that commences with the input variable of the head and concludes with the output variable of the head cannot be encoded in mode declarations. If one uses an unmodified refinement operator, such checks are by necessity implemented by syntactic checks *after* invalid clauses have been constructed.

Another question that naturally arises is why this can not be done by rewriting to propositionalize the inductive step, at least wrt. the threaded variables: threaded variables are used to accumulate state information, and not simply achieve a certain order of execution. Body literals might depend on the binding of these variables, as is the case in our DL learning experiments below (cf. Section 5).

2.2 Custom Refinement Operators

Besides its application to constructing definite Horn clauses, ILP has also been applied to other fragments of first-order logic. This is usually achieved by modifying the refinement operator so that a different hypotheses space is explored instead of that of definite Horn clauses. *Description logics* (DL) are such a family of logics that have been targeted as a non-Horn domain of application of ILP. DLs are ubiquitous in the Semantic Web, where they have found many applications in conceptual and semantic modeling.

DL statements formally describe *concepts* (classes, unary predicates) using complex class expressions involving the primitive class descriptions supported by each given DL. The inventory of such primitives varies, yielding logics with different expressivity and inference efficiency.

Badea and Nienhuys-Cheng (2000) propose a top-down refinement operator for DLs. Although it addresses the over-specificity of the clauses learnt by previous bottom-up approaches (Cohen and Hirsh, 1994), it is restricted to a simple DL without negation, numerical quantification, and relation composition. Lehmann and Hitzler (2008) and Iannone et al (2007) address negation, but do not learn all descriptions possible in $\mathcal{SROIQ}$, the DL that covers the widely-used Semantic Web language OWL (W3C OWL Working Group, 2009).

A further possible extension is the ability to synthesise knowledge under *many-valued semantics*, where an individual's membership in a set gets associated with one of many (instead of two) possible values, typically within the real unit interval. The logical connectives are, accordingly, interpreted by numerical operators.

Previous approaches to many-valued ILP typically use thresholds to create crisp representations of many-valued logics and then apply crisp DL-learning methodologies. This approach has been proven to work well, but is based on the assumption that the domain can be quantized. Bobillo et al (2007), for instance, successfully applied this approach to learning many-valued description under Gödel (min-max) semantics.

Konstantopoulos and Charalambidis (2010) have also demonstrated how expressive DL constructs (under both many-valued and crisp semantics) can be expressed in a threaded-variable Logic Programming setting, effectively using Prolog meta-predicates to define DL primitives that fall outside the expressivity of pure LP. In this article we are re-visiting this work in order to demonstrate how this approach can be optimized by applying our variable threading refinement operator.

3 The Variable Threading Operator

3.1 Task Definition

As stated in the previous section, there are use cases where we want to induce hypotheses that are only meaningful when the clauses respect the single-thread constraint. In that case, the search spaces can be drastically reduced if guided correctly to only valid clauses. We will proceed by giving the definition of the single variable thread constraint.

Consider a literal L of form $l(a_1, \cdots, a_n)$ and $n \geq 2$. We will assume that two arguments of literal L are denoted especially as input and output arguments. We will use the operators $in(L) = a_i$ and $out(L) = a_j$ to refer to the input and output arguments of literal L. Consider now a sequence B of literals $L_1, \cdots, L_k$. Let $\mathcal{I}_B$ and $\mathcal{O}_B$ be the sets where for every literal $L \in B$ it holds that $in(L) \in \mathcal{I}_B$ and $out(L) \in \mathcal{O}_B$.

We will say that B consists a single variable thread of input i and output o if for every literal $L \in B$ $in(L)$ and $out(L)$ are variables and for the input variable it holds that $in(L) \in \mathcal{O}_B \cup \{i\}$ and $in(L) \notin \mathcal{I}_{B-\{L\}}$ and for the output variable it holds that $out(L) \in \mathcal{I}_B \cup \{o\}$ and $out(L) \notin \mathcal{O}_{B-\{L\}}$. We will also denote the input and the output of the thread as $in(B) = i$ and $out(B) = o$.

Consider the definite clause $C = H \leftarrow B$. We will say that the clause C respects the single variable-thread constraint if B is a single variable thread and $in(H) \equiv in(B)$ and $out(H) \equiv out(B)$.

The input/output threading requirement can be encoded as mode declarations, by restricting that the newly added literal will have input variable one of the output variables that are already exist. This encoding will also construct clauses where the variables will be used as input in other literal, therefore violating the single variable-thread constraint.

3.2   Operator Definition

We have to devise a custom refinement operator in order to produce clauses where the body literals consists a single thread. A custom refinement operator is usually user defined machinery that generates a set of refinements for the ILP procedure and its logic is more sophisticated than the standard tools of an ILP system can handle.

Since the clauses that do not comply with that rule are considered invalid by design, there will be eventually rejected and therefore the construction and evaluation of such clauses can be omitted. In other words, the refinement operator will optimize the search by visiting only the single-thread clauses. Eventually, though, the whole valid search space will be visited.

The operator applies a top-to-bottom breadth-first search, starting from the most general clause, namely the clause that has an empty body and gradually producing refinements that are more specific than the original clause. In each step of the search the operator, given the current valid node, will produce a set of clauses where the body is expanded by a single literal and it is ensured that all the body literals satisfy the single-thread constraint.

Let the refinement operator ρ be a function that given a clause C it returns one-step refinements $r_i \in \rho(C)$ where r_i is more specific from C and respects the single variable-thread constraint. More specifically, let C be a clause $H \leftarrow B$ and L a *legitimate literal*, then $\rho(C) = \{H \leftarrow B' \wedge L\}$ where B' is exactly like B except $out(B) \equiv in(L)$ and $out(L) \equiv out(H)$.

A literal is *legitimate* if it is sufficiently instantiated with respect to the domain of application: its arguments contain appropriate values and the input and output arguments are variables. A legitimate literal can be either produced by a literal generator or be selected by the bottom clause. In the latter case, the refinement operator will only produce refinements that are subset of the bottom clause, since its literal of the refinement is also in the bottom clause.

In practice, there are cases where it is desired to avoid the construction of the bottom clause. In that cases, the refinement procedure will continue by generating literals using either mode definitions supplied by the user, or a user-defined generator that has direct access to the structure and the background of the case.

The generator is mainly used when it is inefficient to construct a bottom clause. A generator can produce literals by consulting the supplied mode definitions. Moreover, a more sophisticated generator can be supplied that is aware of the semantic restrictions of the background knowledge.

The generator is used for inducing definite clause grammar (see Section 4), while the bottom clause is used for inducing Description Logic theories described in Section 5. In this technique, both description logic concepts and relations are occurring as arguments of a literal. The reified concept and relation names used in these literals are chosen after proving them against relation domain and range axioms, the concept hierarchy, and disjointness axioms.

3.3   Technical Details

As mentioned above, our experiments is based on ALEPH. [1]. The refinement operator is designed to work along with the ILP system as a user-defined refinement providing the top-level predicate `refine/2` that accepts two arguments: the current node of the search and unifies non-deterministically the second argument with the candidate refinements. This predicate is used directly by ALEPH to produce new refinements in order to continue the search.

The refinement operator can be configured to use either a generator of legitimate literals, or generate literals using the user defined modes definitions. Moreover, the operator can select a literal that exists on the bottom clause, if it has been already constructed. In this case, the produced clause is a subset of the body of the bottom clause and in addition this subset must obey the syntactic rules.

3.4   Analysis

The refinement operator described in the previous sections constructively generates valid clauses of increasing body literals starting from the more generic to the more specific hypotheses. All clauses are compliant with the single-thread constraint.

There are use cases where the clauses of any form other than the ones that are compliant with this constraint are meaningless and it is guaranteed that eventually when evaluated the ILP system will reject them. The expected behaviour of the ILP system is to visit eventually all the valid nodes of the search space. In that sense, using the refinement operator does not imply any loss of valid answers, since every valid hypothesis will be eventually constructed.

Let us assume that the clause that is constructed is of length $n - 1$, meaning that has a head literal and $n - 1$ body literals and let k the choices of the different literals that can be produced using the mode declarations.

Consider now that we want to produce a new refinement of a clause of length n. Using the standard refinement, there must be at least k different choices, since there are k different literal instantiations. Assume now that each literal has at least an input and an output variable. This implies that the input and the output variable choices of the newly added literal can be n^2. The input variable of the newly added literal is either a new variable or one of the $n - 1$ output variables of the previous literals. Moreover, using the same reasoning, the output variable of the new literal has n candidates. The immediate candidate set is now totaling to $n^2 k$.

On the other hand, the refinement operator will produce k' clauses where $k' \leq k$ are the different instantiations of the added literal. Notice that if we use a standard instantiation strategy then $k' = k$.

---

[1] The implementation of the refinement operator that cooperates with ALEPH is in `pl/refine.pl` and can be downloaded from `http://sourceforge.net/projects/yadlr`

Let us assume that the search space that produces clauses of maximum length n. Then, the total nodes of the tree in the first case will be $k^n n!^2$ while the refinement operator only constructs a space of k'^n nodes where k and k' is now the average different instantiations of each literal.

4 Inducing Definite Clause Grammars

The syntax of an utterance is the way in which words combine to form grammatical phrases and sentences and the way in which the semantics of the individual words combine to give rise to the semantics of phrases and sentences. It is, in other words, the structure hidden behind the (flat) utterance heard and seen on the surface.

It is the fundamental assumption of linguistics that this structure has the form of a tree, where the terminal symbols are the actual word-forms and the non-terminal symbols abstractions of words or multi-word phrases. It is the task of a *computational grammar* to describe the mapping between syntactic trees structures and flat utterances. Computational grammars are typically Definite Clause Grammars,[2] which describe syntactic structures in terms of derivations. Such grammars are used to control a *parser*, the piece of computational machinery that builds the *parse tree* of the phrase from the derivations necessary to successfully recognise the phrase.

Definite Clause Grammars can be naturally represented as Definite Clause Programs and their derivations map directly to Definite Logic Programming proofs. This does not, however, mean that grammar construction can be trivially formulated as an ILP task: this is for various reasons, but most importantly because ILP algorithms will typically perform single-predicate learning without attempting background knowledge revision.

Because of these difficulties, the alternative of refining an incomplete original grammar has been explored. In this approach learning is broken down in an abductive example generation and an inductive example generalization step. In the first step, examples are generated from a linguistic corpus as follows: the incomplete parser is applied to the corpus and, each time a parse fails, the most specific missing piece of the proof tree is identified. In other words, for all sentences that fail to parse, a ground, phrase-specific rule is identified that—if incorporated in the grammar—would have allowed that phrase's parse to succeed.

For the inductive step the parsing mechanism and the original grammar are included in the background knowledge and an ILP algorithm generalizes the

---

[2] Many grammatical formalisms of greater computational complexity, for example, Head Phrase-driven Structure Grammar (HPSG), have also been proposed in the literature and gained wide acceptance. It is, however, the general consensus that grammatical formalisms beyond DCGs are justified on grounds of linguistic-theoretical conformance and compression on the grammar and that definite-clause grammars are powerful enough to capture most of the actual linguistic phenomena.

Table 1. The number of nodes visited and the execution time using the original settings and the VT operator for inducing definite clause grammars using chart parsing

Configuration	Nodes visited	Execution time (sec)	Time per node (ms)
Standard refinement	15246	155.97	10.23
Thread refinement	2310	51.07	22.10
Gain	84.85%	67.26%	-116%

phrase-specific rule examples into general rules that are appended to the original grammar to yield a complete grammar of the language of the corpus.

This abductive-inductive methodology has been successfully applied to Definite Clause grammars of English in two different parsing frameworks: Cussens and Pulman (2000) use it to to extend the coverage of a chart parser (Pereira and Warren, 1983) by learning chart-licensing rules and Zelle and Mooney (1996) to restrict the coverage of an overly general shift-reduce parser (Tomita, 1986) by learning control rules for the parser's shift and reduce operators.

4.1   Experimental Results

We have executed an experiment for inducing definite clause grammars using chart parsing. We have run the experiment with two different configurations: (a) using the original setup used in (Cussens and Pulman, 2000) and (b) by integrating the VT operator. In the first case, the original setup has a saturation phase that produces a bottom clause for each positive example and then consults the mode declarations for the search phase. Each clause has been evaluated using a cost function by counting the successful parsings. Moreover, no pruning optimizations have been used in this run. In the second case, the VT operator has been used for refinements that respect both the constructed bottom clause and the variable-threading nature of the clauses. Both configurations have eventually constructed identical theories, namely the same set of clauses. The experimental results are depicted in Table 1.

The VT operator reduces the number of the nodes constructed by a significant factor. This is due to the fact that the mode declarations cannot capture the structure of single threading, and thus many nodes have a poor evaluation and eventually rejected.

In terms of execution time, the VT operator also reduces the overall execution of the induction search. Notice though that the average execution time per reduction is more expensive. Firstly, for each refinement more calculations must be executed in order to create a single thread clause. Secondly, the clauses that are not a valid parser, will fail early in the cost evaluation function in contrast to the clauses produced by the VT operator which are all valid. Although this affects the average execution time per reduction, the overall execution time is greatly reduced since the visited nodes are also reduced by a significant factor.

4.2 Discussion

In this section we have applied the variable-threading operator in order to optimize the induction of definite clause grammars. Since definite clause grammars are by design rules that respect the variable threading principle, the VT operator is a perfect candidate for optimizing the induction procedure by restricting the search space only to the variable threading clauses.

There are already related parsing frameworks for inducing definite clause grammars. We have tested the operator to the framework described in (Cussens and Pulman, 2000) without changing the overall methodology. The experimental results show that the visited nodes have been greatly reduced compared to the original setup. As a result the overall execution time of the induction search has also been reduced by a significant factor.

5   Inducing Semantic Knowledge

Description logics (DL) are a family of logics that has found many applications in conceptual and semantic modeling, and is one of the key technologies behind *Semantic Web* applications. Despite, however, the rapid progress in inference methods, there has been very limited success—as briefly discussed in Section 2—in defining a refinement operator for the more expressive members of the family (such as those covering OWL) that are routinely used in web intelligence applications.

Coming from the opposite direction, one can also express DL constructs within Logic Programming, rather than adapting ILP systems to cover DL constructs that fall outside Logic Programming (Konstantopoulos and Charalambidis, 2010). This approach uses Prolog meta-predicates to define DL constructs that fall outside the expressivity of pure LP. In other words, the full Prolog language (including meta-predicates) is used within the definitions of background predicates that implement DL semantics. In this manner the ILP algorithm has the building blocks needed to construct and evaluate DL statements *without* being given unrestricted access to meta-predicates for constructing clauses so that no fundamental assumption about the ILP setting is violated.

Using this base definitions, complex DL expressions can be expressed by clauses where body literals 'threading' In and Out variables. The In and Out are essentially lists of instances together with the a membership degree: namely a real number that depicts how much that instance is considered member of a class. Fuzziness then is represented by the membership degrees of the instances.

The resulting logic program is, effectively, an instance-based DL reasoning service implemented as a transformation of DL expressions to Prolog clauses that implement the semantics of DL operators. The In argument is the initial members of the class together with their membership degrees and the Out is the list of instances that remain after the transformation is applied.

```
concept_select(In, 'RoundCar', Out) :-
    concept_select(In, 'Round', C),
```

Table 2. Predicates implementing the conjunction of a set of instances with a DL class description. The second column shows the DL description that corresponds to each predicate and the third column the semantics of the predicate. C is a class name, R a relation name, and n a non-negative integer; $\cdot^I$ is the interpretation function, and card$(\cdot)$ is set cardinality; $\mathbf{X}$ and $\mathbf{Y}$ are sets of instances.

Predicate	Descr.	Set-theoretic semantics
`concept_select/3`	C	$\{(\mathbf{X}, C, \mathbf{Y}) \mid \mathbf{Y} = \mathbf{X} \cap C^I\}$
`forall_select/4`	$\forall R.C$	$\{(\mathbf{X}, R, C, \mathbf{Y}) \mid \mathbf{Y} = \mathbf{X} \cap \{u \mid \forall v.\,(u,v) \in R^I \to v \in C^I\}\}$
`atleast_select/5`	$\geq nR.C$	$\{(\mathbf{X}, n, R, C, \mathbf{Y}) \mid \mathbf{Y} = \mathbf{X} \cap \{u \mid \mathrm{card}(\{v \mid (u,v) \in R^I \wedge v \in C^I\}) \geq n\}\}$
`atmost_select/5`	$\leq nR.C$	$\{(\mathbf{X}, n, R, C, \mathbf{Y}) \mid \mathbf{Y} = \mathbf{X} \cap \{u \mid \mathrm{card}(\{v \mid (u,v) \in R^I \wedge v \in C^I\}) \leq n\}\}$
`self_select/3`	$\exists R.\mathrm{Self}$	$\{(\mathbf{X}, R, \mathbf{Y}) \mid \mathbf{Y} = \mathbf{X} \cap \{u \mid (u,u) \in R^I\}\}$

```
    concept_select(C, 'Car', Out),

concept_select(In, 'LongRoundTrain', Out) :-
    concept_select(In, 'Train', C),
    atleast_select(C, 5, hasCar, 'RoundCar', Out).
```

to mean:
$$\text{Train} \sqcap\, \geq 5\ \text{hasCar. (Round} \sqcap \text{Car)} \sqsubseteq \text{LongRoundTrain}$$

Complementization is expressed by the `compl/1` operator, which creates a new class where membership degree is calculated by applying the negation norm to the membership degree in the original class. This mechanism lies in the implementation of the `_select` predicates in Table 2.

5.1   Establishing Priors

At the core of the Progol algorithm, and ILP approaches in general, is a search through the space of admissible hypotheses. This space is never made explicit, but is defined by (a) a *refinement operator* which traverses the space by repeatedly generating clauses to be evaluated by the search heuristics; and (b) other pieces of *prior knowledge* such as the syntactic and semantic constraints that valid hypotheses must satisfy.

Such restrictions reflect prior knowledge about the domain of application, including the input and output types of predicate arguments, disjointness axioms, etc. These guide the search away from solutions that are known to be unsatisfiable or otherwise unwarranted, providing mechanisms for optimization and for enforcing conformance with a prior theoretical framework.

In our approach, there are two sources of prior knowledge: (a) DL syntax and semantics, so that all constructed hypotheses correspond to valid DL propositions, and (b) the axiomatization of the domain within which a theory is to be constructed, providing semantic restrictions that can optimize the search.

Constructing Syntactically Valid DL Clauses. Description Logics are mainly characterized as the dyadic fragment of *first-order logic (FOL)*, that is the fragment that only includes one and two-place predicates.[3]

DLs typically employ a variable-free syntax where the logical connectives are used to connect variable-free literals; all literals share an implicit, universally-quantified first argument; where *relations* are involved (i.e., two-argument predicates), the second argument is an existentially or universally-quantified variable, locally scoped to each literal. Finally, although *general concept inclusion (GCI)* is allowed by the syntax (that is, complex constructs may appear on either side of the implication sign), inference engines typically require that a simple concept name appears in at least one side of the implication sign.

Consider, as an example, this statement equivalently formulated in DL, FOL, and our Horn notation:

$$\exists \mathsf{hasLocomotive} \sqcap \forall \mathsf{hasCar}.\mathsf{Green} \sqsubseteq \mathsf{GreenTrain}$$

$$\forall x.\,(\exists y.\mathrm{hasLocomotive}(x,y) \wedge \forall y.\,(\mathrm{hasCar}(x,y) \to \mathrm{Green}(y)) \to \mathrm{GreenTrain}(x))$$

```
concept_select(X,'GreenTrain',Y) :-
    atleast_select(X,1,hasLocomotive,thing,Z),
    forall_select(Z,hasCar,'Green',Y).
```

As one can immediately see by comparing the three notations, the variable-free DL syntax provides a very natural way of restricting what can be written in it to the dyadic fragment of FOL, whereas extra checks are necessary in both FOL and our Horn syntax.

In the case of our Horn syntax, in particular, it must be checked that the first and last arguments of the body literals form a single in/out thread.

Consistency with Background Semantics. Besides learning hypotheses that can be cast into DL statements, it is also desired that the refinement operator constructs clauses that are consistent with the relations that hold between the domain-specific background predicates.

Background predicates in ILP are typically seen as the building blocks for constructing hypotheses, as they define the literals that are used in the bodies

---

[3] With some further restrictions that guarantee the efficiency of inference over the formalism, and that are not directly relevant to the current discussion.

of the constructed clauses. Our approach partially departs from this tradition, as clauses are made up of the `*_select` predicates that define the framework or meta-theory within which theories are constructed. The first-order predicates of the theory itself (the concepts and relations of the DL-expressed domain) are *reified* into arguments for these predicates, and are treated as instances.

As such, the background theory is simultaneously represented as a first-order theory and as relations between the (reified) classes and relations of the theory. As an example, asserting that C is a sub-class of D asserts that all members of C are also members of D, but it also asserts a ground clause in the extension of the `concept_sub/2` predicate that keeps the class hierarchy:

```
concept_select(X,'Train',Y) :- concept_select(X,'GreenTrain',Y).
concept_sub('GreenTrain', 'Train').
```

Although not strictly necessary, `concept_sub/2` and related predicates affords our refinement operation direct access to the DL axioms that define the domain without having to parse the clauses of the `concept_select/3` predicate.

The refinement operator exploits this information to restrict the generated clauses to those that are consistent with these axioms. To make this more concrete, consider the existence of the classes Locomotive, Car, and Train in our domain and the knowledge that these are all disjoint as well as the knowledge that our target class, GreenTrain, is a sub-class of Train and the class Green is a sub-class of Car. Also consider that there are relations hasCar and hasLocomotive with domain Train and range Car and Thing respectively. In that case, the refinement operator will generate clauses with literals that respect the background constraints. For example, the clause

```
concept_select(A, 'GreenTrain', B) :-
    forall_select(A, 'hasCar', 'Green', B).
```

is a valid generated clause. On the other hand, the literal `forall_select(A, 'hasCar', Train, B)` must be avoided by the refinement operator because has-Car has a range that is disjoint with the class Train.

5.2   Experimental Results

We demonstrate our system on classic and well-studied machine learning problems: the smaller, artificially constructed, trainspotting problem and the larger, naturally observed Iris dataset. The train-spotting problem is rich in relations and relation chains, demonstrating the method's adequacy in constructing complex expressions. The Iris dataset is structurally simpler, but demonstrates the method's ability to handle numerical valuations beyond thresholding and the min-max algebra.

We tested our approach on a variation (Konstantopoulos and Charalambidis, 2010) of the famous trainspotting machine-learning problem (Larson and Michalski, 1977) and on the Iris dataset (Asuncion and Newman, 2007). The Train-spotting dataset has run multiple times with several optimizations. The standard configuration is used as a baseline for the comparison of the

Table 3. The number of nodes visited and the total execution time of different configurations for the trainspotting dataset. We also give the gain of each optimization compared to the standard.

Trains Configuration	Nodes Visited	Gain	Total Time (ms)	Gain
Standard	41	0%	32	0%
Pruning	18	56%	28	12%
VT Operator	15	63%	20	38%
VT Operator and DL	11	73%	12	63%

Table 4. The number of nodes visited before constructing identical theories using the VT operator and the pruning-based implementation. In pruning, we show the number of constructed and the number of those that have been pruned.

	Nodes visited				Avg. Time per node (ms)	
Priors enforced	Constructed	Pruned	VT Op	Gain	Pruning	VT Op
Iris Setosa	16886	12246	696	95.88%	2.10	4.00
Iris Versicolour	24698	20180	742	96.99%	1.28	2.81
Iris Virginica	11202	9118	351	96.86%	1.32	2.62

several optimizations. The pruning-based approach prunes the invalid clauses, so we avoid evaluating hopeless clauses. Both aforemented configurations uses mode declarations to restricted the generation of the clauses, yet they produce clauses that are not single variable threads. The VT operator is used with the standard literal generator and with the DL-aware literal generator. On the Iris dataset the experiment has successively run in order to learn three different target predicates: setosa, versicolor and virginica. In both datasets the background contains instances that are fuzzily members of-classes. The induction run statistics are collected in Tables 3 and 4.

It is straightforward to observe from the results that in all experiments the custom refinement operator constructs a dramatically smaller number of clauses: the standard operator generates a large number of clauses that either do not comply with the syntactic restrictions (Section 5.1) and are pruned after construction or do not comply with the semantic restrictions of the domain (Section 5.1) and are known in advance to not cover any positive examples.

In Table 3 are depicted the effects of the different optimization settings in the amount of nodes visited and the total execution of the search until all runs constructs identical theories. Since the train-spotting problem contains constraints about the domain and the range of the relations, the VT operator is further improving the results when is using the DL-aware literal generator.

In terms of execution time, the total execution time is also reduced in both datasets in the same amount as the reduction of the visited nodes. We have derived the average time of a single reduction from the total visited nodes and the total execution time. The reduction time includes the construction of a new refinement, the evaluation of the clause and the application of the clause to the positive and negative examples.

In the results presented in Table 4 the average times when we are using the custom operator are higher on the Iris datasets. The reason for this is that in the Iris dataset the priors only involve syntactic constraints and there are no prior semantic grounds for rejecting a syntactically valid clause. For this reason, the more complex calculations in the custom refinement clause make each refinement step more expensive, although the overall induction time is greatly improved using the custom refinement operator due to the smaller number of constructed and evaluated nodes.

The learning task and experimental setup is as described previously by Konstantopoulos and Charalambidis (2010), the only difference being the application of the refinement operator introduced here. In that work, conformance to DL syntax and semantics was only partially achieved via mode declarations and had to be complemented by pruning, whereas here we are able to altogether avoid refining into ill-formed clauses.

5.3 Discussion

In this section we pursue a line of research on expressing DL class descriptions *within the Logic programming framework,* allowing for the direct application of well-tried and highly-optimized ILP systems to the task of synthesising class descriptions. More specifically, we investigate a principled way of extracting more ILP bias from DL models in order to increase the efficiency of ILP runs.

This investigation led to the definition and implementation of a new refinement operator that is empirically shown to dramatically narrow the search space by being aware of restrictions on the Horn representation stemming from both the syntax and semantics of DLs as well as from axioms present in the specific background of each experiment.

6 Conclusions

In this paper we present a new refinement operator that restricts the ILP search space to that of clauses that follow the variable-threading pattern. We also present a series of experiments in settings where only such clauses are meaningful, thus introducing a stronger bias that allows the learner to visit a dramatically smaller number of hypotheses.

In the DL experiments, in particular, we have further expored how our oprimization can be combined with *lazy bottom construction,* an important optimization that has the refinement operator directly access the background knowledge rather than relying on the bottom clause in order to delimit the search. Although in its current implementation, our lazy-bottom operator cannot be applied to arbitrary backgrounds without customization, one of our first future research priorities is to carefully work out the interaction between our VT optimization and the lazy-bottom optimization over an arbitrary background.

Further plans also include the extension of a DL-specific strain of the VT optimization that can access further domain axioms that can contribute to ILP prior bias, besides the class hierarchy and relation domain and range axioms currently used. In this manner we can further restrict the search space, as well as contribute to the line of investigation that explores evaluation functions appropriate for DL hypothese, in the vein of recent research by Lisi and Straccia (2011) and Fanizzi (2011).

Acknowledgements. We would like to thank our anonymous reviewers for their helpful and substantial comments. We would also like to thank James Cussens and Stephen Pulman for sharing the data and the experimental setup from their inductive chart parsing experiments. Naturally, any errors and omissions in our usage of their data and experimental setup are solely our own.

The research leading to these results has received funding from the European Union's Seventh Framework Programme (FP7/2007-2013) under grant agreement no 288532. For more details, please see the website of the USEFIL project, http://www.usefil.eu

A. Charalambidis also acknowledges the partial support of the European Union's Seventh Framework Programme (FP7/2007-2013) under the PASCAL2 Network of Excellence, grant agreement no 216886.

References

Asuncion, A., Newman, D.J.: UCI machine learning repository. University of California, Irvine, School of Information and Computer Sciences (2007), http://www.ics.uci.edu/~mlearn/MLRepository.html

Badea, L., Nienhuys-Cheng, S.-H.: A refinement operator for description logics. In: Cussens, J., Frisch, A.M. (eds.) ILP 2000. LNCS (LNAI), vol. 1866, pp. 40–59. Springer, Heidelberg (2000)

Bobillo, F., Delgado, M., Gómez-Romero, J.: Optimizing the crisp representation of the fuzzy description logic $\mathcal{SROIQ}$. In: da Costa, P.C.G., d'Amato, C., Fanizzi, N., Laskey, K.B., Laskey, K.J., Lukasiewicz, T., Nickles, M., Pool, M. (eds.) URSW 2005-2007. LNCS (LNAI), vol. 5327, pp. 189–206. Springer, Heidelberg (2008)

Cohen, W.W., Hirsh, H.: Learning the CLASSIC description logic: Theoretical and experimental results. In: Proceedings 4th Intl. Conf. on Principles of Knowledge Representation and Reasoning, pp. 121–133 (1994)

Cussens, J., Pulman, S.: Experiments in inductive chart parsing. In: Cussens, J., Džeroski, S. (eds.) LLL 1999. LNCS (LNAI), vol. 1925, pp. 143–156. Springer, Heidelberg (2000)

Fanizzi, N.: Concept induction in Description Logics using information-theoretic heuristics. Intl. J. on Semantic Web and Information Systems 7(2) (2011)

Iannone, L., Palmisano, I., Fanizzi, N.: An algorithm based on counterfactuals for concept learning in the semantic web. Journal of Applied Intelligence 26(2), 139–159 (2007)

Konstantopoulos, S., Charalambidis, A.: Formulating description logic learning as an inductive logic programming task. In: Proceedings of Intl Conference on Fuzzy Systems (FUZZ-IEEE 2010), July 18-23. IEEE, Barcelona (2010)

Larson, J., Michalski, R.S.: Inductive inference of VL decision rules. ACM SIGART Bulletin 63, 38–44 (1977)

Lehmann, J., Hitzler, P.: A refinement operator based learning algorithm for the $\mathcal{ALC}$ description logic. In: Blockeel, H., Ramon, J., Shavlik, J., Tadepalli, P. (eds.) ILP 2007. LNCS (LNAI), vol. 4894, pp. 147–160. Springer, Heidelberg (2008)

Lisi, F.A., Straccia, U.: Can ILP deal with incomplete and vague structured knowledge? In: Proceedings ILP, Windsor Great Park, U.K. (2011)

Muggleton, S.: Inverse entailment and Progol. New Generation Computing 13, 245–286 (1995)

Pereira, F., Warren, D.S.: Parsing as deduction. In: Dunno (ed.) Proceedings of the 21st Conference of the ACL, pp. 137–144 (1983)

Srinivasan, A.: The Aleph Manual (2004), `http://www.comlab.ox.ac.uk/activities/machinelearning/Aleph/`

Tomita, M.: Efficient Parsing for Natural Language. Kluwer Academic Publishers, Boston (1986)

W3C OWL Working Group, OWL 2 web ontology language. W3C Recommendation (October 27, 2009), `http://www.w3.org/TR/owl2-overview`

Zelle, J.M., Mooney, R.J.: Learning to parse database queries using Inductive Logic Programming. In: Proceedings of the 13th National Conference on Artificial Intelligence, Portland, USA (1996)

Propositionalisation of Continuous Attributes beyond Simple Aggregation

Soufiane El Jelali, Agnès Braud, and Nicolas Lachiche

University of Strasbourg, LSIIT
Pôle API, Bd Brant, 67400 Illkirch, France
{eljelali,agnes.braud,nicolas.lachiche}@unistra.fr

Abstract. Existing propositionalisation approaches mainly deal with categorical attributes. Few approaches deal with continuous attributes. A first solution is then to discretise numeric attributes to transform them into categorical ones. Alternative approaches dealing with numeric attributes consist in aggregating them with simple functions such as average, minimum, maximum, etc. We propose an approach dual to discretisation that reverses the processing of objects and thresholds, and whose discretisation corresponds to quantiles. Our approach is evaluated thoroughly on artificial data to characterize its behaviour with respect to two attribute-value learners, and on real datasets.

Keywords: Propositionalisation, Continuous attributes, Aggregation.

1 Introduction

Relational Data Mining [7] considers data stored in at least two tables linked by a one-to-many relationship, as for example in the case of customers and their purchases, or molecules and their atoms. A way of mining these data consists in transforming them into a single attribute-value table. This transformation is called propositionalisation [14]. This paper focuses on propositionalisation of relational data involving continuous attributes.

A geographical problem motivated this work. This problem consists in predicting the class of urban blocks (see Fig. 1). The experts have defined 7 classes [17]: Continuous urban fabric (city center), Discontinuous urban fabric with individual houses, Discontinuous urban fabric with housing blocks (blocks of flats), Mixed urban fabric (including individual housing and housing blocks), Mixed areas, High density of specialised areas (including industrial, commercial, hospital or scholar buildings), and Low density of specialised areas (containing few or no building). A urban block is characterised only by the geometrical properties of its polygon: area, elongation and convexity. The buildings contained in the urban block are represented as polygons characterised by the same geometrical properties. Density is an additional property of the urban block.

Example 1. *Below is a sample of the data, that we will use as a running example in the article. It contains the two urban blocks of Fig. 1:*

F. Riguzzi and F. Železný (Eds.): ILP 2012, LNAI 7842, pp. 32–44, 2013.
© Springer-Verlag Berlin Heidelberg 2013

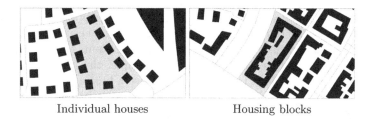

| Individual houses | Housing blocks |

Fig. 1. Geographical example: prediction of the class of a urban block

idblock	density	area	elong.	convex.	class
9601	0.194	6812	0.772	0.921	indiv
9602	0.455	11119	0.470	0.916	hous
⋮	⋮	⋮	⋮	⋮	⋮

and their buildings:

idbuild	area	elong.	convex.	idblock
4519	122	0.765	1.00	9601
4521	122	0.752	1.00	9601
4528	119	0.948	1.00	9601
4537	112	0.918	1.00	9601
4545	121	0.829	1.00	9601
4556	136	0.739	0.999	9601
4564	115	0.755	1.00	9601
4568	134	0.829	0.999	9601

idbuild	area	elong.	convex.	idblock
4579	125	0.745	1.00	9601
4583	98	0.935	0.999	9601
4589	113	0.909	1.000	9601
4231	1669	0.955	0.680	9602
4866	2239	0.772	0.595	9602
4867	229	0.818	0.999	9602
4868	164	0.795	0.936	9602
4869	559	0.451	0.894	9602
4870	205	0.271	0.999	9602

Discussions with the experts showed that the class depends on conditions about the geometry of buildings and the number, or proportion, of buildings satisfying those conditions. For example, the class "individual houses" mainly depends on the presence of small buildings. So the learning task consists in determining relevant attributes and their thresholds, as well as the number of those buildings. Existing approaches are not optimised to search at the same time a threshold on an attribute and a threshold on the number of objects satisfying this condition, as we will see in the next section. We propose a dedicated approach, called cardinalisation. It is introduced in Section 3. Section 4 presents an experimental validation on artificial data and on real data.

2 Related Works

Relational data mining approaches mainly come from Inductive Logic Programming (ILP) [16]. These approaches deal with categorical attributes rather than with numeric ones, and generally rely on a discretisation to transform numeric attributes into categorical ones. Other approaches use relational databases and propose to employ aggregate functions available in SQL. Finally, some approaches in full-fledged relational data mining systems integrate aggregate operators to the hypothesis construction. We detail those three approaches in this section.

In what follows, the main table is the table that contains the target column. Its rows are the individuals. In our running example, the individuals are the urban blocks. The secondary table describes objects linked to, or components of, the individual. E denotes the objects linked to an individual i, for example the buildings of a urban block. The cardinality of this set is denoted by $|E|$. In our example, it is the number of buildings. For reason of clarity, the attributes built by propositionalisation are called *features* to distinguish them from the columns of the original tables.

Discretisation. The search space of relational data mining is larger than the one, already exponential, of the attribute-value learning. Therefore, most works in ILP concern the efficient exploration of this search space. Some ILP works focus on handling numeric attributes. Main approaches [3][5][1] introduce constraints on the numeric attributes of related objects, but they consider the existential quantifier only. None of them evaluate how many such objects there are. Other ILP algorithms often deal with numerical attributes as if they were categorical. For example, recent propositionalisation algorithms such as RSD [23], Hifi [12] and RelF [13] do not deal explicitly with continuous attributes.

Thus, in order to deal with numeric attributes, a first approach consists in discretising them. This discretisation is made *a priori*, regardless of the model construction. The advantage is that the whole set of objects is considered, *i.e.* components of all individuals are taken together. For example, the areas of the buildings of all urban blocks can be discretised at the same time into small buildings (individual houses), medium (housing blocks) and big (industrial, commercial). Appropriate thresholds might be found more easily than with smaller samples, in particular than with a single urban block. The drawback is that the choice of the thresholds is not optimised with respect to a subset of the examples, as in the case of decision trees [20]. In this direction, let us cite the work of [15] where the authors propose a supervised technique for discretisation based on multi-instance learning: Lahbib et al. apply a pre-process to discretise numerical variables of the secondary table, and after that, they look for optimal thresholds using a MAP criterion.

Example 2. *The areas of buildings are discretised into 20 intervals of equal frequency. So each interval contains one twentieth of the buildings. This produces 19 thresholds and thus 38 features of the form "number of buildings whose area is lower (respectively greater) than such threshold":*

idblock	...	area<66	area>66	area<101	area>101	area<110	...
9601	...	0	11	0	11	1	...
9602	...	0	6	0	6	0	...
⋮	⋮	⋮	⋮	⋮	⋮	⋮	⋮

Features are generated for the elongation and convexity in the same way.

Once the continuous attribute is turned into categorical ones (two for each threshold), propositionalisation generates features. The simpler features built by a state-of-the-art system such as Relaggs [10,11] are generated: the number of buildings for each elementary condition. An elementary condition is a single constraint on one attribute. Let us notice that other propositionalisation systems dealing with categorical attributes, such as Hifi [12], would be able to generate a conjunction of attribute-value constraints, using an existential quantifier, and thus to build more complex features than the propositionalisation systems dedicated to numeric attributes do. Nevertheless the number of features increases exponentially, and it is difficult in practice to generate exhaustively all the features.

Aggregation. The systems Polka [9] and Relaggs [10,11] explicitly deal with numeric attributes. They apply the usual SQL aggregate functions to the values v_A of each numeric attribute A for the set E of tuples linked to the current individual,

$$\text{aggregation}(A, f, E) = f(\{v_A(t), t \in E\})$$

where f is an aggregate function. The available functions are minimum, maximum, average, standard deviation, sum, count, first quartile, median, third quartile, and the difference between the maximum and the minimum.

The difference between Polka and Relaggs can just be seen for problems with a secondary table linked to a third table by a one-to-many relationship. The sequence of tables linked by one-to-many relationships is called nested tables. In this case, Polka applies successively the aggregate operators in a depth-first order. Relaggs makes joins between the secondary table and the tables depending on it, and then apply the aggregate operators to the columns obtained.

Example 3. *Relaggs that computes the sum, the average, the minimum, the maximum and the standard deviation are applied to each numeric attribute:*

idblock	...	sum(area)	avg(area)	min(area)	max(area)	stddev(area)	...
9601	...	1317	119	98	136	10.5	...
9602	...	5065	844	164	2239	889	...
⋮	⋮	⋮	⋮	⋮	⋮	⋮	⋮

Those approaches are interesting because they summarize the data and they are not subject to combinatorial explosion, even when applied to nested tables.

Complex Aggregates. Full-fledged relational data mining systems do not dissociate the propositionalisation step from the construction of the model. Most of the full-fledged relational data mining systems, as for example Progol [18], do not explicitly deal with numeric attributes. Nevertheless, Tilde [4] has been modified to construct complex aggregates [21]. A complex aggregate is the application of an aggregate function to a conjunction of elementary conditions, for example the number of buildings whose areas are lower than 180 and their elongations are lower than 0.8. A simple aggregate is the application of an aggregate function to the objects selected by a single elementary condition, such as the number of buildings with an area lower than 180. The features considered in this paper are simple aggregates. Their generation with Tilde is not easy because the language bias has to be defined by hand, in particular the values to test for the thresholds on the number of buildings must be enumerated.

In theory, Tilde is able to generate complex aggregates. In practice, the combinatorial number of complex aggregates exceeds in a few minutes the memory capacity of the program. The expressible features are thus restricted to an aggregate and an inequality, for example "at least 2 buildings with an area greater than 180". The functions minimum, maximum, average, sum, mode, count and count-distinct are implemented, but standard deviation, first quartile, median and third quartile are missing. Tilde can also build features implying a proportion, such as "at least 30% of small buildings", but it is not optimised to do so. So, we will use a dedicated implementation in a propositionalisation approach rather than Tilde.

The contribution of this paper consists in proposing new ways to relate a continuous attribute of a secondary table to the main table. It is easier to present and evaluate in a propositionalisation approach, but it could be used in full-fledged relational data mining systems as well.

3   Cardinalisation

We propose a new approach for propositionalising in presence of numeric attributes, without discretising in advance, and without restricting ourselves to a fixed number of features limited by the number of aggregate functions. We call this approach cardinalisation to highlight the fact that it sets the cardinality between the main table and the secondary one, on the contrary to the existing approaches based on a discretisation that sets a threshold on the domain of the numeric attribute.

This duality is the subject of the first part of this section. In the second part, we will see that discretising the cardinalities correspond to quantiles. So it is a generalisation of aggregate functions, allowing to generate a greater number of features.

An Approach Dual to Discretisation. Aggregate functions provide a simple solution for propositionalising numeric attributes. Nevertheless, the number of functions proposed is small, and does not allow choosing both a threshold on the numeric attributes and a minimum number of the corresponding objects.

An alternative approach consists in discretising the numeric attribute so as to transform it into a categorical attribute. Thus thresholds are introduced on the numeric attribute and they will not be changed during the model construction. Then, the propositionalisation is made by applying an aggregate function to count the corresponding number of objects per individual. Given a numeric attribute A of the secondary table, a feature is built for each threshold s coming from the discretisation. The value of this feature for an individual linked to a set of tuples E of the secondary table is the number of tuples t whose value $v_A(t)$ for the attribute A is lower than the threshold s:

$$\text{aggregationAfterDiscretisation}(A, s, E) = |t \in E \text{ such that } v_A(t) \leq s|$$

This produces a numeric feature that the attribute-value learner can use. The value of this feature is a number of objects. During its contruction process, a classifier such as a decision tree learner will only have the possibility to choose the threshold on a numeric attribute by choosing the most discriminant feature, since the feature is defined by the threshold.

Cardinalisation tries to reverse the roles of the thresholds set on the numeric attribute and on the cardinality. It sets a threshold on the cardinality, and then lets the attribute-value learner choose the threshold on the numeric attribute. Actually, given a numeric attribute A of the secondary table, a feature is built for each possible threshold k of the cardinality, between 1 and the maximal number of objects per individual. The value of this feature is the minimal value of the threshold on the numeric attribute such that at least k tuples t have a value $v_A(t)$ for the numeric attribute A less than this threshold:

$$\begin{aligned} &\text{cardinalisation}(A, k, E) \\ &= min(s \in Domain(A) \text{ such that } |t \in E \text{ such that } v_A(t) \leq s| \geq k) \\ &= min(s \in Domain(A) \text{ such that aggregationAfterDiscretisation}(A, s, E) \geq k) \end{aligned}$$

When a urban block has less than k objects, an infinite threshold is assigned to the feature.

Example 4. *For each attribute, new features compute the minimal thresholds to select k buildings, k varying between 1 and the maximal number of buildings per urban block. For example, min_3_area indicates the minimal area such that the urban block contains at least 3 buildings with an area lower than or equal to that threshold.*

idblock	...	min_1_area	min_2_area	min_3_area	min_4_area	min_5_area	...
9601	...	98	112	113	115	119	...
9602	...	164	205	229	559	1669	...
⋮	⋮	⋮	⋮	⋮	⋮	⋮	⋮

We note that the thresholds depend on each urban block, in contrary to the discretisation that chooses each threshold independently from the urban blocks.

The interest of cardinalisation is to not set the threshold on the numeric attribute during propositionalisation, thus allowing the attribute-value program to choose the relevant threshold on the numeric attribute, in particular by taking into account the context as in the branches of a decision tree for example.

Cardinalisation is an approach dual to discretisation followed by an aggregation: a discretisation theoretically equivalent to cardinalisation consists in introducing a threshold between each value of the numeric attribute. The drawback is that the number of thresholds produced by this discretisation (in intervals containing a single object each) is the maximal number of values taken by the attribute on all objects from all individuals together. In the worst case, the number of distinct values is of the order of the total number of objects of the secondary table, for each of its numeric attributes. Moreover the aggregation produces two features per threshold: the number of objects below and above each threshold. On the other hand, the cardinalisation produces a number of features that is exactly the maximal number of objects per individual. In general, the maximal number of objects per individual is smaller than the double of the number of distinct values of a continuous attribute. Therefore, the discretisation that is theoretically equivalent to the cardinalisation, *i.e.* an interval per value, is often not tractable: the number of columns exceeds the maximal number of columns of the RDBMS (1600 columns in PostgreSQL and 4096 in MySQL, respectively). So, the cardinalisation is a more tractable way to let the attribute-value learner see all distinct numeric values.

If the number of features is too large in the case of a discretisation followed by an aggregation, it is possible to discretise the numeric attribute in larger intervals. Nevertheless, the drawback remains that the thresholds are set with respect to all the objects, all individuals together, for example the areas of all the buildings from all the urban blocks together. Thus, this setting is not as expressive as the cardinalisation.

If the cardinalisation generates too many features as well, the cardinality can be discretised.

Quantiles. Instead of building a feature per possible value of the cardinality, it is possible to discretise the cardinality in order to obtain a fixed number of features, in a similar way to the discretisation of the numeric attribute itself followed by the aggregation. Instead of setting an absolute number for the threshold on the cardinality, the discretised cardinalisation uses a relative number. Actually, this corresponds to quantiles. If four intervals are chosen, the features correspond to the first quartile, the median, and the third quartile. The kth n-quantile is the threshold such that a proportion of at least k/n objects are selected:

$$\text{quantile}(A, k, n, E) = min(s \in Domain(A) \text{ such that}$$
$$\text{agregationAfterDiscretisation}(A, s, E) \geq k \times |E|/n)$$

Example 5. *For each attribute of the buildings table, the features compute the minimal thresholds to select at least k twentieth of the buildings. The number of intervals, 20 in this case, is a parameter of the program.*

idblock	...	min_1_20_area	min_2_20_area	min_3_20_area	min_4_20_area	...
9601	...	98	98	112	112	...
9602	...	164	164	164	164	...
⋮	⋮	⋮	⋮	⋮	⋮	⋮

While quartiles are implemented in some propositionalisation approaches, none of them generalise quartiles to quantiles. Moreover, no full-fledged relational system but Tilde accepts a language bias allowing to construct aggregates corresponding to quantiles, and they are not really intended to. We propose to use quantiles in propositionalisation, so as to be able to configure the number of generated features, and thus to obtain a greater expressivity than in the existing propositionalisation approaches.

4   Experimental Evaluation

Four propositionalisation approaches, aggregationAfterDiscretisation[1], simple numeric aggregates using Relaggs, cardinalisation, and quantiles, are evaluated with respect to two attribute-value learners (classifiers).

The attribute value learners are the decision tree learner J48, and the support vector machine with a gaussian kernel using weka [22]. The width of the gaussian, γ, and the complexity, C, are tuned by a grid search on the training data: We select the combination of $\gamma \in \{10^{-5}, 10^{-3}, 10^{-1}\}$ and $C \in \{10^{-4}, 10^{-2}, 1, 10^2, 10^4\}$ that gets the best accuracy on the training set.

The experimental evaluation consists of two parts. The first part concerns artificial data where the class is chosen according to two known target functions. The second test consists in comparing the approaches on real data.

The 20-quantiles were used and the agregationAfterDiscretisation used 20 equal-frequency intervals, in order to use the same number of intervals.

Artificial Data. Artificial data have a structure similar to the geographical data that motivated this work. The main table describes some urban blocks with a Boolean class and one numeric attribute, the area. The building table also has a single numeric attribute, the area, and each building belongs to one and only one urban block. The values of the numeric attributes have been generated according to the distribution estimated on the real data: an exponential distribution with mean 28962 for the areas of urban blocks, an exponential distribution with mean 351 for the areas of buildings, and an exponential distribution with mean 12 for the number of buildings.

The number of training instances was increased from 100 to 1000 by step of 100, according to the same exponential distributions. An independent test set of 2000 urban blocks was generated. 10 training sets were generated for each number of training instances. The figures report the average of the accuracies on the test set obtained by models built from the 10 training sets, with error-bars corresponding to the standard deviation. Experiments on the real geographical data [19] showed that each class is described by a disjunction of rules with supports of the same order. Therefore two target functions involving a disjunction with equal supports were chosen in order to characterise the influence of the different propositionalisations with respect to the attribute-value learners:

Absolute Number: urban blocks that have more than 12 buildings with areas lower than 500, or more than 2 buildings with areas lower than 1200 and

---
[1] The aggregationAfterDiscretisation is denoted Discretisation in the figures.

less than 7 buildings with areas lower than 100 and less than 12 buildings with areas lower than 500.

Relative Number: urban blocks that have less than 60% buildings with areas lower than 400, or less than 10% buildings with areas lower than 30 and more than 20% buildings with areas lower than 100 and more than 60% buildings with areas lower than 400.

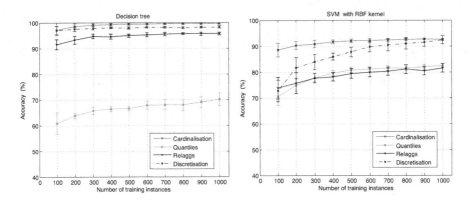

Fig. 2. Accuracy of two attribute-value learners on artificial data for a target involving absolute numbers

A first round of experiments concerns the absolute number target function defined above. Figure 2 reports the accuracies of a decision tree learner and of a support vector machine with a gaussian kernel, with respect to the four propositionalisation approaches. We observe that the cardinalisation enables both the decision tree and the support vector machine to get a more accurate model than the agregationAfterDiscretisation. Both of those propositionalisations get a higher accuracy than the simple aggregates of Relaggs. The quantiles are not designed to learn such a target function involving absolute numbers of objects.

A second round o experiments concerns the relative number target function defined above. Figure 3 reports the accuracies of the two attribute-value learners with respect to the four propositionalisation approaches. We observe that the quantiles enable the decision tree to learn perfectly a target function involving proportions of objects. The decision trees based on the other three propositionalisations are less accurate. We notice that the effectiveness of the propositionalisation depends on the attribute-value learners applied afterward. In particular, the support vector machine produces less accurate models than the decision tree on this target function involving relative numbers, and shows a clear ordering between the accuracies resulting from the four propositionalisations: aggregation-AfterDiscretisation, quantiles, simple aggregates of Relaggs, and cardinalisation, in decreasing order.

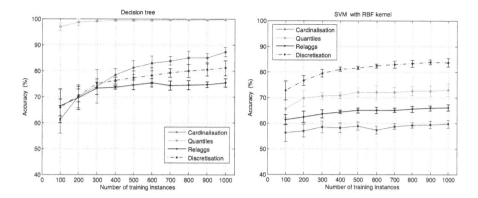

Fig. 3. Accuracy of two attribute-value learners on artificial data for a target involving relative numbers

These experiments show that the different combinations of propositionalisation and attribute-value learner get different accuracies, depending on the target problem. Thus, the different propositionalisations provide different inductive biases that can be beneficial on some datasets, but not necessarily on all datasets.

Real Data. Let us evaluate how the different propositionalisation approaches perform on real datasets. Few ILP datasets mainly involve continuous attributes. We consider six benchmarks from three application domains. Musk1 and Musk2 deal with properties of molecules [6]. Each molecule is described by a set of conformations, each conformation is described by 166 continuous attributes. Tiger, Elephant and Fox are content-based image retrieval problems [2]. Each image is described by a set of segments, each segment is described by 230 continuous attributes about color, texture and shape. Continuous attributes of images with low standard deviation have been filtered out. 91, 105 and 108 attributes were kept for Elephant, Fox and Tiger, respectively. The geographical data correspond to suburbs of the city of Strasbourg. The aim is to predict the classes of urban blocks from their geometrical properties (density, area, elongation, convexity) and the geometrical properties (area, elongation, convexity) of their buildings. The five datasets on musk and images are available on http://www.uco.es/grupos/kdis/mil/dataset.html. Those datasets have no dedicated test set. The reported accuracy is the average of 10 runs of 10-fold cross-validations. For the geographical dataset, all urban areas corresponding to year 2002 were used for the training set, and all urban areas from other years (1956, 1966, 1976, 1989 and 2008) for the test set. Let us emphasize that the suburbs changed a lot in the last fifty years, and nearly all urban blocks changed from one date to the next one, as can be seen on Figure 4 of the geographical study published in [19]. The number of quantiles and of intervals for the agregationAfterDiscretisation were restricted to 4 for the musk and the images datasets in order to keep the number of features below 1600. Cardinalisation produced

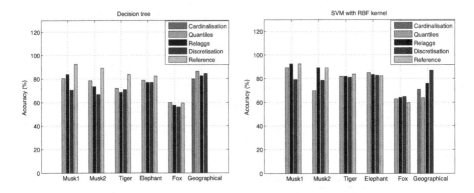

Fig. 4. Accuracy of two attribute-value learners on real data

more than 1600 features on those datasets. So, cardinalisation was only run on the geographical data.

The accuracies of the two standard attribute-value learners with respect to the four propositionalisation approaches are reported in Figure 4 and compared to the accuracy, denoted Reference, of the original dedicated multi-instance learning algorithms for the musk [6] and image [2] datasets.

First of all, we observe that the accuracies vary with respect to the propositionalisation and to the attribute-value learner, in a similar way to the results on the artificial data. No single propositionalisation and attribute-value learner wins on all datasets. On the opposite, each propositionalisation leads to the maximal accuracy on some dataset: quantiles on all datasets but Musk1 for J48, and on Tiger and Elephant for the SVM, simple aggregates of Relaggs on Musk1 and Musk2 for the SVM, and the agregationAfterDiscretisation on Fox and Geographical for the SVM. Indeed, the different propositionalisation approaches provide different biases for different attribute-value learners on different domains.

Moreover, the best accuracies of two standard attribute-value learners with those propositionalisation approaches reach, and in some cases outperform, the accuracies of the original dedicated learning algorithms.

Finally, we notice that the quantiles enable the attribute-value decision tree learner to produce an accurate model and solve the real seven classes geographical problem that motivated this work. The experts preferred the decision tree learner to the support vector machine because the models are easier to interpret [19].

5   Conclusion

This paper proposed two new propositionalisation techniques dealing with continuous attributes. Experiments on artificial and on real data showed that those propositionalisation techniques offer a different expressivity than existing propositionalisation approaches. This new expressivity can be beneficial on some target problems, according to the used attribute-value learner. Since the accuracy depends on the propositionalisation and on the applied attribute-value learner, we plan in future work to select the best combination on the training set.

The image and musk datasets contain too many columns to generate more than 4 quantiles in a propositionalisation approach. Indeed, propositionalisation is not tractable on all datasets. Therefore another perspective concerns the generation of complex aggregates on-the-fly, *i.e.* in a full-fledged relational data mining system, such as BET [8].

Acknowledgments. The geographical data were prepared by the LIVE laboratory of the University of Strasbourg and the COGIT laboratory of the French National Geographic Institute. Alain Shakour and Jonathan Haehnel contributed to the implementation.

References

1. Alphonse, E., Girschick, T., Buchwald, F., Kramer, S.: A numerical refinement operator based on multi-instance learning. In: Frasconi, P., Lisi, F.A. (eds.) ILP 2010. LNCS (LNAI), vol. 6489, pp. 14–21. Springer, Heidelberg (2011)
2. Andrews, S., Tsochantaridis, I., Hofmann, T.: Support vector machines for multiple-instance learning. In: Becker, S., Thrun, S., Obermayer, K. (eds.) NIPS, pp. 561–568. MIT Press (2002)
3. Anthony, S., Frisch, A.M.: Generating numerical literals during refinement. In: Džeroski, S., Lavrač, N. (eds.) ILP 1997. LNCS (LNAI), vol. 1297, pp. 61–76. Springer, Heidelberg (1997)
4. Blockeel, H., De Raedt, L.: Top-down induction of first-order logical decision trees. Artif. Intell. 101(1-2), 285–297 (1998)
5. Botta, M., Piola, R.: Refining numerical constants in first order logic theories. Mach. Learn. 38(1-2), 109–131 (2000)
6. Dietterich, T.G., Lathrop, R.H., Lozano-Pérez, T.: Solving the multiple instance problem with axis-parallel rectangles. Artif. Intell. 89(1-2), 31–71 (1997)
7. Džeroski, S., Lavrač, N. (eds.): Relational data mining. Springer (2001)
8. Kalgi, S., Gosar, C., Gawde, P., Ramakrishnan, G., Gada, K., Iyer, C., Kiran, T.V.S., Srinivasan, A.: BET: An inductive logic programming workbench. In: Frasconi, P., Lisi, F.A. (eds.) ILP 2010. LNCS (LNAI), vol. 6489, pp. 130–137. Springer, Heidelberg (2011)
9. Knobbe, A.J., de Haas, M., Siebes, A.: Propositionalisation and aggregates. In: De Raedt, L., Siebes, A. (eds.) PKDD 2001. LNCS (LNAI), vol. 2168, pp. 277–288. Springer, Heidelberg (2001)
10. Krogel, M.-A., Wrobel, S.: Transformation-based learning using multirelational aggregation. In: Rouveirol, C., Sebag, M. (eds.) ILP 2001. LNCS (LNAI), vol. 2157, pp. 142–155. Springer, Heidelberg (2001)
11. Krogel, M.A., Wrobel, S.: Facets of aggregation approaches to propositionalization. Work-in-Progress Session of the 13th Int. Conf. on ILP (2003)
12. Kuželka, O., Železný, F.: Hifi: Tractable propositionalization through hierarchical feature construction. In: Železný, F., Lavrač, N. (eds.) Late Breaking Papers, the 18th Int. Conf. on ILP (2008)

13. Kuželka, O., Železný, F.: Block-wise construction of acyclic relational features with monotone irreducibility and relevancy properties. In: Danyluk, A.P., Bottou, L., Littman, M.L. (eds.) ICML. ACM Int. Conf. Proceeding Series, vol. 382, p. 72. ACM (2009)
14. Lachiche, N.: Propositionalization. In: Sammut, C., Webb, G.I. (eds.) Encyclopedia of Machine Learning. Springer (2010)
15. Lahbib, D., Boullé, M., Laurent, D.: Prétraitement supervisé des variables numériques pour la fouille de données multi-tables. In: Lechevallier, Y., Melançon, G., Pinaud, B. (eds.) EGC. Revue des Nouvelles Technologies de l'Information, vol. RNTI-E-23, pp. 501–512. Hermann-Éditions (2012)
16. Lavrač, N., Džeroski, S.: Inductive Logic Programming: Techniques and Applications. Ellis Horwood (1994)
17. Lesbegueries, J., Lachiche, N., Braud, A., Puissant, A., Skupinski, G., Perret, J.: A platform for spatial data labelling in an urban context. In: Bocher, E., Neteler, M. (eds.) Geospatial Free and Open Source Software in the 21st Century. Lecture Notes in Geoinformation and Cartography, pp. 49–61. Springer (2012)
18. Muggleton, S.: Inverse entailment and progol. New Generation Computing 13(3-4), 245–286 (1995)
19. Puissant, A., Skupinski, G., Lachiche, N., Braud, A., Perret, J.: Classification et évolution des tissus urbains à partir de données vectorielles. Revue Internationale de Géomatique 21(4), 513–532 (2011)
20. Quinlan, J.R.: C4.5: Programs for Machine Learning. Morgan Kaufmann (1993)
21. Vens, C., Ramon, J., Blockeel, H.: Refining aggregate conditions in relational learning. In: Fürnkranz, J., Scheffer, T., Spiliopoulou, M. (eds.) PKDD 2006. LNCS (LNAI), vol. 4213, pp. 383–394. Springer, Heidelberg (2006)
22. Witten, I.H., Frank, E.: Data Mining: Practical machine learning tools and techniques, 2nd edn. Morgan Kaufmann (2005)
23. Zelezný, F., Lavrac, N.: Propositionalization-based relational subgroup discovery with RSD. Machine Learning 62(1-2), 33–63 (2006)

Topic Models with Relational Features for Drug Design

Tanveer A. Faruquie[1], Ashwin Srinivasan[2,*], and Ross D. King[3]

[1] IBM Research—India, Block 4-C, Vasant Institutional Area, New Delhi, India
[2] Indraprastha Institute of Information Technology, Delhi (IIIT-D) New Delhi, India
[3] School of Computer Science, University of Manchester, United Kingdom

Abstract. To date, ILP models in drug design have largely focussed on models in first-order logic that relate two- or three-dimensional molecular structure of a potential drug (a ligand) to its activity (for example, inhibition of some protein). In modelling terms: (a) the models have largely been logic-based (although there have been some attempts at probabilistic models); (b) the models have been mostly of a discriminatory nature (they have been mainly used for classification tasks); and (c) data for concepts to be learned are usually provided explicitly: "hidden" or latent concept learning is rare. Each of these aspects imposes certain limitations on the use of such models for drug design. Here, we propose the use of "topic models"—correctly, hierarchical Bayesian models—as a general and powerful modelling technique for drug design. Specifically, we use the feature-construction cabilities of a general-purpose ILP system to incorporate complex relational information into topic models for drug-like molecules. Our main interest in this paper is to describe computational tools to assist the discovery of drugs for malaria. To this end, we describe the construction of topic models using the GlaxoSmithKline Tres Cantos Antimalarial TCAMS dataset. This consists of about 13,000 inhibitors of the 3D7 strain of *P. falciparum* in human erythrocytes, obtained by screening of approximately 2 million compounds. We investigate the discrimination of molecules into groups (for example, "more active" and "less active"). For this task, we present evidence that suggests that when it is important to maximise the detection of molecules with high activity ("hits"), topic-based classifiers may be better than those that operate directly on the feature-space representation of the molecules. Besides the applicability for modelling anti-malarials, an obvious utility of topic-modelling as a technique of reducing the dimensionality of ILP-constructed feature spaces is also apparent.

1 Introduction

Malaria is one of the world's worst diseases. The WHO's *World Malaria Report 2011* estimates that there were about 216 million cases of malaria in 2010 with

---

* Corresponding Author.

F. Riguzzi and F. Železný (Eds.): ILP 2012, LNAI 7842, pp. 45–57, 2013.

about 655,000 fatalities.[1] Despite some decrease in the incidence of malaria over the last decade (about 17% since 2000), the disease remains widespread in tropical regions of Africa, Asia and the Americas. The primary parasite involved in the most severe cases is *Plasmodium falciparum*, and the drugs currently possessing the greatest efficacy against this parasite are artemisinin (empirical formula $C_{15}H_{22}O_5$) and its derivatives. The primary reason for the bioactivity for this group of drugs and the actual molecular targets that they attack is still not completely known [10], but the peroxide (two oxygen atoms bonded together) in a seven-membered ring is believed to play a significant role (see Fig. 1).

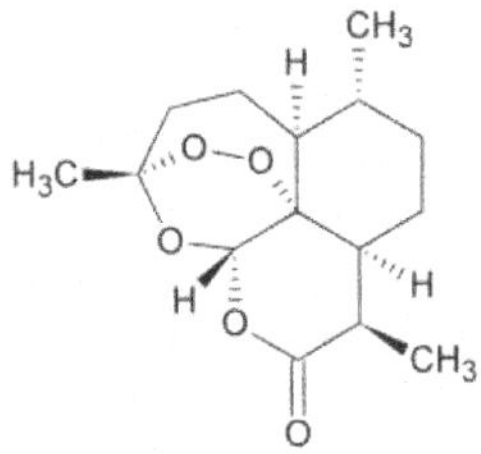

Fig. 1. The molecular structure of artemisinin (from `en.wikipedia.org/wiki/Artemisinin`). Artemisinin and its derivatives are currently the most effective treatment against the malaria parasite *P. falciparum*. The principal structural feature responsible for the activity of the molecule is the peroxide "bridge" in the 7-membered ring. This ring in artemisinin is connected to a lactone ring that contains 2 additional oxygen atoms.

A source of great concern is that a resistance to artemisinins has been reported in a growing number of countries [16]. While this resistance has been controlled to some degree by the use of therapies that combine artemisinins with other drugs, there is now an urgency to develop new anti-malarial drugs that can match the efficacy of artemisinin-based drugs. The development of such drugs is a key recommendation in the WHO's *Global Plan on Artemisinin Resistance Containment* [15]. It is probable that compounds that share significant structural similarities to artemisinin may not be effective against artemisinin-resistant parasites. Developing new anti-malarials therefore necessarily has to account for the presence or absence of certain kinds of structural features, and it is important that any tool assisting this development also be capable of representing such features, and construct models that account for their importance in some principled manner.

Inductive Logic Programming (ILP) systems have shown in the past that they are capable of representing and constructing models using structures such as those shown in Fig. 1 [8]. It is conceivable, with appropriate background knowledge and data, an ILP system could have found the logical expression of the following rule:

---

[1] The Institute for Health Medicine and Evaluation (IHME) at the University of Washington, Seattle (Vogel, G. (2012) How do you count the dead?, Science, 336. 1372-1374.) estimate a death toll twice as large.

A molecule m is active if:
> m has a 7-membered ring r_1 and
> r_1 has a peroxide bridge and
> m has a lactone ring r_2 and
> r_1 and r_2 are connected

To find such a rule an ILP system would need at least general definitions of ring-structures, peroxide bridges, and predicates to decide when one or more rings are connected to each other. It is evident that this particular rule could be used to confirm that the artemisinin molecule is active. More could be done: the rule could be used to identify all molecules in a database that satisfy these constraints. Also, with some slight effort, a "molecular generator" could be constructed that was guaranteed to satisfy these constraints. But there are clear limitations. The rule requires the joint presence of the peroxide bridge and the lactone ring. It is therefore not possible to consider these separately in the search or generation of new molecules, although it might in fact be the case that the two structural features may be responsible for different aspects of the behaviour of artemisinin. That is, if we want to look for molecules that contain at least one of the two structural features:

A molecule m is active if:
> m has a 7-membered ring r_1 and
> r_1 has a peroxide bridge

A molecule m is active if:
> m has a 7-membered ring r_1 and
> m has a lactone ring r_2 and
> r_1 and r_2 are connected

then this cannot be inferred directly from molecules that satisfy the original rule. The rule of inference used by ILP systems also do not allow for the generation of molecules using a stochastic mechanism that "weights" these two features differently.[2] Yet these are all clearly important tasks in the search for new drugs. Therefore what we would like is to combine the usual advantages of an ILP system: the use of background knowledge and the discovery of complex features first-order logic—with that of modelling technique that discovers automatically partitions on the features to represent sub-concepts, and allows the search and generation of new molecules that uses these concepts. We propose the probabilistic setting of hierarchical Bayesian modelling, sometimes called "topic" models, as the appropriate one for fulfilling the modelling requirements outlined here.

2 Topic Models for Molecules

Topic modelling, originally used in the analysis of text documents, is concerned with three principal entities: documents, topics, and words. Documents consist of one or more topics, which in turn consist of one or more words. We adapt the original topic modelling approach to modelling molecules in the following way.

---

[2] Although it is conceivable that recent work on probabilistic ILP, or PILP, systems will allow this.

Molecules ("documents") will be taken to consist of one or more concepts ("topics"). Examples of concepts relevant to drug-design are "activity" and "toxicity". Concepts will be taken to consist of one or more features ("words", although more like phrases in the molecular setting). For example, an active molecule may consist of the following features: the presence of a 7-membered ring, the presence of a lactone ring connected to a 7-membered ring, and the presence of a peroxide bridge in a 7-membered ring. It is understood that some mechanism exists for deciding whether such features are present or absent in any molecule. The question of how the features are to be obtained in the first place is discussed in the next section.

Tools for constructing topic models for molecules accept as input feature-vector descriptions of the molecules. In this paper, we will assume these descriptions to be Boolean vectors, in which a "1" signifies the presence of the corresponding feature in a molecule, and the phrase "molecule m has feature f" should be taken to mean that feature f has the value 1 for molecule m. The output is a set of concepts, each characterised by a probability distribution over the features. Each molecule in turn is characterised by a probability distribution over the concepts. We elaborate on these further, using a standard technique for topic modelling (Latent Dirichlet Allocation, or LDA [2]).

Suppose there is a set of features $\mathcal{F}$, and a set of molecules $\mathcal{M}$. Molecule m has a set features that is a subset of $\mathcal{F}$. We will take $|\mathcal{F}| = V$ and $|\mathcal{M}| = M$. We further assume that the the total number of concepts is K (usually, K is much smaller than V). Then, the LDA model asssumes the following:

1. Associated with each molecule m is a multinomial distribution over the entire set of K concepts. The parameter θ_m of this distribution gives the probability of observing the concepts for the molecule m. It is assumed that the molecule-specific concept-distributions are drawn using some prior distribution over multinomial parameters. It is convenient to assume that this prior distribution is a (K-dimensional) Dirichlet distribution with parameter α.

2. Similarly, associated with each concept k is a multinomial distribution over the entire set of V features. The parameter ϕ_k of this gives the probability of observing the features for concept k. Again, it is assumed that these feature-specific distributions are in fact drawn using some prior distribution over V-dimensional multinomial parameters. A mathematically convenient prior distribution for the parameters is a (V-dimensional) Dirichlet distribution with parameter β.

Once the prior distributions, and the multinomials ϕ and θ are fully specified, it is possible to compute the (posterior) probability of observing the feature-values of any molecule. The difficulty however is this: the only observables are the molecule-specific feature values. All other quantities are hidden or latent. Estimating the parameters of the distributions from the observable data is a problem of Bayesian inference.

Once parameters have been estimated the probabilistic model can be used in a number of different ways. Given any molecule it can compute the probability of each concept. The data are thus automatically reduced from the original

V-dimensional representation to a smaller K-dimensional one. This reduced representation can then be used to construct models that classify molecules into one of several known categories, or to retrieve molecules that have similar concept-probabilities to a known ligand. More interesting to a synthetic chemist is that the model can be used to generate fragments of new molecules by drawing features from the underlying probability distributions. This can be done using the stochastic process that defines the generation of the molecule and features. To generate a new molecule, its composition is determined using the "concepts". The concepts are generated by sampling a multinomial distribution from a Dirichlet distribution. Alternatively, the concept distribution can be fixed to a multinomial distribution based on the desired composition. The features are generated by sampling the concepts and then sampling the feature from the multinomial distribution corresponding to the selected concept.

2.1 Relational Features for Molecules

The probabilistic modelling technique requires that the objects being modelled be represented by feature-vectors. When modelling documents in a natural language, an obvious choice for features is the counts of words from a vocabulary. What about objects like molecules? The "characters" making up these objects are the usual chemical atoms (along perhaps with some "punctuation" symbols, to denote some special bonds and so on). These are connected together by bonds of various kinds to form various cyclic structures (rings), functional groups and the like. However a representation of molecules that simply counts certain kinds of atoms, rings, or functional groups is unlikely to be useful. A key assumption in the structure-activity modelling is that it is the relationships amongst these structures that gives rise to its chemical activity. Thus, it is not just that molecule has 3 benzene rings, but that it has 3 fused benzene rings that is actually important. This suggests a representation in which features are relations constructed from the basic molecular structural building blocks. This is somewhat akin to representing documents by phrases rather than words, and just as there may be a very large number of possible phrases, there may be a very large number of such relational features for molecules. A significant effort has been invested in ILP on how to obtain features for molecular data [11,13]. We will not reproduce these techniques here, but simply show in Fig. 2 the process used to obtain structural fragments from a set of molecules using an ILP engine.

3   Empirical Evaluation

We intend to investigate the use of topic models for drug-design by using tasks relevant to the discovery of new anti-malarials. Within the broad area of structure-activity relations, we consider the discrimination of molecules into categories (like more or less active), based on their two-dimensional structure and some bulk properties (molecular weight and hydrophobicity). Specifically, we investigate the comparative performance of a classifier that uses the probabilities computed by an

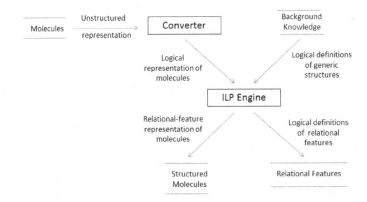

Fig. 2. The process of extracting relational features for molecules from data using an ILP engine. The ILP engine can use any of the "propositionalisation" techniques reviewed in [11]: the the specific one employed here is in the "Materials" section. The database of "structured molecules" will be used to construct topic models.

LDA model using relational features against some standard classifiers operating directly on the feature-space.

3.1   Materials

Data. The Tres Cantos Antimalarial TCAMS dataset is available at the ChEMBL Neglected Tropical Disease archive (www.ebi.ac.uk/chemblntd). This archive provides free access to screening and chemical data relevant to a number of tropical diseases affecting countries in Asia, Africa and the Americas. The TCAMS database is a result of screening GlaxoSmithKline's library of approximately 2 million compounds. The database consists of 13,000 of the chemicals that were found, on screening, to inhibit significantly the growth of the 3D7 strain of *P. falciparum* in human erythrocytes [5]. Data made available include some bulk-properties of the compounds (like molecular weight and hydrophobicity), along with the SMILES representation of the structure of the molecules. The distribution of inhibition activities of the molecules is shown in Fig. 3

Background Knowledge. The ILP engine uses definitions of a number of standard cyclic structures and functional groups that have been used in structure-activity applications before [8,9]. For reasons of space, we do not reproduce these here.

Algorithms and Machines. The algorithms used are the following:

1. The conversion of the representation of molecules from their SMILES representation to a logical form is done by a combination of Open Babel [3], that converts from SMILES to a "mol2" format, and a converter that converts the "mol2" representation to a logical form.

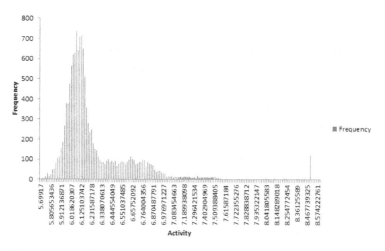

Fig. 3. Distribution of inhibition activity of the 3D7 strain of *P. falciparum* by molecules in the TCAMS database (pXC50_3D7 in the data).

2. The ILP engine used to generate relational features from the logical representation of the molecules is Aleph [12].
3. The topic models are constructed by a package provided with the R statistical package [6].
4. All classification is done by classifiers within the WEKA data mining software [7].

The topic models were constructed on a 2 GHz processor 4 core laptop with 4GB main memory running Mac OS v10.6.8. The ILP engine and classification were performed on a Pentium Core i5 laptop with access to 4GB of main memory running Fedora 13, and using the Yap Prolog engine.

3.2 Method

Our method is straightforward:

1. Two-dimensional structure information in terms of a logical representation of their atom and bond structure is obtained for the 13,403 molecules using the converters.
2. Partition the data into "training" and "test" subsets (the former for constructing models and the latter for testing them).
3. With the training data:
    (a) The ILP engine is used to construct relational features using the logical representation of the molecules and the background knowlege provided.
    (b) Each molecule is converted into a feature-vector representation in which an entry of "1" denotes that the corresponding relational feature is present in the molecule. This will be called the "feature-space representation" of the training data.

(c) The topic-modeller is given molecules in the feature-vector representation to obtain a topic-model with exactly K topics

(d) Molecules are re-expressed as K-dimensional probability vectors, in which the ith entry for molecule m represents the probability that m contains topic i. This will be called the "topic-space representation of the training data"

4. With the test data:
   (a) Using the features constructed with the training data, obtain the feature-space representation of the molecules.
   (b) With the topic model constructed with the training data, obtain the topic-space representation of the molecules.
5. Obtain classification labels for all the molecules.
6. With both training and test data:
   (a) Add label information to the feature-space representation. We will call this the "labelled feature-space representation" of the data.
   (b) Add label information to the topic-space representation. We will call this the "labelled topic-space representation" of the data.
7. Construct the best classifier possible with the labelled feature-space representation of the training data. We will call this the "feature-based model": let its performance on the corresponding test data be P_F.
8. Construct the best classifier possible with the labelled topic-space representation of the training data. We will call this the "topic-based model": let its performance on the corresponding test data be P_T.
9. Compare P_F and P_T.

The following details are relevant:

1. 30% of the total number of molecules are set aside for testing the models obtained. This results in 9382 molecules in the training set and 4021 molecules in the test set.
2. The ILP engine Aleph is used to obtain features. We restrict these to be from at most $10,000$ definite clauses with no more than 5 literals. The features are constructed using Aleph's random-search strategy, independent of each other, and without access to any class labels. The minimum number of molecules to be "covered" by a feature is set very low to allow a large number of frequent features to be found.
3. There is no theoretically optimal number of topics. We construct models with $K = 10, 25$ and 50 topics. We note also that if classification is the only task of interest, then better variants of topic models exist. We would expect that an approach similar to that described in [14] would be better than the unsupervised topic-modelling we perform here.
4. In the classification task addressed here, we attempt to discriminate between molecules in the right-hand tail of the distribution in Fig.3. By "right-hand tail" we will mean the molecules in the 85-percentile or above, which translates to an activity threshold of approximately 6.6. We call these molecules "highly active", and the rest simply "active". The details of the resulting datasets are as follows:

Class	Train	Test
Highly Active	1486	666
Active	7896	3355
Total	9382	4021

5. All molecules in the data are inhibitors of *P. falciparum*. The focus on the right-hand tail results in a substantially skewed class distribution. There are differential costs associated with misclassification. Specifically, the cost of missing "highly active" molecules is substantially greater. Our experience suggests 10-fold difference. That is, denoting "highly active" as positives for learning and the rest as negatives, the cost of false negatives is likely to be 10 times that of false positives. A more complete picture is obtained by considering a distribution of cost values: in [1] a triangular distribution with a mode at the likely value is suggested, using a re-scaled version of costs in which the cost of false negatives (c_1 in [1]) is in the range $0 \ldots 1$ and the sum of costs of false negatives and false positives (c_0) is 1. We will consider a variation of cost ratios from 1:1 to 1:20. That is, c_0/c_1 is in the range $[0.05, 1]$ with the most likely value corresponding to $c_1 = 0.91$ (that is, a cost ratio of 1:10). Following [1], the distribution of c_1 values is taken to be a triangle with end-points at $c_1 = 0.67$ and $c_1 = 0.95$, with a peak at $c_1 = 0.91$. The height of the triangle is $2/(0.95 - 0.67) = 7.12$ (obtained so that the area of the triangle is 1). We will use this distribution to compute the LC-index.

6. The classifiers used here are all implementations in the WEKA toolbox. We examine techniques that span a range of different ways of constructing a discrimination function. These are: decision-tree classifiers; classifiers based on posterior probability estimates (naive Bayes and logistic regression); thresholds on linear functions using weights on features or topics (Winnow); perceptron-based classification; and support-vector machines. A meta-classifier is used to minimise expected misclassification cost. This requires posterior probability estimates from the classifiers, and it is known that heavily pruned trees give poor estimates of such probabilities [4]. We therefore use unpruned trees, with a Laplacian smoothing on the posterior probabilities. For Naive Bayes, we reiterate that feature-construction by the ILP engine is performed randomly, without using information from other features or the class label. We would expect conditional independence assumptions of naive Bayes to hold when using the feature-based representation. The extent to which the topic-space attributes satisfy these assumptions is however less clear. The Winnow algorithm requires Boolean feature vectors, which poses a difficulty when using data produced by the topic models. To convert the probabilities produced by the topic models into Boolean values all "significant topics" are set to the value 1. These are topics such that their sum covers atleast 80% of the probability mass.

3.3   Results

The ILP engine identifies about 2100 features. The results obtained using topic models derived from these features are shown in Fig. 4. We note that for all the classification models shown, the best topic model exhibits superior performance as the cost of false negatives approaches the likely value of 10, or is greater. In some cases (for example, Naive Bayes or Logistic Regression), topic models are better across the full range of costs implied by the cost-distribution.

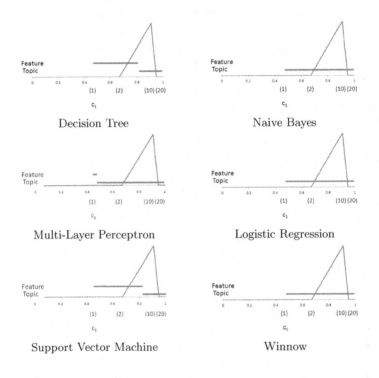

Fig. 4. A cost distribution and the performance of classificatory models in feature- and topic-space. In each case, "Feature" represents the classifier model in the feature-space, and "Topic" represents the model in the topic-space. For simplicity, the latter only shows the best model obtained with 10, 25 and 50 topics (the "best" model is taken to be one with the highest LC-index: see text for details). For the Multi-layer Perceptron and Support Vector Machine models, feature-selection was required to allow construction of models in feature-space (again, see text for details). The horizontal axis (c_1) is the cost of false negatives, re-scaled to the interval $[0, 1]$ (the number in parentheses gives the actual costs of some points). The mode of the triangular distribution is at $c_1 = 0.91$ (unscaled value = 10). The bold segments against each classifier denotes the range of c_1 values for which that model has lower expected cost.

A quantitative measure of comparative performance can be obtained using the LC-index.[3] Figure 5 tabulates the LC-index obtained by pairwise comparison of the different kinds of topic models against the corresponding feature-based model. We note here that it was not possible to obtain feature-based models with the multi-layer perceptron or the SVM implementation using all the features. The comparison here is therefore restricted to a comparison against a model using the 50 best features (selected using an information-gain criterion). This tabulation suggests that building discriminatory models in topic-space is better than building discriminatory models in feature-space.

Topic Model	Feature-based Model					
	Tree	NBayes	Percept	Logit	SVM	Winnow
10-Topic	−0.71	−1.00	−1.00	−1.00	−0.49	−1.00
25-Topic	−0.42	−1.00	−1.00	−1.00	−0.49	−1.00
50-Topic	+1.00	−1.00	−0.89	−1.00	−0.49	−1.00

Fig. 5. A comparison of classifier models constructed using features and topics. The comparison tabulates the LC-index [1] obtained by comparing the performance of a feature-based model against a topic-based model, across a range of costs. A positive value of the index means that the feature-based model performs better than the topic model. A negative value means that the topic model is better. Index values lie between −1 and +1, and the proximity of a number to +1 or −1 is a quantitative indicator of how much better or worse the feature-based model performs than its topic-model counterpart. For reasons of efficiency, for "Percept" and "SVM" LC-indices are computed only against a model that uses the 50 best features. In all other cases, all features found by the ILP system were used.

There are some additional sanity checks that are needed before we are able to claim good reasons to use topic models. First, the features constructed by the ILP system need to be conditionally independent, given the molecule-specific multinomial. While the ILP system was not provided with this constraint *a priori*, we have verified that the features are at least largely dis-similar to each other using a simple Jaccard-index calculation. Second, are topic-models simply performing a kind of feature selection? A comparative evaluation of topic models against models obtained with standard feature-selection methods could provide an answer to this question. In our experiments, we have found that the expected cost of a classifier constructed with K topics is almost always lower than that of a classifier constructed with the K best features (selected using an information-

---

[3] The LC-index ranges between −1 and +1, and is a (distribution-)weighted summation of the value of a function $L(c_1)$ that takes the value +1 when the feature-based model is better than the topic-model being compared against. A positive LC-index indicates the feature-based model performs better (the closer to +1, the better it is): see [1] for more details.

gain strategy). The reader would correctly point out that this result may change with a different cost distribution; or a different feature-selection method.

The quantitative results appear to suggest that 10-topic models perform best across all classification methods. However, we find that models with 25-topics are more informative: models with 10-topics often have low expected costs by predicting all molecules to be highly active. The 25-topic models in contrast are less trivial.

4   Concluding Remarks

In this paper, we have introduced the application of hierarchical Bayesian models—commonly called topic models—that use relational features for tasks in drug-design. Our goals are specifically to assist in the discovery of antimalarials, and in this paper we have focussed on discriminating amongst molecules with two different qualitative levels of activity. The results obtained are promising: if the costs of missing "hits" are high, then the use of topic-based modelling appears to perform better than the usual approach adopted in ILP-based propositionalisation (that is: constructing models over a space of features identified by an ILP engine). The use of ILP to construct relational features for molecules retains some important aspects of those systems namely: the use of background knowledge and the benefits of a first-order representation to capture complex structural features. We believe that combining this with hierarchical probabilistic modelling allows us further advantages in terms of discovery of hidden concepts (with some connection to the invention of new predicates in ILP), handling uncertainty, and providing a generative model for molecules. In the future, we propose to extend the approach presented here to investigating the use of this form of hybrid modelling for molecule retrieval and synthesis. A shortcoming of the approach should already be evident, namely, the need to specify the appropriate number of topics beforehand. Our results suggest that irrespective of the classification method, a 10- or 25-topic model performs better than using all the features. In practice, we would usually not know this, and we may be forced to use techniques that determine the number of topics automatically. One possibility is the more sophisticated technique of non-parameteric hierarchical Bayesian modelling, in which the number of topics is determined automatically from the data.

Acknowledgements. A.S. holds a Ramanujan Fellowship from the Government of India. He is a Visiting Professor at the Department of Computing Science, University of Oxford; and a Visiting Professorial Fellow at the University of New South Wales, Sydney.

References

1. Adams, N.M., Hand, D.J.: Comparing classifiers when misallocation costs are uncertain. Pattern Recognition 32, 1139–1147 (1999)

2. Blei, D.M., Ng, A.Y., Jordan, M.I.: Latent dirichlet allocation. Journal of Machine Learning Research 3, 993–1022 (2003)
3. Boyle, N.M., Banck, M., James, C.A., Morley, C., Vandermeersch, T., Hutchison, D.R.: Open Babel: an open chemical toolbox. Chemoinformatics 3, 33 (2011), http://www.openbabel.org
4. Elkan, C.: The foundations of cost-sensitive learning. In: Proceedings of the Seventeenth International Joint Conference on Artificial Intelligence, pp. 973–978 (2001)
5. Gamo, F., Sanz, L.M., Vidal, J., de Cozar, C., Alvarez, E., Lavandera, J., Vanderwall, D.E., Green, D.V.S., Kumar, V., Hasan, S., Brown, J.R., Peishoff, C.E., Cardon, L.R., Garcia-Bustos, J.F.: Thousands of chemical starting points for antimalarial lead identification. Nature 465(7296), 305–310 (2010)
6. Grun, B., Hornik, K.: topicmodels: An R Package for fitting Topic Models. Journal of Statistical Software 40(13), 1–30 (2011)
7. Hall, M., Frank, E., Holmes, G., Pfahringer, B., Reutemann, P., Witten, I.H.: The WEKA Data Mining Software: An Update. SIGKDD Explorations 11(1) (2009)
8. King, R.D., Muggleton, S.H., Srinivasan, A., Sternberg, M.J.E.: Structure-activity relationships derived by machine learning: The use of atoms and their bond connectivities to predict mutagenicity by inductive logic programming. Proc. of the National Academy of Sciences 93, 438–442 (1996)
9. King, R.D., Srinivasan, A.: Prediction of rodent carcinogenicity bioassays from molecular structure using inductive logic programming. Environmental Health Perspectives 104(5), 1031–1040 (1996)
10. O'Neill, P.M., Barton, V.E., Ward, S.A.: The Molecular Mechanism of Action of Artemisinin—The Debate Continues. Molecules 15, 1705–1721 (2010)
11. Kramer, S., Lavrac, N., Flach, P.: Propositionalization approaches to relational data mining. In: Relational Data Mining, pp. 262–286. Springer, New York (2001)
12. Srinivasan, A.: The Aleph Manual (1999), http://www.comlab.ox.ac.uk/oucl/research/areas/machlearn/Aleph/
13. Srinivasan, A., King, R.D.: Feature construction with Inductive Logic Programming: a study of quantitative predictions of biological activity aided by structural attributes. In: Muggleton, S. (ed.) ILP 1996. LNCS (LNAI), vol. 1314, pp. 89–104. Springer, Heidelberg (1997)
14. Taranto, C., Di Mauro, N., Esposito, F.: rsLDA: A Bayesian heirarchical model for relational learning. In: ICDKE, pp. 68–74 (2011)
15. WHO. Global plan for artemisinin resistance containment, GPARC (2011), http://www.who.int/malaria/publications/atoz/9789241500838/en/index.html
16. WHO. World Malaria Report 2011 (2011), http://www.who.int/malaria/world_malaria_report_2011/en/

Pairwise Markov Logic

Daan Fierens[1], Kristian Kersting[2,3], Jesse Davis[1], Jian Chen[1],
and Martin Mladenov[3]

[1] Dept. of Computer Science, KULeuven, Belgium
fierens.daan@gmail.com,
jesse.davis@cs.kuleuven.be
[2] Fraunhofer IAIS, Germany
kristian.kersting@iais.fraunhofer.de
[3] University of Bonn, Germany
mladenov@igg.uni-bonn.de

Abstract. For many tasks in fields like computer vision, computational
biology and information extraction, popular probabilistic inference methods have been devised mainly for propositional models that contain only
unary and pairwise clique potentials. In contrast, statistical relational
approaches typically do not restrict a model's representational power
and use high-order potentials to capture the rich structure of relational
domains. This paper aims to bring both worlds closer together.

We introduce pairwise Markov Logic, a subset of Markov Logic where
each formula contains at most two atoms. We show that every non-
pairwise Markov Logic Network (MLN) can be transformed or 'reduced'
to a pairwise MLN. Thus, existing, highly efficient probabilistic inference
methods can be employed for pairwise MLNs without the overhead of
devising or implementing high-order variants. Experiments on two relational datasets confirm the usefulness of this reduction approach.

1 Introduction

In the probabilistic graphical models literature, many inference algorithms have
been designed specifically for *low-order models*. The term *order* is used here
in the probabilistic sense (as opposed to the logical sense of 'first-order' logic).
Concretely, the order of a factor graph is defined as the maximum number of
arguments of a factor [1]; the order of a Markov Random Field (MRF) is the
size of the largest clique. A probabilistic model is called *pairwise* if its order is
(at most) two.

Pairwise graphical models first became popular in the field of statistical physics
and are nowadays used for many applications, in diverse fields like computer
vision, computational biology and information extraction [1]. The typical 'pairwise approach' is to write a probabilistic model with unary and pairwise clique
potentials, understand the Bayesian priors it incorporates, and then perform
inference [2]. This is especially common in the case of *MAP inference*. MAP
is the task of, given the observed state of some random variables, finding the

F. Riguzzi and F. Železný (Eds.): ILP 2012, LNAI 7842, pp. 58–73, 2013.
© Springer-Verlag Berlin Heidelberg 2013

most likely state of all other random variables occurring in the model.[1] Many state-of-the-art methods for MAP inference were mainly developed for pairwise models. High-order variants of such methods (capable of handling non-pairwise models) often do not exist and if they do, they are typically more complex and lack implementations. This is particularly true for MAP methods based on Linear Programming [3], Quadratic Programming [4], graph cuts [5,6], etc. While most of these methods can in principle work on non-pairwise models, pairwise models receive the most attention because this facilitates implementation and theoretical analysis (e.g., convergence analysis for iterative methods like belief propagation [3]). Table 1 provides an overview of some inference methods and their ability to handle non-pairwise models.

Table 1. An (incomplete) overview of MAP inference methods and their support for non-pairwise models

Method	Supports non-pairwise models
Max Product Variable Elimination	✓
Max Product Belief Propagation	✓
Linear programming methods [3]	✓/−
Graph cut methods [6] (QPBO [5], ...)	−
Quadratic programming [4]	−
...	

In statistical relational learning (SRL), the situation is different. Typical SRL approaches, like Markov Logic [7], do not restrict a model's representational power and use high-order potentials to capture the rich structure of relational domains. The notion of pairwise models has so far not been considered in the SRL literature. With this paper, we aim to bridge this gap between SRL and probabilistic graphical models. To this end, we introduce *pairwise Markov logic*. We show that any high-order Markov logic network can be reduced to a model in pairwise Markov logic, where we can then bring to bear the powerful inference techniques for low-order models. Section 3 explains the reduction for propositional MLNs and Section 4 explains the lifted reduction for first-order MLNs. Section 5 empirically demonstrates the utilty of the reduction approach for both ground and lifted inference on some common relational datasets.

2   Pairwise Markov Logic

A Markov Logic Network (MLN) is a set of pairs (φ_i, w_i), where φ_i is a formula in first-order logic and w_i a real-valued weight. Following the propositional case, we define the *order* of an MLN as the maximum number of atoms in a formula,

---

[1] This is sometimes also called *MPE (Most Probable Explanation)* inference, or *full MAP* inference (to distinguish it from *partial MAP*, in which some of the random variables need to be summed out).

i.e., the maximum 'length' of a formula in the MLN. We call an MLN (or a single MLN formula) *pairwise* if its order is two. For instance, the formula $Smokes(x) \Rightarrow Asthma(x)$ is pairwise, but $Friends(x, y) \Rightarrow (Smokes(x) \Leftrightarrow Smokes(y))$ is not. We call the latter a *triplewise* formula since it has length three.

Pairwise MLNs have advantages for both ground and lifted inference. Ground inference typically applies methods from the graphical models literature, many of which focus on pairwise models. A similar argument holds for lifted inference, where some recent work uses graph-theoretical notions that assume that the network is pairwise [8] (as it can then be represented as a simple graph rather than a hypergraph).

Despite the advantages of pairwise MLNs, one should not discard non-pairwise MLNs. It is difficult, if not impossible, to capture the rich structure of many relational domains using a pairwise model. For example, when performing structure learning, restricting the hypothesis space to pairwise MLNs would simply ignore too many relevant patterns in relational datasets. Many typical relational patterns require triplewise formulas. Common examples are found in collective classification (e.g., $Class(x, c) \wedge Link(x, y) \Rightarrow Class(y, c)$), in link prediction (e.g., $Property(x) \wedge Property(y) \Rightarrow Link(x, y)$), in social networks (e.g., the above Smokes formula), etc.

In summary: we would like to use non-pairwise MLNs during modelling and learning but use only pairwise MLNs during inference. In the graphical models literature, this apparent contradiction is typically solved by *reducing* the non-pairwise model to an equivalent pairwise model when performing inference. In this paper we (for the first time) show that this can also be done for Markov Logic: any MLN can be reduced to a pairwise MLN.

3   Reduction for Propositional MLNs

A propositional MLN is a set of pairs (φ_i, w_i) where φ_i is a propositional logic formula (using connectives $\neg, \wedge, \vee, \Rightarrow$ and $\Leftrightarrow$) and $w_i \in \mathbb{R}$. A propositional MLN defines a probability distribution on the set of possible worlds (interpretations): the probability of a world ω is $P(\omega) = \frac{1}{Z} exp(\sum_i w_i \delta_i(\omega))$, with Z a normalization constant and $\delta_i(\omega)$ the indicator function being 1 if formula φ_i is true in world ω and 0 otherwise.

Below we show how to reduce a given propositional non-pairwise MLN to a pairwise MLN. In graphical models, reduction is typically done by converting the model (or its energy function) to a pseudo-Boolean function or multi-linear polynomial [1]. While our reduction for propositional MLNs can also be phrased in this terminology, we instead chose for a more self-contained formulation, referring only to Markov Logic. This will allow us to extend our approach from propositional MLNs to first-order MLNs (lifted reduction, Section 4).

3.1   Outline of the Reduction Algorithm

The reduction is done by means of a rewriting process that modifies the MLN knowledge base, i.e., the set of weighted formulas. This is a two-step process.

The first step is an enabling step that brings the MLN into a certain normal form. The second step does the actual reduction to pairwise form.

Algorithm 1 provides an outline of the procedure, which we explain in detail in the following sections. We will also discuss in what sense the obtained pairwise MLN is 'equivalent' to the original MLN. To simplify the discussion, we first show how to reduce *triplewise* MLNs (i.e., with maximum formula length 3). Section 3.5 explains how to reduce MLNs with order higher than 3.

3.2 Step 1: Write Triplewise Formulas in Positive Normal Form (PNF)

Since we for now only consider triplewise MLNs, each formula in the MLN is either triplewise, pairwise or unary. The goal of Step 1 is to bring all triplewise formulas into what we call positive normal form.[2] This will simplify the actual reduction to pairwise form in Step 2. A formula is in *positive normal form* or *PNF* if it is a conjunction of atoms (not involving negation), e.g., $P \wedge Q \wedge R$.

Step 1 loops over all weighted formulas (φ, w) in the given MLN M. If φ is triplewise and not in PNF, we execute an *atomic rewriting step* on φ.

Atomic Rewriting Step. An atomic rewriting step executed on a non-PNF triplewise formula φ with weight w removes φ from the MLN and replaces it by a set of equivalent PNF formulas. In Algorithm 1a, this is denoted by the line $M := M \cup RewriteToPNF(\varphi, w) \setminus \{(\varphi, w)\}$. To do so, we call the function $RewriteToPNF(\varphi, w)$ (see Algorithm 1b) which returns an equivalent set of seven weighted PNF formulas that we add in place of φ in the MLN. The equations for the weights of these formulas, w_1 to w_7, depend on the truth table of φ, which is denoted $v_{...}$ in Algorithm 1b. Concretely, let P, Q and R be the atoms in φ, then v_{pqr} denotes the Boolean truth value (1 or 0) of φ under the interpretation $P = p$, $Q = q$ and $R = r$. For instance, v_{ttf} is the truth value of φ when P and Q are true and R is false. In general the output of $RewriteToPNF()$ consists of seven formulas, but in practice it is often the case that the weights of some of these formulas equal zero, so they can be omitted.

Example. Consider the triplewise non-PNF formula $P \wedge Q \Rightarrow R$ with weight w. This formula is satisfied unless P and Q are true and R is false. Hence, the truth table of this formula is: v_{ttf} equals 0, all other $v_{...}$'s equal 1. Applying the function $RewriteToPNF()$ to this formula, we get an equivalent set of two PNF formulas. The first formula is $P \wedge Q \wedge R$ with weight w (since $w_1 = w(1 - 0 - 1 + 1 - 1 + 1 + 1 - 1) = w$). The second formula is $P \wedge Q$ with weight $-w$ (since $w_2 = w(0 - 1 - 1 + 1) = -w$). The five other formulas get weight zero and hence can be omitted.

Equivalence of MLNs. Let M_0 be the original MLN with triplewise formulas and M_1 be the MLN obtained after applying Step 1 to M_0. M_0 and M_1 are *equivalent* in the sense that they determine the same probability distribution over possible worlds. To prove this, we show that executing one atomic rewriting

---

[2] Our use of the term positive normal form is unrelated to its use in other fields.

Algorithm 1. The algorithm for reducing a triplewise MLN: **1a)** outer loop, **1b)** Step 1, writing to positive normal form or 'PNF', **1c)** Step 2, reduction to pairwise form.
Lines in gray do not apply to the propositional reduction but only to the lifted reduction for first-order MLNs.

1a)
> **procedure** *ReduceMLN*(M)
> **in:** a triplewise MLN M
>
> // Step 1: rewrite to PNF
> **for each** triplewise weighted formula $(\varphi, w) \in M$ that is not in PNF
> $M := M \cup RewriteToPNF(\varphi, w) \setminus \{(\varphi, w)\}$ // an atomic rewriting step
>
> // Step 2: reduce to pairwise
> **for each** triplewise weighted (PNF) formula $(\varphi, w) \in M$
> $M := M \cup ReduceToPairwise(\varphi, w) \setminus \{(\varphi, w)\}$ // an atomic reduction step

---

1b)
> **function** *RewriteToPNF*(φ, w)
> **in:** a triplewise formula φ with weight w
>   let P, Q and R denote the atoms in φ
>   let $v_{...}$ denote the truth table of φ (see text)
>
> **return** { $(w_1, P \wedge Q \wedge R)$, $(w_2, P \wedge Q)$, $(w_3, P \wedge R)$,
> $(w_4, Q \wedge R)$, (w_5, P), (w_6, Q), (w_7, R) }
>   with
>
> $$w_1 = w(v_{ttt} - v_{ttf} - v_{tft} + v_{tff} - v_{ftt} + v_{ftf} + v_{fft} - v_{fff})$$
> $$w_2 = w(v_{ttf} - v_{tff} - v_{ftf} + v_{fff})$$
> $$w_3 = w(v_{tft} - v_{tff} - v_{fft} + v_{fff})$$
> $$w_4 = w(v_{ftt} - v_{ftf} - v_{fft} + v_{fff})$$
> $$w_5 = w(v_{tff} - v_{fff})$$
> $$w_6 = w(v_{ftf} - v_{fff})$$
> $$w_7 = w(v_{fft} - v_{fff})$$
>
> For the 2nd up to 7th formula (formulas with weights $w_2, \ldots, w_7$):
> multiply the weight with a correction factor for lost logvars (see text)

1c)
> **function** *ReduceToPairwise*(φ, w)
> **in:** a triplewise PNF formula φ of the form $P \wedge Q \wedge R$ with weight w
>
> Introduce a <u>new</u> auxiliary atom A Logvars of A:
>   union of all logvars in P, Q and R
> **if** $w > 0$
>   **return** { $(w, A \wedge P)$, $(w, A \wedge Q)$, $(w, A \wedge R)$, $(-2w, A)$ }
> **else**
>   **return** { $(w, P \wedge Q)$, $(w, P \wedge R)$, $(w, Q \wedge R)$,
> $(-w, A \wedge P)$, $(-w, A \wedge Q)$, $(-w, A \wedge R)$, (w, A) }
>   For the first three formulas ($P \wedge Q$, $P \wedge R$ and $Q \wedge R$):
>   multiply the weight with a correction factor for lost logvars (see text)

step preserves the distribution of the MLN. As Step 1 is just a sequence of atomic rewriting steps, it follows by transitivity of equality that M_1 is equivalent to M_0.

Recall that the probability of a world ω is $P(\omega) = \frac{1}{Z}exp(\sum_i w_i \delta_i(\omega))$ where $\sum_i w_i \delta_i(\omega)$ is the sum of weights of satisfied formulas in the world ω and Z is the normalization constant, i.e., $Z = \sum_\omega exp(\sum_i w_i \delta_i(\omega))$. Hence, any rewriting step that leaves the *sum of weights of satisfied formulas* or *SWSF* of every world unchanged, preserves the distribution of the MLN. In fact, this is also the case if the rewriting step adds a constant number to the SWSF of every world, since this constant will disappear into the normalization constant, i.e., the resulting MLN will have a different normalization constant but will still define the same distribution over possible worlds.

An atomic rewriting step only has a *local* effect: it replaces one triplewise non-PNF formula φ, involving only three atoms P, Q and R, by a set of at most seven new formulas involving P, Q and R. Because of this locality, the SWSF of each world remains largely unchanged under an atomic rewriting step: we only need to show that the contribution of the seven new formulas to the SWSF in each world is the same (up to a constant) as the contribution of φ. This can be seen from the following table over the possible worlds of P, Q and R.

P Q R	φ	new formulas
t t t	$w\ v_{ttt}$	$w(v_{ttt} - v_{fff})$
t t f	$w\ v_{ttf}$	$w(v_{ttf} - v_{fff})$
t f t	$w\ v_{tft}$	$w(v_{tft} - v_{fff})$
t f f	$w\ v_{tff}$	$w(v_{tff} - v_{fff})$
f t t	$w\ v_{ftt}$	$w(v_{ftt} - v_{fff})$
f t f	$w\ v_{ftf}$	$w(v_{ftf} - v_{fff})$
f f t	$w\ v_{fft}$	$w(v_{fft} - v_{fff})$
f f f	$w\ v_{fff}$	$w(v_{fff} - v_{fff})$

This table is read as follows. The first three columns determine the possible world ('t' is *true*, 'f' is *false*). The fourth column gives the contribution of the original formula φ to the SWSF for that world, which by definition is the weight w multiplied by either 1 or 0, depending on the truth value $v_{...}$ of φ in the considered world. The last column gives the summed contribution of the seven new formulas to the SWSF, this can be calculated from the equations given in Algorithm 1b. Let us illustrate this for the first row in the table, i.e., for possible world in which P, Q and R are all true. In this world, one can verify that all seven new formulas are satisfied. Hence the contribution of these formulas to the SWSF is the sum of their weights, $\sum_{i=1}^{7} w_i$. Calculating this sum using the equalities in Algorithm 1b, and simplifying the resulting expression, we obtain $w(v_{ttt} - v_{fff})$, as indicated in the first row of the table. In the same way, the reader can verify the other rows in the table.

As the table shows, the contribution to the SWSF of the seven new formulas is the same as that of φ, up to a constant (namely $-w\, v_{fff}$). Hence, although the normalization constant is different, the distribution of the MLN is preserved.[3]

3.3   Step 2: Reduce Triplewise Formulas to Pairwise Form

Step 2, shown in Algorithm 1a, reduces all triplewise formulas in the MLN to pairwise form. It loops over all formulas in the MLN. When encountering a triplewise formula φ, we execute an *atomic reduction step* on φ.

Atomic Reduction Step. An atomic reduction step executed on a triplewise PNF formula φ with weigth w removes φ from the MLN and replaces it by an equivalent set of pairwise (or unary) formulas. To find this set, we call *ReduceToPairwise*(φ, w), as defined in Algorithm 1c. If w is positive, this returns four formulas; if w is negative, seven formulas are needed.

The key part of the reduction is that we introduce an *auxiliary atom* (*aux-atom*) into the vocabulary. This is analogous to an auxiliary random variable from the traditional reduction in the graphical models literature [1]. In the function *ReduceToPairwise*(), this aux-atom is denoted A.

Example. We continue our previous example. After Step 1, we had an MLN with a triplewise formula $P \wedge Q \wedge R$ with weight w, and a pairwise formula $P \wedge Q$ with weight $-w$. In Step 2, the pairwise formula is left unchanged, while (assuming $w > 0$) the triplewise formula is replaced by its equivalent set of four pairwise or unary formulas, as given by Algorithm 1c. The end result is hence a set a five formulas: the pairwise formula $P \wedge Q$ with weight $-w$, the pairwise formulas $A \wedge P$, $A \wedge Q$ and $A \wedge R$ each with weight w, and the unary formula A with weight $-2w$.

One important technicality that is not illustrated by this small example is that, for the reduction to be correct, we need to introduce a *separate* aux-atom (i.e., with a new, unique name) for every triplewise formula being reduced. For instance, for the i-th triplewise formula, we can introduce an aux-atom named A_i.

3.4   Equivalence of the Reduced MLN (max-equivalence)

Let us call the original MLN M_0, the MLN obtained after Step 1 M_1, and the MLN after Step 2 M_2. Let P_0, P_1 and P_2 denote the probability distributions specified by respectively M_0, M_1 and M_2.

We have already shown (Section 3.2) that M_1 is equivalent to M_0. Now the question is: how does M_2 relate to M_0 and M_1? Because Step 2 introduced aux-atoms, M_2 defines a probability distribution over possible worlds described in terms of a larger vocabulary than M_0 and M_1. We show below that M_2 is *max-equivalent* to M_0 and M_1, i.e., when *maxing-out* all aux-atoms from the probability distribution of M_2, the obtained distribution is the same as that of M_0

---

[3] We could add the trivial formula *true* (which is satisfied in all worlds) with weight $w\, v_{fff}$ in order to also make the normalization constants equal. However, from the perspective of equality of distributions, there is no need to do this.

and M_1. Maxing-out an atom (or random variable), as used in all max-product algorithms in graphical models, is the counterpart of summing-out (marginalization).

Example (maxing-out). In our running example, the MLN M_2 obtained after Step 2 contains the atoms A, P, Q, and R and hence defines a distribution $P_2(A, P, Q, R)$ over $2^4=16$ possible worlds. Maxing-out A from this distribution yields a distribution $P_2'(P, Q, R)$ over 8 possible worlds, where each 'entry' of the distribution $P_2'()$ is defined as the maximum of the two corresponding entries of the distribution $P_2()$. Concretely, $\forall a, p, q, r \in \{true, false\}$:

$$P_2'(P = p, Q = q, R = r)$$
$$\stackrel{def}{=} max(P_2(A = \underline{true}, P = p, Q = q, R = r), P_2(A = \underline{false}, P = p, Q = q, R = r)).$$

We write this more concisely as:

$$P_2'(P, Q, R) \stackrel{def}{=} max_A P_2(A, P, Q, R).$$

Example (max-equivalence). The MLNs M_0 and M_1 for our running example define a distribution $P_0(P, Q, R) = P_1(P, Q, R)$. Max-equivalence means:

$$max_A P_2(A, P, Q, R) = P_1(P, Q, R) = P_0(P, Q, R).$$

As mentioned before, the example illustrates a simple case with only one triplewise formula in the original MLN, and hence one aux-atom. In general, multiple aux-atoms are needed (one per triplewise formula). In such cases, max-equivalence is satisfied after successively maxing-out *all* aux-atoms.

Before we prove that max-equivalence indeed holds, we first clarify why max-equivalence is meaningful.

Inference on the Reduced MLN. As mentioned in the introduction, pairwise reductions are mostly used for MAP inference. This is the task of, given the observed truth value of some atoms in the MLN, finding the most likely truth value of all others atoms (the *MAP solution*). Given a non-pairwise MLN M, there are two possible approaches. The *direct approach* runs MAP inference on M. The *pairwise approach* first reduces M to pairwise form and then applies MAP inference. This approach returns the most likely truth value for all non-observed atoms in M, as well as for all aux-atoms. The aux-atoms' truth values can simply be discarded as they are not part of the original inference task. Max-equivalence implies that the pairwise approach returns the same MAP solution as the direct approach (assuming we perform *exact* MAP inference and that M has a single MAP optimum). This is because MAP inferences maximizes over all non-observed atoms including the aux-atoms, and maximizing over all aux-atoms is exactly what the notion of max-equivalence assumes. This explains why max-equivalence is useful in the context of MAP inference.

Proof of Max-Equivalence. For any given (propositional triplewise) MLN M_0 it holds that the MLN M_2 obtained after Step 2 of our reduction is max-equivalent to M_0. Step 2 repeatedly applies atomic reduction steps. Each step introduces a separate aux-atom, which is independent of any previous aux-atoms. Hence, maxing-out one aux-atom does not influence maxing-out any other aux-atom. Thus, to show that max-equivalence holds, it suffices to show that a single atomic reduction step applied to some MLN leads to a max-equivalent MLN. We show this below.

Each atomic reduction step takes a triplewise PNF formula $P \wedge Q \wedge R$ with weight w and replaces it by a set of four (if $w > 0$) or seven (if $w < 0$) pairwise or unary formulas. As we did for Step 1, we will again focus on the contribution of the involved formulas to the SWSF of each world. We again use a table over possible worlds to show this. We first consider the case $w > 0$. According to Algorithm 1c, $P \wedge Q \wedge R$ is replaced by four new formulas: $A \wedge P$, $A \wedge Q$ and $A \wedge R$ each with weight w, and A with weight $-2w$. The left table below show the contribution to the SWSF for (some of) the 16 possible worlds of A, P, Q and R. The right table shows the contribution after maxing-out A.

P	Q	R	A	SWSF	
t	t	t	t	w	$\}\ max = w$
t	t	t	f	0	
t	t	f	t	0	$\}\ max = 0$
t	t	f	f	0	
$\ldots$				$\ldots$	
f	f	f	t	$-2w$	$\}\ max = 0$
f	f	f	f	0	

$\Rightarrow$

P	Q	R	SWSF maxed-out
t	t	t	w
t	t	f	0
$\ldots$			$\ldots$
f	f	f	0

These tables should be read as follows. Consider the first row of the left table, which is for the possible world in which P, Q, R and A are all true. The four new formulas are all satisfied in this world, so their total contribution to the SWSF is the sum of their weights, namely $w + w + w - 2w = w$. For the other worlds, the contribution can be computed similarly. If we max-out A, the table over 16 possible worlds (left) collapses into one over 8 possible worlds (right). For instance, the first two rows of the left table are both for the case where P, Q and R are all true. Taking the maximum of their SWSF contributions, we obtain $max(w, 0) = w$, which is the value in the first row of the right table. The other entries of the maxed-out table (right) can be computed similarly. The end result is that the SWSF contribution is w for the first world, and zero for all other worlds.

The goal of the atomic rewriting step is to replace the formula $P \wedge Q \wedge R$. The contribution of this formula is exactly the same as that in the maxed-out table: w for the first world, and zero for the other worlds. This establishes max-equivalence.

This is for $w > 0$; the reader can verify in the same way that max-equivalence also holds if $w < 0$ (using the seven formulas of Algorithm 1c).

3.5   Beyond Triplewise MLNs

The above shows how to reduce triplewise propositional MLNs to pairwise form. When given a non-triplewise MLN (i.e., when the maximum formula length is larger than 3), we apply a *preprocessing step* that converts the MLN to triplewise form. This can be done using existing techniques: we first convert every non-triplewise formula to clausal form [7], then we reduce each non-triplewise clause to a set of triplewise formulas using the classic method with auxiliary atoms of Karp [9]. For instance, consider the clausal formula $P \vee \neg Q \vee R \vee \neg S$, with weight w. This can be converted to two triplewise formulas, namely a formula $P \vee \neg Q \vee T$ with weight w, and a hard formula $T \Leftrightarrow R \vee \neg S$, with T an auxiliary atom.[4] Note that this method can be used to bring any MLN into triplewise form, but not to reduce it further to pairwise form. For this, our reduction algorithm is needed.

4 Lifted Reduction for First-Order MLNs

So far, we only considered *propositional* MLNs. A *first-order* MLN is a set of pairs (φ_i, w_i) where each φ_i is a formula in first-order logic (we do not allow existential quantifiers or functors).[5] The resulting probability distribution is: $P(\omega) = \frac{1}{Z} exp(\sum_i w_i n_i(\omega))$, with $n_i(\omega)$ the number of satisfied groundings of formula φ_i in world ω. We refer to logical variables as *logvars* and write them in lowercase[6], e.g., x.

Reducing a first-order MLN can be done at the *ground level* or at the *lifted level*. Reduction at the ground level simply consists of using existing methods to ground the MLN and then carrying out the reduction as in the previous section, treating each ground atom as a proposition (this requires using a separate auxiliary proposition for each grounding of each triplewise formula). More interestingly, we can also do the reduction at the first-order level, i.e., in a lifted way.

Lifted Reduction. This reduction is useful when performing lifted inference [8]. The approach is very similar to the propositional case, below we focus on the differences.

Given a non-triplewise MLN, we apply a preprocessing step to make it triplewise, as in the propositional case. Concretely, we convert it to first-order clausal form and then reduce non-triplewise clauses by introducing auxiliary variables using Karp's method on the first-order level [9]. For instance, consider the clausal formula $P(x) \vee Q(x, y) \vee R(y, z) \vee S(z)$, with weight w. We replace this by two triplewise formulas, namely $P(x) \vee Q(x, y) \vee T(y, z)$ with weight w, and a hard formula $T(y, z) \Leftrightarrow R(y, z) \vee S(z)$, with T an auxiliary predicate.

Once we have a triplewise MLN, the outline of the reduction algorithm is the same as in the propositional case, although there are some differences in the actual steps, see the colored comments in Algorithm 1. Step 1 is the same as before, except for the complication of 'lost logvars', which we discuss later in this section. Step 2 is also similar, the difference being that in the propositional

---

[4] A 'hard' formula in an MLN has infinite weight [7].

[5] This is a common restriction in most of the work on Markov Logic.

[6] This is the MLN (and first-order logic) convention, and is the opposite of the Prolog convention.

case the introduced aux-atom is a proposition A, while in the lifted case it is an atom containing the necessary logvars, as illustrated in the following example.

Example. Consider the first-order formula $P(x) \wedge Q(x, y) \Rightarrow R(y)$. This type of formula is used often in SRL, for instance in collective classification: $Class_i(x) \wedge Link(x, y) \Rightarrow Class_j(y)$. Note that this formula has exactly the same structure (connectives) as our earlier propositional example $P \wedge Q \Rightarrow R$. The reduction is thus also very similar. In Step 1, we rewrite the formula to PNF, using Algorithm 1b. This yields two formulas: $P(x) \wedge Q(x, y) \wedge R(y)$ with weight w and $P(x) \wedge Q(x, y)$ with weight $-w$. In Step 2, we reduce the first formula to pairwise form using Algorithm 1c, i.e., we replace it by four formulas: the pairwise formulas $A(x, y) \wedge P(x)$, $A(x, y) \wedge Q(x, y)$, $A(x, y) \wedge R(y)$ each with weight w, and the unary formula $A(x, y)$ with weight $-2w$. The only actual difference with the propositional example is that, as an aux-atom, we need to use $A(x, y)$, with A an auxiliary predicate. It is necessary that this atom contains all the logvars that occur in the original triplewise formula, i.e., x and y in this case. This ensures that, if we ground the obtained pairwise MLN, it will contain a separate aux-atom for every instantiation of (x, y), just as it would if we would have done the reduction at the ground level. If we would not use a separate aux-atom for each instantiation, the ground reductions would become inter-dependent and max-equivalence would no longer hold. For this reason, it is also necessary to use a separate aux-predicate for every first-order triplewise formula in the MLN.

Lifted Versus Ground Reduction. The above lifted reduction is equivalent to the reduction on the ground level in the sense that first reducing an MLN on the lifted level and then grounding it will yield exactly the same result as first grounding the MLN and then reducing it with our ground/propositional reduction. This implies as a corollary that the max-equivalence that we had for our propositional reduction carries over to the lifted level: when reducing a first-order MLN to first-order pairwise form, the result will be max-equivalent, provided that we max-out *all instantiations* of all aux-atoms.

Correction Factors for Lost Logvars. Finally, there is one technical complication that we have not discussed yet. This complication occurs only in special cases; it does for instance not occur in any of our previous examples, nor in any of our experiments (Section 5). For completeness, we illustrate it on another example. Consider the triplewise formula $P(x) \wedge Q(x) \Leftrightarrow R$ with weight w. If we apply Step 1 (Algorithm 1b), we obtain, after some calculations, three PNF formulas: $P(x) \wedge Q(x) \wedge R$ with weight $2w$, $P(x) \wedge Q(x)$ with weight $-w$, and R with weight $-w$. The problem is with the last formula (namely R): we have '*lost*' the logvar x in the reduction (i.e., x does not occur anymore in this formula). This is incorrect. If we would do the reduction on the ground level, we would separately rewrite every instantiation of the original formula to PNF, and we would (correctly) get one occurrence of the formula R (with weight $-w$) *for each possible instantiation* of x. Such duplicate occurrences are equivalent to a single occurrence of R with weight $-wN_x$, where N_x is the *domain size* (number of constants) of logvar x. However, in the above lifted reduction, we get only one occurrence of R with weight $-w$. The solution is simple: we correct for the lost logvar x in our lifted reduction by setting the weight of the formula R not to $-w$ but to $-wN_x$, as it would be in the ground case.

In more general terms, whenever a formula in our lifted reduction has a lost logvar, we need to compensate by multiplying the weight with the domain size of the lost logvar. Note that lost logvars can arise in Step 1 (as in the above example) or in Step 2. The solution with the correction factor applies to both cases.

5   Experiments

To evaluate the usefulness of our reduction, we used it in experiments with MAP inference. We used ground inference algorithms (with our propositional reduction) as well as lifted algorithms (with our lifted reduction).

5.1   Ground Inference: Setup

MLNs and Datasets. We used two triplewise MLNs. The first is the MLN for the Smokers domain [7]. On the first-order level, this contains one triplewise formula, namely $Friends(x, y) \Rightarrow (Smokes(x) \Leftrightarrow Smokes(y))$, and several pairwise or unary formulas. While this is a synthetic domain, it is of interest because its formulas are similar to those in real-world MLNs (e.g., the above formula follows the common pattern $Link(x, y) \Rightarrow (Property(x) \Leftrightarrow Property(y))$). We generated ground MLNs for eight different domain sizes (number of people): 10, 20, 40, 60, 80, 100, 120 and 150 (beyond that, the experiments timed-out). For each domain size, we defined MAP tasks by selecting the required number of people, sampling the truth value for all ground atoms, and randomly selecting 80% of all resulting ground atoms as evidence. The MAP task was to max-out the remaining 20% of the atoms. We repeated this construction 20 times, yielding 20 different MAP tasks per domain size, so 160 in total.

The second MLN is for WebKB (http://www.cs.cmu.edu/~webkb/), a dataset about collective classification. On the first-order level, it contains 49 triplewise formulas of the form $Class_i(x) \wedge Link(x, y) \Rightarrow Class_j(y)$ (with i and j one of 7 classes), and a number of pairwise and unary formulas. We varied the domain size, by subsampling the set of webpages, from 10 to 70 in multiples of 10 (beyond that, the experiments timed-out). We learned the parameters of the MLNs from data, using 4-fold cross validation over the different data-subsets. We defined MAP tasks for each test fold by randomly selecting 80% of all ground atoms in the data as evidence and maxing-out the other 20%. We constructed 5 different MAP tasks per fold, so 20 per domain size, so 140 in total.

Algorithms. We used two ground approximate MAP algorithms, MaxWalkSAT (MWS) and MPLP. MWS [7] works in the same way irrespective of whether the MLN is pairwise or not. MPLP [3] is a Max Product variant based on Linear Programming. While the MPLP paper only discusses the pairwise case, we obtained an implementation from the authors that supports also non-pairwise models. We ran MPLP on the original triplewise MLNs (*MPLP-t*) as well as the pairwise MLNs returned by our reduction algorithm (*MPLP-p*), with identical settings. We also ran MWS on both types of MLNs (*MWS-t* and *MWS-p*). Since MWS is an anytime algorithm, we had to choose the time-budget: for each MAP task, we set the budget equal to the runtime of MPLP-p on that task.

Evaluation Measure. We evaluated the quality of the MAP assignment returned by an algorithm by computing the sum of weights of satisfied formulas in the original triplewise MLN under this assignment, as this sum is proportional to the probability of the MAP assignment, which we want to maximize [7].

5.2   Ground Inference: Results

MLN Characteristics. We recorded the characteristics of the triplewise MLNs and their pairwise reductions. There is of course a certain blow up when doing the reduction, e.g., the aux-atoms are introduced. How big this effect is depends on the formulas involved, on the evidence, etc. On Smokers, the effect is very small: the ratio of the number of ground atoms in the pairwise MLN versus in the triplewise MLN is always in the interval $[1.04, 1.07]$. On WebKB, the effect is larger, with ratios in the interval $[2.22, 2.87]$. The fact that these ratios are larger than 2 means that, on WebKB, the majority of all atoms in the pairwise MLNs are aux-atoms.

Algorithm Comparison. When comparing the quality of the MAP solution returned by the 4 methods, we found that MPLP-p is the best. On Smokers, MPLP-p is the best method on 93% of all tasks, versus 62% for MPLP-t, 59% for MWS-t, and 3% for MWS-p. These percentages sum up to more than 100% because of ties: often multiple methods give the same MAP solution and are jointly the best. On WebKB, MPLP-p is the best on 99% of all tasks, versus 82% for MPLP-t, and 15% for both MWS-t and MWS-p. Hence, overall MPLP-p is the preferred method in terms of solution quality. However, in terms of time till convergence, MPLP-p is significantly slower than MPLP-t: a factor 4.07 slower on Smokers and a factor 1.62 slower on WebKB. This is because the pairwise MLNs are larger than the triplewise ones. We now analyze closer the difference between the triplewise and pairwise approaches.

MPLP-p vs MPLP-t. Figure 1 (top row) shows the wins/ties/losses for MPLP-p versus MPLP-t. A win (green) means that MPLP-p, i.e. the pairwise approach, is better; a loss (red) means that the triplewise approach is better. The pairwise approach clearly gives the best results, especially on larger domains. It seems that MPLP cannot deal well with many triplewise formulas, and the solution quality suffers. This proves the usefulness of our reduction to pairwise form.

MWS-p vs MWS-t. Figure 1 (bottom row) shows the results for MWS. Here the opposite trend holds: the triplewise approach gives the best results. Unlike MPLP, MWS does local search to find the MAP optimum. For the pairwise MLNs, the search space is significantly larger due to the extra aux-atoms. The results show that MWS suffers from this enlarged search space and hence performs poorly on the pairwise MLNs.

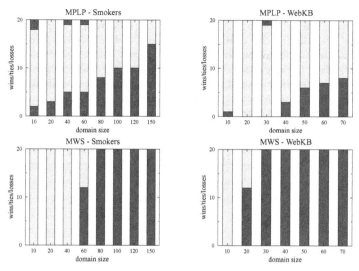

Fig. 1. *(Best viewed in color.)* Comparison of solution quality for pairwise versus triple-wise MLNs. Each bar shows, for the corresponding domain size, how many of the 20 MAP tasks are 'wins' (green; pairwise is better), 'losses' (red; triplewise is better) and 'ties' (yellow).

5.3   Lifted Inference: Setup

We also tested our reduction in combination with *lifted inference*. For this, we used the lifted version of our reduction. As inference algorithm, we used MAP via *Lifted Linear Programming* (using the sparse version of CVXOPT as the underlying solver) [8]. We used the Smokers MLN with domain sizes 50, 100 and 150 (beyond that, some of the experiments timed-out).

5.4   Lifted Inference: Results

Triplewise Versus Pairwise. Table 2 shows the results. Lifted Linear Programming (LLP) performs approximate MAP inference. To indicate the quality of the solution, it computes an upper bound on the MAP objective. Since the MAP objective needs to be maximized, it holds that the lower the upper bound is, the 'tighter' LLP's approximation is. Hence, for the upper bound, lower is better. As the last column of Table 2 shows, the results are slightly better for pairwise MLNs than for the original triplewise MLNs. This shows that also in the lifted case our reduction to pairwise form is useful.

Lifting. The measured speed-ups obtained due to lifting go up to a factor 1.53 for the triplewise MLNs and 1.27 for the pairwise MLNs. The fact that these speed-ups are rather modest, might say more about the practical implementation than about the theoretical potential for lifting: the reduction of the number of variables (number of unknowns) in the LPs is very drastic, as Table 2 shows. Hence, while the actual benefit of lifting in this experiment is modest, we see

Table 2. Lifted inference results. The 1st and 2nd column (*MLN* and *domain size*) specify the input. The 3rd column (*speed-up*) gives the relative speed-up factor achieved due to lifting. The 4th and 5th column (*vars-ground* and *vars-lifted*) give the number of variables in the LP for respectively the ground and lifted case. The 6th column (*upper bound*) gives the upper bound on the LP objective, lower is better (see text); this bound is, by construction of LLP, identical for the ground and lifted case.

MLN	Domain size	Speed-up	Vars-ground	Vars-lifted	Upper bound
	50	1.22	25,200	19	7.83e03
Triplewise	100	1.39	100,400	19	3.12e04
	150	1.53	225,600	19	7.00e04
	50	1.07	60,400	24	7.77e03
Pairwise	100	1.22	240,800	24	3.10e04
	150	1.27	541,200	24	6.96e04

this mainly as a proof-of-concept that our reduction can also be combined with lifted inference.

6   Conclusion

We introduced Pairwise Markov Logic, a new subset of Markov Logic. This subset is of special interest since working with pairwise MLNs has advantages in the context of MAP/MPE inference. Our experiments with the MPLP algorithm confirm this. Since allowing only pairwise MLNs is too restrictive during modelling and learning, we have shown how to reduce a non-pairwise MLN to an equivalent pairwise MLN for running inference, both on the ground level and lifted. While we focussed on inference here, also learning will benefit from this work, as MAP/MPE inference is a sub-procedure in many learning tasks (e.g., discriminative weight learning for MLNs [7]).

We presented some first experiments on combining our lifted reduction with lifted inference. Further research in this direction is interesting future work.

Acknowledgements. We thank the reviewers for useful comments and suggestions. DF is a post-doctoral fellow of the Research Foundation-Flanders (FWO-Vlaanderen). KK was supported by the Fraunhofer ATTRACT fellowship STREAM and by the EC (FP7-248258-First-MM). MM and KK were supported by the German Research Foundation DFG (KE 1686/2-1) within the SPP 1527. JD is supported by the Research Fund K.U.Leuven (CREA/11/015 and OT/11/051), EU FP7 Marie Curie Career Integration Grant (#294068) and FWO-Vlaanderen (G.0356.12).

References

1. Gallagher, A.C., Batra, D., Parikh, D.: Inference for order reduction in Markov random fields. In: IEEE Computer Society Conference on Computer Vision and Pattern Recognition, pp. 1857–1864. IEEE Computer Society, Los Alamitos (2011)

2. Kumar, M., Kolmogorov, V., Torr, P.: An analysis of convex relaxations for MAP estimation of discrete MRFs. Journal of Machine Learning Research 10, 71–106 (2009)
3. Sontag, D., Meltzer, T., Globerson, A., Jaakkola, T., Weiss, Y.: Tightening LP relaxations for MAP using message passing. In: Proceedings of the 24th Annual Conference on Uncertainty in Artificial Intelligence, pp. 503–510. AUAI Press, Corvallis (2008)
4. Cour, T., Shi, J.: Solving Markov random fields with spectral relaxation. Journal of Machine Learning Research - Proceedings Track 2, 75–82 (2007)
5. Rother, C., Kolmogorov, V., Lempitsky, V.S., Szummer, M.: Optimizing binary MRFs via extended roof duality. In: IEEE Computer Society Conference on Computer Vision and Pattern Recognition (2007)
6. Kolmogorov, V., Rother, C.: Minimizing nonsubmodular functions with graph cuts - A review. IEEE Transactions on Pattern Analysis and Machine Intelligence 29(7), 1274–1279 (2007)
7. Domingos, P., Kok, S., Lowd, D., Poon, H., Richardson, M., Singla, P.: Markov Logic. In: De Raedt, L., Frasconi, P., Kersting, K., Muggleton, S.H. (eds.) Probabilistic ILP 2007. LNCS (LNAI), vol. 4911, pp. 92–117. Springer, Heidelberg (2008)
8. Mladenov, M., Ahmadi, B., Kersting, K.: Lifted linear programming. In: 15th International Conference on Artificial Intelligence and Statistics, La Palma, Canary Islands, Spain. Journal of Machine Learning Research: Workshop & Conference Proceedings, vol. 22 (2012)
9. Karp, R.: Reducibility among combinatorial problems. In: Complexity of Computer Computations, pp. 85–103. Plenum Press (1972)

Evaluating Inference Algorithms for the Prolog Factor Language

Tiago Gomes and Vítor Santos Costa

CRACS & INESC TEC, Faculty of Sciences, University of Porto
Rua do Campo Alegre, 1021/1055, 4169-007 Porto, Portugal
{tgomes,vsc}@fc.up.pt

Abstract. Over the last years there has been some interest in models that combine first-order logic and probabilistic graphical models to describe large scale domains, and in efficient ways to perform inference on these domains. Prolog Factor Language (PFL) is a extension of the Prolog language that allows a natural representation of these first-order probabilistic models (either directed or undirected). PFL is also capable of solving probabilistic queries on these models through the implementation of four inference algorithms: variable elimination, belief propagation, lifted variable elimination and lifted belief propagation. We show how these models can be easily represented using PFL and then we perform a comparative study between the different inference algorithms in four artificial problems.

1 Introduction

Over the last years there has been some interest in models that combine first-order logic and probabilistic graphical models to describe large scale domains, and in efficient ways to perform inference and learning using these models [1,2]. A significant number of languages and systems has been proposed and made available, such as Independent Choice Logic [3], PRISM [4,5], Stochastic Logic Programs (ICL) [6], Markov Logic Networks (MLNs) [7], CLP($\mathcal{BN}$) [8,9], ProbLog [10,11], and LPADs [12], to only mention a few. These languages differ widely, both on the formalism they use to represent structured knowledge, on the graphical model they encode, and on how they encode it. Languages such as ICL, Prism, or ProbLog use the distribution semantics to encode probability distributions. MLNs encode relationships through first-order formulas, such that the strength of a true relationship serves as parameter to a corresponding ground markov network. Last, languages such as CLP($\mathcal{BN}$) approach the problem in a more straightforward way, by using the flexibility of logic programming as a way to quickly encode graphical models.

Research in Probabilistic Logic Languages has made it very clear that it is crucial to design models that can support efficient inference. One of the most exciting developments toward this goal has been the notion of lifted inference [13,14]. The idea is to take advantage of the regularities in structured models and perform a number of operations in a fell swoop. The idea was first proposed as an

F. Riguzzi and F. Železný (Eds.): ILP 2012, LNAI 7842, pp. 74–85, 2013.
© Springer-Verlag Berlin Heidelberg 2013

extension of variable elimination [13,14], and has since been applied to belief propagation [15,16] and in the context of theorem proving and model counting [17,18].

Most work in probabilistic inference computes statistics from a sum of products representation, where each element is known as a *factor*. Lifted inference approaches the problem by generalizing factors through the notion of *parametric factor*, commonly called *parfactor*. Parfactors can be seen as templates, or classes, for the actual factors found in the inference process. Lifted inference is based on the idea of manipulating these parfactors, thus creating intermediate parfactors in the process, and in general delaying as much as possible the use of fully instantiated factors.

Parfactors are a compact way to encode distributions and can be seen as a natural way to express complex distributions. This was recognized in the Bayesian Logic Inference Engine (BLOG) [19], and more recently has been the basis for proposals such as Relational Continuous Models [20], and parametrized randvars (random variables) [21]. In this same vein, and in the spirit of CLP($\mathcal{BN}$) we propose an extension of Prolog designed to support parfactors, the *Prolog Factor Language (PFL)*.

The PFL aims at two goals. First, we would like to use the PFL to understand the general usefulness of lifted inference and how it plays with logical and probabilistic inference. Second, we would like to use the PFL as a tool for multi-relational learning. In this work, we focus on the first task. The work therefore introduces two contributions: a new language for probabilistic logical inference, and an experimental evaluation of a number of state-of-the-art inference techniques.

The paper is organized as follows. First, we present the main ideas of the PFL. Second, we discuss how the PFL is implemented. Third, we present a first experimental evaluation of the PFL and draw some conclusions.

2   The Prolog Factor Language (PFL)

First, we briefly review parfactors. We define a parfactor as a tuple of the form:

$$< A, C, \phi >$$

where A is a set of atoms (atomic formulas), C is a set of constraints, and ϕ defines a potential function on $\mathbb{R}_0^+$. Intuitively, the atoms in A describe a set of random variables, the constraints in the C describe the possible instances for those random variables, and the ϕ describe the potential values. The constraints in C apply over a set of logical variables L in A.

As an example, a parfactor for the MLN language could be written as:

$$2.33 : Smokes(X) \wedge X \in \{\texttt{John}, \texttt{Mary}, \texttt{William}\}$$

in the example, $A = \{Smokes(X)\}$, and $L = \{X\}$. Each MLN factor requires a single parameter, in this case 2.33. The constraint $X \in \{\texttt{John}, \texttt{Mary}, \texttt{William}\}$ defines the possible instances of this parfactor, in the example the three factors:

$$2.33 : Smokes(\texttt{John}) \wedge 2.33 : Smokes(\texttt{Mary}) \wedge 2.33 : Smokes(\texttt{William})$$

Notice that all factors share the same potential values.

The main goal of the PFL is to enable one to use this compact representation as an extension of a logic program. The PFL inherits from previous work in CLP($\mathcal{BN}$), which in turn was motivated by prior work on probabilistic relational models (PRMs) [22]. A PRM uses a Bayesian network to represent the joint probability distribution over fields in a relational database. Then, this Bayesian network can be used to make inferences about missing values in the database. In Datalog, missing values are represented by Skolem constants; more generally, in logic programming missing values, or existentially-quantified variables, can be represented by terms built from Skolem functors. CLP($\mathcal{BN}$) represents the joint probability distribution as a function over terms constructed from the Skolem functors in a logic program. Thus, in CLP($\mathcal{BN}$), we see random variables as a special interpretation over a set V of function symbols, the skolem variables.

The PFL. The first insight of the PFL is that CLP($\mathcal{BN}$) skolem functions naturally map to atoms in parfactor formulae. That is, in a parfactor the formula $a(X) \wedge b(X)$ can be seen as identifying two different skolem functions $X \to a(X)$ and $X \to b(X)$. The second observation is that both the constraints and the potential values can be obtained from a logic program. Thus, an example of a parfactor described by PFL is simply:

```
bayes ability(K)::[high,medium,low] ;
    [0.50, 0.40, 0.10] ;
    [professor(K)].
```

This parfactor defines an atom `ability(K)`, within the context of a directed network, with $K \in \{professor(K)\}$ as the constraint. The random variables instantiated by `ability(K)` will have `high`, `medium` and `low` as their domain.

More precisely, the PFL syntax for a factor is

$$Type\ F;\ \phi;\ C.$$

Thus, a PFL factor has four components:

- *Type* refers the type of the network over the parfactor is defined. It can be **bayes**, for directed networks, or **markov**, for undirected ones.
- F is a sequence of Prolog terms that define sets of random variables under the constraints in C. Each term can be seen as the signature of a skolem function whose arguments are given by the unbound logical variables. The set of all logical variables in F is named L. The example includes a single term, $ability(K)$; the only logical variable is K.
- The table ϕ is either a list of potential values or a call to a Prolog goal that will unify its last argument with a list of potential values.
- C is a list of Prolog goals that will impose bindings on the logical variables in L. In other words, the successful substitutions for the goals in C are the valid values for the variables in L. In the example, the goals constrain K to match a professor.

A more complex example is a parfactor for a student's grade in a course:

```
bayes grade(C,S)::[a,b,c,d], intelligence(S), difficulty(C) ;
    grade_table ;
    [registration(_,C,S)].
```

In the example, the `registration/3` relation is part of the extensional data-base.

The next example shows an encoding for the competing workshops problem [23]:

```
markov attends(P), hot(W) ;
    [0.2, 0.8, 0.8, 0.8] ;
    [c(P,W)].
```

```
markov attends(P), series ;
    [0.501, 0.499, 0.499, 0.499] ;
    [c(P,_)].
```

One can observe that the model in this case is undirected. The encoding defines two parfactors: one connects workshop attendance with the workshop being hot, the other with the workshop being a series. In this case, the domains of the random variables are not being specified. Thus, they will default to boolean.

2.1   Querying and Generating Evidence for the PFL

In our approach, each random variable in PFL is implicitly defined by a predicate with the same name and an extra argument. More formally, the value V for a random variable or skolem function $R(A_1, \ldots, A_n)$, is given by the predicate $R(A_1, \ldots, A_n, V)$. Thus in order to query a professor's ability it is sufficient to ask:

```
?- ability(p0, V).
```

This approach follows in the lines of PRISM [4] and CLP($\mathcal{BN}$).

Evidence can be given as facts for the intentional predicates. Hence,

```
ability(p0,high).
```

can be used to indicate that we have evidence on p0's (high) ability. Conditional evidence can also be given as part of a query, hence:

```
?- ability(p0,high), pop(p0,X).
```

would return the marginal probability distribution for professor p0's popularity given he has a high ability.

3   Inference in the PFL

One of our main motivations in designing the PFL is to research on the interplay between logical and probabilistic inference. Indeed, how to execute a PFL program very much depends on the probabilistic inference method used.

Solving the main inference tasks in first-order probabilistic models can be done by first grounding the network and then applying a ground solver, such as variable elimination or belief propagation. However, the cost of this operation will strongly depend on the domain size (number of objects). It may be much more efficient to solve this problem in a *lifted* way, that is, by exploiting the repeated structure of the model to speed up inference by potentially several orders of magnitude.

The PFL thus supports both lifted and grounded inference methods. In *fully grounded solving* the PFL implementation creates a ground network and then calls the solver to obtain marginals. In *lifted solving* the PFL implementation first finds out a graph of parfactors that are needed to address the current query, an then calls the lifted solver.

The fully grounded algorithm implements a transitive closure algorithm (similar to [24]), and is as follows. First, given a query goal Q, it obtains the corresponding random variable $\{V\}$. It also collects the set of evidence variables E. The algorithm then maintains two sets: an open set of variables, initially $O \leftarrow E \cup \{V\}$, and an explored set of variables X, initially empty ($X \leftarrow \{\}$). It proceeds as follows:

1. It selects a variable V from the open-set (O) and removes it from O;
2. Adds V to X;
3. For all parfactors *defining* the variable V, it computes the constraints as a sequence of Prolog goals, and it adds the other random variables V' to O if $V' \notin X$;
4. It terminates when O empty.

The PFL uses the convention that in Bayesian networks a single parfactor defines each variable, and in markov networks all parfactors define the variable. This implementation is currently used by variable elimination, belief propagation and counting belief propagation, and the CLP($\mathcal{BN}$) solvers [25].

Lifted solving performs a first step of computing transitive closure, but only on the graph of parfactors. No constraints are called, and no grounding is made. A second step then computes the extension of the constraints, so the groundings are kept separate from the graph. This implementation is used by lifted variable elimination.

The probabilistic inference algorithms used in the PFL are discussed next.

4   Probabilistic Inference Algorithms

A typical inference task is to compute the marginal probabilities of a set of random variables given the observed values of others (evidence). Variable elimination (VE) and belief propagation (BP) are two popular ways to solve this problem.

Next we briefly describe these two algorithms as well as their lifted versions.

4.1   Variable Elimination

Variable elimination [26] is one of the simplest exact inference methods. As the name indicates, it works by successively eliminating the random variables that appears in the factors until only the query variable remains.

At each step, each variable X is eliminated by first collecting the factors where X appears, then calculating the product of these factors and finally summing out X. The product of two factors works like a join operation in relational algebra, where the set of random variables of the resulting factor is the union of the random variables of the two operands, and the product is performed such that the values of common random variables match with each other. Summing out a variable X is done through the summation of all probabilities where the values of X varies, while the values of the other variables remains fixed.

Unfortunately, the cost of variable elimination is exponential in relation to the *treewidth* of the graph. The treewidth is the size of the largest factor created during the operation of the algorithm using the best order in which the variables are eliminated. And it gets worse: finding the best elimination order is a NP-Hard problem.

4.2   Belief Propagation

Belief propagation [27] is a very efficient way to solve probabilistic queries in models where exact inference is intractable. Here we describe the implementation of the algorithm over factor graphs. The algorithm consists in iteratively exchanging local messages between variable and factor nodes of a factor graph, until convergence be achieved (the probabilities for each variable stabilize). These messages consists in vectors of probabilities that measures the influence that variables have among others.

Belief propagation is known to converge to the exact answers where the factor graph is acyclic. Otherwise, there is no guarantee on the convergence of the algorithm, or that the results will be good approximations of the exact answers. However, experimental results show that often it converges to good approximations and can finish several times faster than other methods [28].

Next, we describe how the messages are calculated and how to obtain the marginals for each variable. The message from a variable X to a factor f is defined by:

$$u_{X \to f}(x) = \prod_{g \in \text{neighbors}(X) \setminus \{f\}} u_{g \to X}(x) \ , \tag{1}$$

that is, it consists in the product of the messages received from X's other neighbor factors. The message that a factor f sends to a variable X is defined by:

$$u_{f \to X}(x) = \sum_{y_1, \dots, y_k} f(x, y_1, \dots, y_k) \prod_{i=1}^{k} u_{Y_i \to f}(y_i) \ , \tag{2}$$

where $Y_1, \ldots, Y_k$ are other f's neighbor variables. In other words, it is the product of the messages received from other f's neighbor variables, followed by summing out all variables except X.

Finally, we estimate the marginal probabilities for a variable X by computing the product of the messages received by all X's neighbor factors:

$$P(x) \propto \prod_{f \in \text{neighbors}(X)} u_{f \to X}(x) \ . \tag{3}$$

Depending on the graphical model, we may also need to normalize the probabilities. That is, to scale them to sum to 1. In our implementation, initially all messages are initialized in a uniform way.

4.3 Lifted Variable Elimination

Lifted variable elimination exploits the symmetries present in first-order probabilistic models, so that it can apply the same principles behind variable elimination to solve a probabilistic query without grounding the model. However, instead of summing out a random variable at a time, it works out by summing out a whole group of interchangeable random variables.

Initially a shattering operation is applied in all of the parfactors. Shattering consists in splitting the parfactors until all atoms represent under the constraint either identical or disjoint sets of ground random variables. Intuitively, this is necessary to ensure that the same reasoning steps are applied to all ground factors represented by a parfactor. Sometimes it may also be necessary to further split the parfactors for some lifted operation be correct, that is, equal to a set of multiple ground operations.

Work on lifting variable elimination started with Poole [13] and was later extended by de Salvo [29]. Milch [23] increased the scope of lifted inference by introducing counting formulas, that can be seen as a compact way to represent a product of repeated ground factors. The current state of art on lifted variable elimination is GC-FOVE [21]. GC-FOVE extends previous work by allowing more flexibility about how the groups of interchangeable random variables are defined.

4.4 Lifted Belief Propagation

There are currently two main approaches for applying belief propagation in a lifted level: lifted first-order belief propagation [15] and counting belief propagation [16]. While the first requires a first-order probabilistic model as input, the second does not (is a generalization of the first). Here we focus only on counting belief propagation.

Counting belief propagation is defined over factor graphs and employs two steps. Firstly, variable and factor nodes that are indistinguishable in terms of messages sent and received are respectively grouped in clusters of variables and clusters of factors, creating with this process a compressed factor graph. To achieve this, counting belief propagation simulates the use of belief propagation by using a color based scheme, where colors are transmitted instead of real probabilities and identical colors identify identical messages.

Secondly, an adapted version of belief propagation is executed over the compressed factor graph. Adapted because it needs to consider that an edge in the compressed factor graph can represent multiples edges in the original factor graph.

As the size of compressed factor graph can be a lot smaller than the original factor graph, the time required to solve some query can substantially decrease.

5   Experimental Evaluation

We compare the performance of variable elimination, belief propagation and their lifted versions in four benchmark problems described using PFL: *workshop attributes* [23], *competing workshops* [23], *city* [13] and *social domain* [30].

Our implementation of lifted belief propagation is based on counting belief propagation (CBP), while the implementation of lifted variable elimination corresponds to GC-FOVE with the difference that we use a simple tree to represent the constraints instead of a constraint tree [21]. All algorithms are written in C++ and are available with the YAP Prolog system[1] [31].

The test environment was a machine with 2 Intel® Xeon® X5650@2.67GHz processors and 99 Gigabytes of main memory, running a Linux kernel 2.6.34.9-69.fc13.x86_64. The times for solving each query presented here are the minimum times of a series of multiple runs. All tests are deterministic.

Figure 1 displays the running times to solve a query on *series* for both workshop attributes and competing workshops problems with an increased number of individuals [23]. GC-FOVE is the method that performs better, followed by CBP. For CBP, the size of the compressed factor graph remains constant as we increase the number of individuals. Hence its cost is mostly dominated by the time used to compress the factor graph. Variable elimination performs reasonable well even in the presence of large domains, as the treewidth remains constant for all number of individuals. Although it converges in few iterations, belief propagation is clearly the slowest method.

Figure 2(a) shows the running times to solve a query on *guilty* for some individual in the city problem, given the descriptions of the others [13]. Here the performances of CBP and GC-FOVE are close, as well as their non-lifted versions. More interesting is the social domain problem [30] where non-lifted exact inference is intractable, since the treewidth of the graph increases exponentially in relation to the number of individuals. Figure 2(b) displays the performances of belief propagation, GC-FOVE and CBP for a query on *friends*. We found that GC-FOVE can solve very efficiently this problem and in a faster way than CBP.

At last, we are interested in study how evidence causes the parfactors to be split for GC-FOVE and how it influences the size of the compressed factor graph for CBP. For this experiment we took the social domain network, fixed

---

[1] Available from http://yap.git.sourceforge.net/git/gitweb.cgi?p=yap/yap-6.3

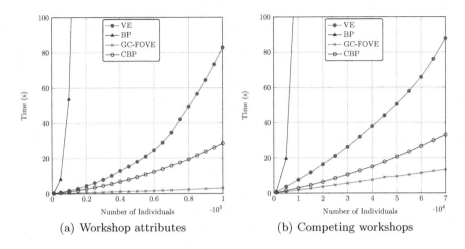

Fig. 1. Performance on workshops attributes and competing workshops problems with an increased number of individuals

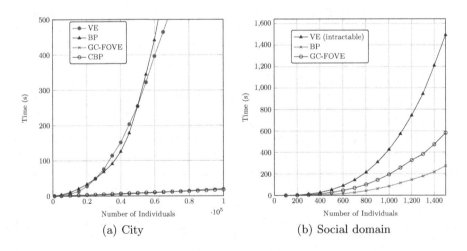

Fig. 2. Performance on city and social domain problems with an increased number of individuals

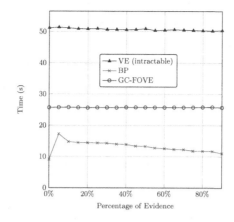

Fig. 3. Performance on social domain varying the percentage of evidence

the number of individuals at 500 and varied the percentage of evidence on random variables produced by *smokes* (observed random variables as well as their observed values were choose randomly). The results are show in Figure 3. The run times for belief propagation and CBP remain stable for all different percentages of evidence. Therefore the size of the compressed factor graph does not considerably increase for CBP. For GC-FOVE, the first set of evidence causes an increase on the run time as some parfactors will be split. However adding more evidence will not cause further splitting and instead the run times will decrease gradually, since the absorption of the evidence will cause some operations used by GC-FOVE to become slightly cheaper.

6   Conclusions and Future Work

We introduce the PFL, an extension of Prolog to manipulate complex distributions represented as products of factors. The PFL has been implemented as an extension to the YAP Prolog system, and is publicly available in YAP's development version. As an initial experiment, we use the PFL to represent several lifted inference benchmarks, and compare a number of inference techniques. Our results show benefit in using first-order variable elimination, even given the complexity of the algorithm.

We see the PFL as a tool in experimenting with ways to combine probabilistic and logical inference, and plan to continue experimenting with different approaches, such as other forms of lifted belief propagation [15]. We also plan to look in more detail to aggregates. Last, and not least, our ultimate focus is to be able to learn PFL programs.

Acknowledgments. This work is funded (or part-funded) by the ERDF - European Regional Development Fund through the COMPETE Programme (operational programme for competitiveness) and by National Funds through the

FCT - Fundação para a Ciência e a Tecnologia (Portuguese Foundation for Science and Technology) within project HORUS (PTDC/EIA/100897/2008) and project LEAP (PTDC/EIA-CCO/112158/2009).

References

1. Getoor, L., Taskar, B.: Introduction to Statistical Relational Learning. MIT Press (2007)
2. De Raedt, L., Frasconi, P., Kersting, K., Muggleton, S. (eds.): Probabilistic Inductive Logic Programming. LNCS (LNAI), vol. 4911. Springer, Heidelberg (2008)
3. Poole, D.: The Independent Choice Logic for Modelling Multiple Agents Under Uncertainty. Artif. Intell. 94(1-2), 7–56 (1997)
4. Sato, T., Kameya, Y.: PRISM: A Language for Symbolic-Statistical Modeling. In: Proceedings of the Fifteenth International Joint Conference on Artificial Intelligence, IJCAI 1997, Nagoya, Japan, August 23-29, vols. 2, pp. 1330–1339. Morgan Kaufmann (1997)
5. Sato, T., Kameya, Y.: New Advances in Logic-Based Probabilistic Modeling by PRISM. In: [2], pp. 118–155
6. Muggleton, S.: Stochastic Logic Programs. In: De Raedt, L. (ed.) Advances in Inductive Logic Programming. Frontiers in Artificial Intelligence and Applications, vol. 32, pp. 254–264. IOS Press, Amsterdam (1996)
7. Richardson, M., Domingos, P.: Markov logic networks. Machine Learning 62(1-2), 107–136 (2006)
8. Santos Costa, V., Page, D., Qazi, M., Cussens, J.: CLP(BN): Constraint Logic Programming for Probabilistic Knowledge. In: Meek, C., Kjærulff, U. (eds.) Proceedings of the 19th Conference in Uncertainty in Artificial Intelligence, UAI 2003, Acapulco, Mexico, August 7-10, pp. 517–524. Morgan Kaufmann (2003)
9. Santos Costa, V., Page, D., Cussens, J.: CLP(BN): Constraint Logic Programming for Probabilistic Knowledge. In: [2], pp. 156–188
10. De Raedt, L., Kimmig, A., Toivonen, H.: ProbLog: A Probabilistic Prolog and Its Application in Link Discovery. In: Veloso, M.M. (ed.) Proceedings of the 20th International Joint Conference on Artificial Intelligence, IJCAI 2007, Hyderabad, India, January 6-12, pp. 2462–2467 (2007)
11. Kimmig, A., Demoen, B., De Raedt, L., Santos Costa, V., Rocha, R.: On the implementation of the probabilistic logic programming language ProbLog. TPLP 11(2-3), 235–262 (2011)
12. Riguzzi, F., Swift, T.: The PITA system: Tabling and answer subsumption for reasoning under uncertainty. TPLP 11(4-5), 433–449 (2011)
13. Poole, D.: First-order probabilistic inference. In: Gottlob, G., Walsh, T. (eds.) IJCAI, pp. 985–991. Morgan Kaufmann (2003)
14. de Salvo Braz, R., Amir, E., Roth, D.: Lifted First-Order Probabilistic Inference. In: Kaelbling, L.P., Saffiotti, A. (eds.) IJCAI, pp. 1319–1325. Professional Book Center (2005)
15. Singla, P., Domingos, P.: Lifted first-order belief propagation. In: Proceedings of the 23rd National Conference on Artificial Intelligence, vol. 2, pp. 1094–1099 (2008)
16. Kersting, K., Ahmadi, B., Natarajan, S.: Counting belief propagation. In: Proceedings of the Twenty-Fifth Conference on Uncertainty in Artificial Intelligence, pp. 277–284. AUAI Press (2009)

17. Gogate, V., Domingos, P.: Probabilistic Theorem Proving. In: Cozman, F.G., Pfeffer, A. (eds.) Proceedings of the Twenty-Seventh Conference on Uncertainty in Artificial Intelligence, UAI 2011, Barcelona, Spain, July 14-17, pp. 256–265. AUAI Press (2011)

18. Van den Broeck, G., Taghipour, N., Meert, W., Davis, J., De Raedt, L.: Lifted Probabilistic Inference by First-Order Knowledge Compilation. In: Walsh, T. (ed.) Proceedings of the 22nd International Joint Conference on Artificial Intelligence, IJCAI 2011, Barcelona, Catalonia, Spain, July 16-22, pp. 2178–2185. IJCAI/AAAI (2011)

19. Milch, B., Marthi, B., Russell, S.J., Sontag, D., Ong, D.L., Kolobov, A.: BLOG: Probabilistic Models with Unknown Objects. In: Kaelbling, L.P., Saffiotti, A. (eds.) Proceedings of the Nineteenth International Joint Conference on Artificial Intelligence, IJCAI 2005, Edinburgh, Scotland, UK, July 30-August 5, pp. 1352–1359. Professional Book Center (2005)

20. Choi, J., Amir, E., Hill, D.J.: Lifted Inference for Relational Continuous Models. In: Grünwald, P., Spirtes, P. (eds.) Proceedings of the Twenty-Sixth Conference on Uncertainty in Artificial Intelligence, UAI 2010, Catalina Island, CA, USA, July 8-11, pp. 126–134. AUAI Press (2010)

21. Taghipour, N., Fierens, D., Davis, J., Blockeel, H.: Lifted variable elimination with arbitrary constraints. In: Proceedings of the Fifteenth International Conference on Artificial Intelligence and Statistics (2012)

22. Getoor, L., Friedman, N., Koller, D., Pfeffer, A.: Learning Probabilistic Relational Models. In: Relational Data Mining, pp. 307–335. Springer (2001)

23. Milch, B., Zettlemoyer, L., Kersting, K., Haimes, M., Kaelbling, L.: Lifted probabilistic inference with counting formulas. In: Proc. 23rd AAAI, pp. 1062–1068 (2008)

24. Kersting, K., De Raedt, L.: Bayesian logic programs. CoRR cs.AI/0111058 (2001)

25. Santos Costa, V.: On the Implementation of the CLP($\mathcal{B}N$) Language. In: Carro, M., Peña, R. (eds.) PADL 2010. LNCS, vol. 5937, pp. 234–248. Springer, Heidelberg (2010)

26. Zhang, N.L., Poole, D.: Exploiting causal independence in bayesian network inference. Journal of Artificial Intelligence Research 5, 301–328 (1996)

27. Kschischang, F., Frey, B., Loeliger, H.: Factor graphs and the sum-product algorithm. IEEE Transactions on Information Theory 47(2), 498–519 (2001)

28. Murphy, K., Weiss, Y., Jordan, M.: Loopy belief propagation for approximate inference: An empirical study. In: Proceedings of the Fifteenth Conference on Uncertainty in Artificial Intelligence, pp. 467–475. Morgan Kaufmann Publishers Inc. (1999)

29. de Salvo Braz, R., Amir, E., Roth, D.: Lifted first-order probabilistic inference. In: Getoor, L., Taskar, B. (eds.) Introduction to Statistical Relational Learning, pp. 433–451. MIT Press (2007)

30. Jha, A.K., Gogate, V., Meliou, A., Suciu, D.: Lifted Inference Seen from the Other Side: The Tractable Features. In: Lafferty, J.D., Williams, C.K.I., Shawe-Taylor, J., Zemel, R.S., Culotta, A. (eds.) NIPS, pp. 973–981. Curran Associates, Inc. (2010)

31. Santos Costa, V., Damas, L., Rocha, R.: The YAP Prolog system. Theory and Practice of Logic Programming 12(Special Issue 1-2), 5–34 (2012)

Polynomial Time Pattern Matching Algorithm for Ordered Graph Patterns

Takahiro Hino[1], Yusuke Suzuki[1], Tomoyuki Uchida[1], and Yuko Itokawa[2]

[1] Department of Intelligent Systems, Hiroshima City University, Japan
{hino@ml.info.,y-suzuki@,uchida@}hiroshima-cu.ac.jp
[2] Faculty of Psychological Science, Hiroshima International University, Japan
y-itoka@he.hirokoku-u.ac.jp

Abstract. Ordered graphs, each of whose vertices has a unique order on edges incident to the vertex, can represent graph structured data such as Web pages, TEX sources, CAD and MAP. In this paper, in order to design computational machine learning for such data, we propose an ordered graph pattern with ordered graph structures and structured variables. We define an ordered graph language for an ordered graph pattern g as the set of all ordered graphs obtained from g by replacing structured variables in g with arbitrary ordered graphs. We present a polynomial time pattern matching algorithm for determining whether or not a given ordered graph is contained in the ordered graph language for a given ordered graph pattern. We also implement the proposed algorithm on a computer and evaluate the algorithm by reporting and discussing experimental results.

1 Introduction

Tree structured data such as Web pages and TEX sources are modelled by ordered trees, each of which contains internal nodes with ordered children. Geometric data such as CAD and MAP can be naturally modelled by planar maps [13] in which a single edge in the outer frame is directed in the clockwise direction. Jiang and Bunke [3] introduced an ordered graph as a more general class of graphs containing ordered trees and planar maps. An ordered graph is a graph in which the edges incident to a vertex are uniquely ordered. Ordered graphs G_1, G_2, F_1 and F_2 are shown in Fig. 1 as examples. In F_1 and F_2, the numbers around each vertex denote the order of edges incident to the vertex. G_1 and G_2 are also planar maps. However, F_1 and F_2 are not planar maps, because F_1 and F_2 are not biconnected.

Suzuki et al. [10] introduced an ordered term tree, which is an ordered tree that has structured variables. Kawamoto et al. [4] proposed a planar map pattern that is a graph structured pattern having a planar map structure and structured variables. In a planar map pattern, structured variables are defined as faces having variable labels. A planar map pattern p is said to match a planar map P if P can be obtained from p by replacing structured variables with arbitrary planar maps. For example, the planar map pattern p in Fig. 1 matches the

F. Riguzzi and F. Železný (Eds.): ILP 2012, LNAI 7842, pp. 86–101, 2013.

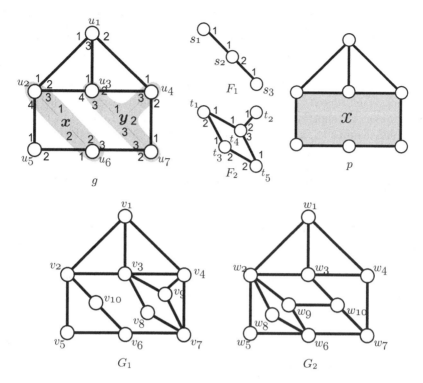

Fig. 1. Ordered graph pattern g, planar map pattern p and ordered graphs F_1, F_2, G_1 and G_2. In our simple drawing of an ordered graph pattern and a planar map pattern, grey regions of an ordered graph pattern or a planar map pattern show structured variables, the order of edges incident to each vertex is arranged in the clockwise direction and edge labels are omitted.

planar maps G_1 and G_2. Namely, p represents common structural features of G_1 and G_2. To represent structural features of ordered graphs, in this paper, we propose a new ordered graph pattern that has structured variables and each of whose vertices has an ordering list on incident edges and structured variables. A structured variable is a hyperedge consisting of distinct vertices and having a variable label. Similar to an ordered term tree and a planar map pattern, an ordered graph pattern f is said to match an ordered graph F if F can be obtained from f by replacing structured variables with arbitrary ordered graphs. In Fig. 1, we show an ordered graph pattern g as an example. g matches the ordered graph G_1 in Fig. 1, because G_1 can be generated by replacing structured variables having variable labels x and y with ordered graphs F_1 and F_2, respectively. However, due to the edge between w_9 and w_{10}, g does not match the ordered graph G_2 in Fig. 1. In g, two variables are in the same face, so there is no planar map pattern corresponding to g. These examples show that an ordered graph pattern provides a richer representation of structural features of ordered graphs than a planar map pattern.

Techniques based on computational machine learning play an important role in the data mining field [5,7]. For a class C, Angluin [1] and Shinohara [8] showed that, if C has finite thickness and two problems for C, a membership problem and a minimal language (MINL) problem, are solvable in polynomial time, then C is polynomial time inductively inferable from positive data. Many classes of graphs such as unordered/ordered trees [9,10], two-terminal series parallel graphs (TTSP graphs, for short) [11], *well-behaved* outerplanar graphs [14], cographs [15] and planar maps [4], are known to be polynomial time inductively inferable from positive data. A well-behaved outerplanar graph is a connected outerplanar graph having only polynomially many simple cycles and a tree-like structure. A TTSP graph and a cograph have unique representations of their parse structures. Hence, in [11,14,15], the pattern matching algorithms solving membership problems for the classes of TTSP graphs, well-behaved outerplanar graphs and cographs have employed polynomial time pattern matching algorithms for unordered/ordered trees. Almost all known algorithms that solve MINL problems for such classes are designed to call polynomial time pattern matching algorithms solving membership problems in many times. For an ordered graph pattern g, we define an ordered graph language, denoted by $GL(g)$, as the set of all ordered graphs generated by replacing all structured variables of g with arbitrarily ordered graphs. The purpose of our research is to show polynomial time learnability of the class of ordered graph languages. As a preliminary step for this purpose, in this paper we present a polynomial time pattern matching algorithm for solving a membership problem which is to determine whether or not the ordered graph language $GL(g)$ for a given ordered graph pattern g contains a given ordered graph G.

The rest of this paper is organized as follows. In Section 2, we formally define an ordered graph pattern as a new graph pattern and its ordered graph language. In Section 3, we propose a polynomial time pattern matching algorithm for determining whether or not a given ordered graph is contained in the ordered graph language for a given ordered graph pattern. In Section 4, we report and discuss the results of experiments using large synthetic ordered graphs that we performed to evaluate the feasibility of the proposed method. Section 5 concludes the paper with a brief summary and a mention of future work.

2   Preliminaries

Jiang and Bunke [3] investigated the class of *ordered graphs*, in which the edges incident to a vertex have a unique order. In this section, we formally define a new graph pattern, called an *ordered graph pattern*, having an ordered graph structure and structured variables. Moreover, we define an *ordered graph language* for an ordered graph pattern.

2.1   Ordered Graph Pattern and Ordered Graph Language

Let Λ and $\mathcal{X}$ be alphabets with $\Lambda \cap \mathcal{X} = \emptyset$. A graph pattern g is a triple (V, E, H), where V is a set of vertices, E is a multiset of edges in $V \times \Lambda \times V$ and H is a set of hyperedges in $V \times \mathcal{X} \times V^+$ such that for any $h \in H$, all vertices in h are distinct. For convenience, to show an undirected edge, whose label is $a \in \Lambda$, between u and v in V, we use the same notation (u, a, v) as a directed edge. Elements in Λ and $\mathcal{X}$ are called *edge labels* and *variable labels*, respectively. A hyperedge in H is called a *variable*. For a variable $h = (u_0, x, u_1, \ldots, u_k) \in H$, each vertex u_i $(0 \leq i \leq k)$ is called a *port* of h. We assume that each variable label x in $\mathcal{X}$ has the *rank*, which is an integer greater than one, and that every variable consisting of k ports has a variable label whose rank is k. The rank of x is denoted by $rank(x)$. The vertex set, edge set and hyperedge set of g are denoted by $V(g)$, $E(g)$ and $H(g)$, respectively.

Definition 1. An *ordered graph pattern* is a five-tuple $g = (V, E, H, \mathcal{OE}, \mathcal{L})$ defined as follows.

(1) (V, E, H) is a graph pattern such that, for any vertex $v \in V$, the number of variables having v as a port is at most one.

(2) $\mathcal{OE} = E \cup H \cup M$ such that M is a multiset of elements in $V \times V$ and the graph $(V, M \cup \{(u, v) \mid (u, a, v) \in E\})$ is connected. An element in $\mathcal{OE}$ is called an *ordering-edge*.

(3) $\mathcal{L} = \{\ell_g(v) \mid v \in V\}$, called an *ordering-set*, is a cyclic list of elements containing v in $\mathcal{OE}$. For a vertex $v \in V$, the element $\ell_g(v)$ of $\mathcal{L}$ is called an *ordering-list* of v on $\mathcal{OE}$.

If E is a multiset of directed edges, g is called a *directed* ordered graph pattern. An *ordered graph* is an ordered graph pattern without variables.

For an ordered graph pattern g, the sets of all ordering-edges and all ordering-lists of g are denoted by $\mathcal{OE}(g)$ and $\mathcal{L}(g)$, respectively. The sets of all ordered graph patterns and all ordered graphs are denoted by $\mathcal{OGP}$ and $\mathcal{OG}$, respectively. Note that $\mathcal{OG} \subset \mathcal{OGP}$ holds.

Example 1. In Fig. 1, we show an ordered graph pattern g and ordered graphs G_1, G_2, F_1 and F_2 as examples, where
$V(g) = \{u_1, \ldots, u_7\}$, $E(g) = \{(u_1, a, u_2), (u_1, a, u_3), \ldots, (u_6, a, u_7)\}$,
$H(g) = \{(u_2, x, u_6), (u_3, y, u_4, u_7)\}$, $\mathcal{OE}(g) = E(g) \cup H(g)$,
$\mathcal{L}(g) = \{\ell_g(u_1), \ell_g(u_2), \ell_g(u_3), \ldots, \ell_g(u_7)\}$,
$\ell_g(u_1) = ((u_1, a, u_2), (u_1, a, u_4), (u_1, a, u_3))$,
$\ell_g(u_2) = ((u_1, a, u_2), (u_2, a, u_3), (u_2, x, u_6), (u_2, a, u_5))$,
$\ell_g(u_3) = ((u_1, a, u_3), (u_3, a, u_4), (u_3, y, u_4, u_7), (u_2, a, u_3)), \ldots$, and
$\ell_g(u_7) = ((u_4, a, u_7), (u_6, a, u_7), (u_3, y, u_4, u_7))$.
We can see that $rank(x) = 2$ and $rank(y) = 3$.

An ordered graph is known to be a more general class of graphs containing ordered trees and plane graphs, because the edges incident to each vertex naturally posses a unique order, either clockwise or counterclockwise [3]. Suzuki et al. [10] introduced an ordered term tree as an ordered tree having structured variables, and Kawamoto et al. [4] proposed a planar map pattern having a planar map

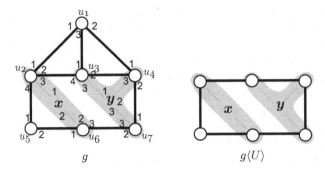

Fig. 2. The ordering-edge induced subgraph pattern $g\langle U \rangle$ of the ordered graph pattern g, where $U = H(g) \cup (E(g) - \{(u_1, a, u_2), (u_1, a, u_3), (u_1, a, u_4)\})$

structure and structured variables. Therefore, ordered term trees and planar map patterns are special types of ordered graph patterns.

For an ordering-list $\ell = (e_1, e_2, \ldots, e_k)$, the number of elements in ℓ is denoted by $|\ell|$, that is, $|\ell| = k$. For an ordering-set $\mathcal{L}(g)$, a *size* of $\mathcal{L}(g)$ is defined as $\|\mathcal{L}(g)\| = \sum_{v \in V(g)} |\ell_g(v)|$. For an ordered graph pattern g and a subset U of $\mathcal{OE}(g)$, an *ordering-edge induced subgraph pattern* of g w.r.t. U, denoted by $g\langle U \rangle$, is the graph pattern f, such that $V(f) = \{u \mid u$ is a vertex contained in $e \in U\}$, $E(f) = E(g) \cap U$, $H(f) = H(g) \cap U$, $\mathcal{OE}(f) = U$, and $\mathcal{L}(f) = \{\ell_f(u) \mid u \in V(f), \ell_f(u)$ is the ordering-list obtained by removing all ordering-edges not in U from $\ell_g(u)\}$.

Example 2. For the ordered graph pattern g in Fig. 2, let U be the set $(E(g) - \{(u_1, a, u_2), (u_1, a, u_3), (u_1, a, u_4)\}) \cup H(g)$ of ordering-edges in g. Then, we show the ordering-edge induced subgraph pattern $g\langle U \rangle$ in Fig. 2 as examples.

Definition 2. An ordered graph pattern g is said to be *ordering isomorphic* to an ordered graph pattern f, denoted by $g \cong f$, if there is a bijection $\varphi : V(g) \to V(f)$ satisfying the following three conditions.

(1) $(u, a, v) \in E(g)$ if and only if $(\varphi(u), a, \varphi(v)) \in E(f)$.
(2) $(u_0, x, u_1, \ldots, u_k) \in H(g)$ if and only if $(\varphi(u_0), x, \varphi(u_1), \ldots, \varphi(u_k)) \in H(f)$.
(3) $\ell_g(v) = (e_1, \ldots, e_k) \in \mathcal{L}(g)$ if and only if $\ell_f(\varphi(v)) = (\varphi(e_1), \ldots, \varphi(e_k)) \in \mathcal{L}(f)$, where for an edge (u_1, a, u_2), $\varphi((u_1, a, u_2)) = (\varphi(u_1), a, \varphi(u_2))$ and for a variable $(w_1, x, w_2, \ldots, w_t)$, $\varphi((w_1, x, w_2, \ldots, w_t)) = (\varphi(w_1), x, \varphi(w_2), \ldots, \varphi(w_t))$.

Such a bijection φ is called an *ordering isomorphism* between g and f.

We can prove the following lemma by modifying an algorithm in Jiang and Bunke's work [3].

Lemma 1. *For two ordered graph patterns g and f, the problem of determining whether or not g is ordering isomorphic to f is solvable in $O(\|\mathcal{L}(g)\|^2)$ time.*

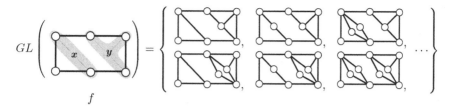

Fig. 3. Ordered graph language $GL(f)$ of the ordered graph pattern f

Especially, given a vertex $u \in V(g)$, an ordering-edge $e \in \ell_g(u)$, a vertex $u' \in V(f)$ and an ordering-edge $e' \in \ell_f(u')$, we can determine whether or not there is an ordering isomorphism φ between g and f, such that $\varphi(u) = u'$ and $\varphi(e) = e'$ in $O(\|\mathcal{L}(g)\|)$ time.

Let f and g be ordered graph patterns and x be a variable label whose rank is r. Let $\sigma = \langle(u_1, p_1), (u_2, p_2), \dots, (u_r, p_r)\rangle$ be a list of r pairs in $V(g) \times \mathcal{OE}(g)$, such that $u_1, u_2, \dots, u_r$ are distinct and for each i $(1 \le i \le r)$, $p_i \in \ell_g(u_i)$ holds. Then, the form $x := [g, \sigma]$ is called a *binding* for x. A new ordered graph pattern f' is obtained by applying the binding $x := [g, \sigma]$ to f in the following way. For each variable $h = (v_1, x, v_2, \dots, v_r)$, we attach a copy g' of g to f by removing the variable h from $H(f)$ and, for each i $(1 \le i \le i)$, by identifying the vertex v_i of f with the vertex u'_i of g', where the u'_i of g' corresponds to the u_i of g. Then, for each i $(1 \le i \le r)$, we update the ordering-list of the v_i of f by replacing $h \in \ell_f(v_i)$ with the list from the element p'_i to the previous element p_i^{-1} of p'_i in the ordering-list $\ell_g(u'_i)$ of u'_i, which is a cyclic list. A *substitution* θ is a finite collection of bindings $\{x_1 := [g_1, \sigma_1], \cdots, x_n := [g_n, \sigma_n]\}$ such that for any i, j $(1 \le i < j \le n)$, the variable labels x_i and x_j are distinct and for each i $(1 \le i \le n)$, the ordered graph pattern g_i has no variable labelled with any variable label in $\{x_1, \dots, x_n\}$. For an ordered graph pattern f, the ordered graph pattern obtained by applying all the bindings $x_i := [g_i, \sigma_i]$ to f is denoted by $f\theta$. For an ordered graph pattern $g \in \mathcal{OGP}$, the **ordered graph language** of g, denoted by $GL(g)$, is defined as the set $\{G \in \mathcal{OG} \mid G \cong g\theta$ for some substitution $\theta\}$.

Example 3. The ordered graph G_1 in Fig. 1 is ordering isomorphic to the ordered graph $g\theta$ obtained by applying the substitution $\theta = \{x := [F_1, \langle(s_1, (s_1, a, s_2)), (s_3, (s_2, a, s_3))\rangle], y := [F_2, \langle(t_1, (t_1, a, t_4)), (t_2, (t_2, a, t_4)), (t_5, (t_4, a, t_5))\rangle]\}$ to g in Fig. 1, that is, $G_1 \cong g\theta$. For the ordered graph pattern g in Fig. 1, we can see that $GL(g)$ contains G_1 but does not G_2, since the edge (w_9, a, w_{10}) exists in G_2. In Fig. 3, we give the ordered graph language $GL(f)$ of the ordered graph pattern f as examples.

2.2   Inductive Inference from Positive Data

From the results of Angluin [1] and Shinohara [8] in a computational machine learning study, in order to show that the class $\mathcal{OGPL} = \{GL(g) \subseteq \mathcal{OG} \mid g \in \mathcal{OGP}\}$

is polynomial time inductively inferable from positive data, it is necessary to prove that the following three propositions hold.

(1) $\mathcal{OGPL}$ has a finite **thickness**, that is, for any nonempty finite set $S \subset \mathcal{OG}$, the cardinality of the set $\{L \in \mathcal{OGPL} \mid S \subseteq L\}$ is finite.
(2) The **membership problem for** $\mathcal{OGPL}$ is solvable in polynomial time. The membership problem for $\mathcal{OGPL}$ is a problem of determining whether or not a given ordered graph in $\mathcal{OG}$ is contained in the ordered graph language of a given ordered graph pattern in $\mathcal{OGP}$.
(3) The **minimal language (MINL) problem** for $\mathcal{OGPL}$ is solvable in polynomial time. The MINL problem for $\mathcal{OGPL}$ is a problem of finding an ordered graph pattern g in $\mathcal{OGP}$ such that, given a set S of ordered graphs in $\mathcal{OG}$, $GL(g)$ is *minimal* among all ordered graph languages that contain all ordered graphs in S, that is, there is no ordered graph pattern $f \in \mathcal{OGP}$ such that $GL(g) \supseteq GL(f) \supseteq S$ holds.

From the definitions of an ordered graph pattern and a substitution, it is easy to show that the following proposition holds.

Proposition 1. *The class $\mathcal{OGPL}$ has a finite thickness.*

We have given polynomial time pattern matching algorithms for solving the membership problems for the classes of ordered trees [10] and planar maps [4]. In the next section, we provide a polynomial time pattern matching algorithm for solving the membership problem for the class $\mathcal{OGPL}$ containing the classes of ordered trees and planar maps as subclasses.

3   Polynomial Time Pattern Matching Algorithm for an Ordered Graph Pattern

For an ordered graph pattern g in $\mathcal{OGP}$ and an ordered graph G in $\mathcal{OG}$, g is said to *match* G if $GL(g)$ contains G. In other words, we can describe the membership problem for the class $\mathcal{OGPL}$ as follows.

Membership problem for $\mathcal{OGPL}$
Instance: An ordered graph pattern $g \in \mathcal{OGP}$ and an ordered graph $G \in \mathcal{OG}$.
Question: Does g match G?

We present a polynomial time algorithm $\mathcal{OGPL}$-MATCHING shown in Algorithm 1 for solving the membership problem for $\mathcal{OGPL}$. In Procedure 2, we present a procedure CODING of $\mathcal{OGPL}$-MATCHING by modifying polynomial time coding algorithms for ordered graphs [3]. Given an ordered graph pattern g, a vertex u of g and an ordering-edge $e \in \ell_g(u)$, the procedure CODING generates a list $List_g$ of ordering-edges of g w.r.t. (u, e) and a sequence w.r.t. $List_g$, called a *code* of g w.r.t. $List_g$. In Procedures 3 and 4, we provide CODEMATCH and STRUCTUREMATCH procedures of $\mathcal{OGPL}$-MATCHING, respectively. Given an ordered graph G, a vertex u of G, an ordering-edge e in $\ell_G(u)$, a code C_g of g and a list $List_g$ of ordering-edges of g, the procedure CODEMATCH generates

---

Algorithm 1. $\mathcal{OGPL}$-MATCHING

Require: g: ordered graph pattern, G:ordered graph
Ensure: true or **false**
1: Select a vertex u of g and an ordering-edge e in $(\ell_g(u) - H(g))$
2: $(C_g, List_g) :=$ CODING(g, u, e)
3: **for** u' in $V(G)$ **do**
4:   **for** e' in $\ell_G(u')$ **do**
5:     $List_G :=$ CODEMATCH$(G, u', e', C_g, List_g)$
6:     **if** $(List_G \neq ())$ **and** STRUCTUREMATCH$(g, G, List_g, List_G)$ **then**
7:       **return true**
8:     **end if**
9:   **end for**
10: **end for**
11: **return false**

---

a list $List_G$ of ordering-edges of G w.r.t. the pair (u, e) and a code of G w.r.t. $List_G$ and checks whether or not the code of G is equal to the code C_g. In the procedures CODING and CODEMATCH, for a sequence C and a character 'a', $C \circ a$ means the sequence obtained by concatenating 'a' at the end of C.

Example 4. In Fig. 1, given the ordered graph pattern g and the ordering-edge (u_1, a, u_2), the procedure CODING$(g, u_1, (u_1, a, u_2))$ works in the following way. First, the procedure assigns the number 1 to u_1. Since the ordering-set of u_1 is $\ell_g(u_1) = ((u_1, a, u_2), (u_1, a, u_4), (u_1, a, u_3))$, the procedure assigns the numbers $2, 3$ and 4 to u_2, u_4 and u_3, respectively. Therefore, the procedure generates the code "#234#". Next, the ordering-set of u_2 is $\ell_g(u_2) = ((u_1, a, u_2), (u_2, a, u_3), (u_2, x, u_6), (u_2, a, u_5))$, and then the procedure assigns the number 5 to u_5 and generates the code "#234#14x5#". The procedure repeatedly assigns the number to each vertex of g and generates the code. Finally, CODING$(g, u_1, (u_1, a, u_2))$ outputs the code $C_g=$"#234#14x5#16x4#13x2#27#37x#5x6#" and the list $List_g = ((u_1, a, u_2), (u_1, a, u_4), (u_1, a, u_3), (u_1, a, u_2), (u_2, a, u_3), (u_2, x, u_6), (u_2, a, u_5), (u_1, a, u_4), \ldots)$. For the ordered graph pattern G_1 in Fig. 1 and the ordering-edge (v_1, a, v_2), CODEMATCH$(G_1, v_1, (v_1, a, v_2), C_g, List_g)$ generates the code $C_{G_1} =$"#234#14x5#16x4#13x2#27#37x#5x6#". Since the code C_g is equal to C_{G_1}, CODEMATCH$(G_1, v_1, (v_1, a, v_2), C_g, List_g)$ outputs the list $List_{G_1} = ((v_1, a, v_2), (v_1, a, v_4), (v_1, a, v_3), (v_1, a, v_2), (v_2, a, v_3), (v_2, a, v_5), (v_1, a, v_4), \ldots)$.

Lemma 2. *For an ordered graph pattern g and an ordered graph G, if g matches G, the procedure* CODEMATCH *returns* $List_G$ *such that* $g\langle \mathcal{OE}(g) - H(g) \rangle \cong G\langle \{e \mid e \in List_G \} \rangle$ *holds.*

Proof. Since g matches G, there exists a set of ordering-edges $W \subseteq \mathcal{OE}(G)$ with $g\langle \mathcal{OE}(g) - H(g) \rangle \cong G\langle W \rangle$. Let ψ be an ordering isomorphism between $g\langle \mathcal{OE}(g) - H(g) \rangle$ and $G\langle W \rangle$. Let $List_g$ and C_g be a list of ordering-edges and a code of g output by CODING(g, u, e) for some $u \in V(g)$ and $e \in \ell_G(v)$, respectively. We show that CODEMATCH$(G, \psi(u), e', C_g, List_g)$ returns $List_G$, such

---

Procedure 2. CODING

Require: g: ordered graph pattern, u: vertex of g, e: ordering-edge in $(\ell_g(u) - H(g))$
Ensure: Code C of g and list of ordering-edges of g

 1: Initially assign the number '1' to u, the number '0' to the other vertices of $V(g)$
 2: Enqueue (u, e) in a queue Q
 3: $List := ()$
 4: $C =$ "#"
 5: $i := 2$
 6: **while** Q is not empty **do**
 7:     Dequeue a pair (v, e') of a vertex and an ordering-edge from Q
 8:     **for** w in $\ell_g(v)$ in order starting from e' **do**
 9:         Append w to the end of $List$
10:         **if** $w \in H(g)$ **then**
11:             $C := C \circ$ 'x'
12:         **else if** w is an ordering-edge between v and a vertex u' assigned by '0' **then**
13:             Assign i to u'
14:             $C := C \circ i$
15:             $i := i + 1$
16:             Enqueue (u', w) in Q
17:         **else**
18:             $C := C \circ j$
                {w is an ordering-edge between v and a vertex u' assigned by j}
19:         **end if**
20:     **end for**
21:     $C := C \circ$ '#'
22: **end while**
23: **return** $(C, List)$

---

that $g\langle \mathcal{OE}(g) - H(g) \rangle \cong G\langle \{e \mid e \in List_G\} \rangle$ holds, where e' corresponds to e on ψ. For any vertex $v \in V(g)$, the number of variables having v as a port is at most one, and the code C_g contains at most one 'x' between each '#'. By comparing $|\ell_g(v)|$ with $|\ell_G(\psi(v))|$, we fix the interval I in $\ell_G(\psi(v))$ such that I corresponds to the variable. Thus, for any vertex $v \in V(g)$, CODEMATCH assigns to $\psi(v)$ the same number of v and generates the same code of C_g. Moreover, CODEMATCH appends each ordering-edge $e \in \ell_G(\psi(v))$ to the $List_G$, such that e does not correspond to the variable. Therefore, for any ordering-edge $e \in List_g$ (except variables), $List_G$ contains the ordering-edge corresponding to e. Hence, $List_G$ satisfies that $g\langle \mathcal{OE}(g) - H(g) \rangle \cong G\langle \{e \mid e \in List_G\} \rangle$ must hold.   □

For an ordered graph pattern g and an ordered graph G, we say that g and G *match on graph structure* if there exists a list $List_G$ of ordering-edges of G such that the following conditions hold. Let $W = \{e \mid e \in \mathcal{OE}(G), e \notin List_G\}$.

(1) ψ is an ordering isomorphism between $g\langle \mathcal{OE}(g) - H(g) \rangle$ and $G\langle \{e \mid e \in List_G\} \rangle$.

(2) $(u_0, x, u_1, \ldots, u_k) \in H(g)$ if and only if all vertices $\psi(u_0), \psi(u_1), \ldots, \psi(u_k)$ in $V(G)$ are in the same component of $G\langle W \rangle$.

(3) For any two distinct variables $(u_0, x, u_1, \ldots, u_k), (v_0, x, v_1, \ldots, v_k) \in H(g)$ having the same variable label x, the component in $G\langle W \rangle$ having $\psi(u_0), \psi(u_1)$, $\ldots, \psi(u_k)$ is ordering isomorphic to the component in $G\langle W \rangle$ having $\psi(v_0)$, $\psi(v_1), \ldots, \psi(v_k)$.

For an ordered graph pattern $g \in \mathcal{OGP}$ and an ordered graph $G \in \mathcal{OG}$, the procedure STRUCTUREMATCH of $\mathcal{OGPL}$-MATCHING decides whether or not the list, which CODEMATCH outputs, satisfies the above conditions (2) and (3). Examples of STRUCTUREMATCH are shown in Fig 4. g_1 and G_1 satisfy Condition (2), but g_2 and G_2 do not.

Lemma 3. *For an ordered graph pattern $g \in \mathcal{OGP}$ and an ordered graph $G \in \mathcal{OG}$, if g and G match on graph structure, g matches G.*

Proof. Let $List_G$ be the ordering-edges of G such that g and G match on graph structure. Let $W = \{e \mid e \in \mathcal{OE}(G), e \notin List_G\}$. We show that there exist a substitution θ and a bijection ϕ such that ϕ is an ordering isomorphism between $g\theta$ and G. From Condition (1), ψ is an ordering isomorphism between $g\langle \mathcal{OE}(g) - H(g)\rangle$ and $G\langle \{e \mid e \in List_G\}\rangle$, and for each vertex v of $g\langle \mathcal{OE}(g) - H(g)\rangle$, $\phi(v) = \psi(v)$. From Conditions (2) and (3), for each variable $h = (u_0, x, u_1, \ldots, u_k)$ in $H(g)$, let $F(h)$ be the component containing $\psi(u_0), \psi(u_1), \ldots, \psi(u_k)$ in $G\langle W \rangle$. Let $\sigma(h) = \langle (\psi(u_0), p_0), (\psi(u_1), p_1), \ldots, (\psi(u_k), p_k)\rangle$, where for each i $(0 \leq i \leq k)$, p_i is the ordering-edge of $\ell_{F(h)}(\psi(u_i))$ fixed by $List_G$. Let $H'(g)$ be a subset of $H(g)$ such that variable labels of any two variables in $H'(g)$ are different. Let θ be the substitution $\bigcup_{h=(u_0,x,u_1,\ldots,u_k) \in H'(g)} \{x := [F(h), \sigma(h)]\}$. For each variable $h \in \{H(g) - H'(g)\}$, there exists a variable $h' \in H'(g)$ such that the variable labels of h and h' are the same. From Condition (3), $F(h)$ is ordering isomorphic to $F(h')$. From the construction of $g\theta$, for each vertex v of $g\theta\langle \mathcal{OE}(g\theta) - \mathcal{OE}(g)\rangle$, we can fix $\phi(v)$. Therefore, if the above three conditions hold, g matches G. $\square$

Finally, we show the main theorem.

Theorem 1. *The membership problem for $\mathcal{OGPL}$ is solvable in polynomial time.*

Proof. We can see that the termination of $\mathcal{OGPL}$-MATCHING is certainly held. The correctness follows from Lemmas 2 and 3. Given an ordered graph pattern $g \in \mathcal{OGP}$ and an ordered graph $G \in \mathcal{OG}$, $\mathcal{OGPL}$-MATCHING correctly determines whether or not $GL(g)$ contains G. For a particular ordering-edge $e = (u, a, v)$ of g, CODING(g, u, e) generates the code in $O(\|\mathcal{L}(g)\|)$ time. For each of the $O(\|\mathcal{L}(G)\|)$ ordering-edges of G, CODEMATCH needs $O(\|\mathcal{L}(g)\|)$ time, and STRUCTUREMATCH needs $O(|V(g)| \times \|\mathcal{L}(G)\|)$ time from Lemma 1. Hence the total time for all executions in the algorithm is $O(|V(g)| \times \|\mathcal{L}(G)\|^2)$. $\square$

4   Experimental Results

In order to evaluate the algorithm $\mathcal{OGPL}$-MATCHING proposed in the previous section, we implemented $\mathcal{OGPL}$-MATCHING in JAVA on a computer equipped

---

Procedure 3. CodeMatch
_____

Require: G: ordered graph, u: vertex of G, e: ordering-edge in $\ell_G(u)$, C_g: code, $List_g$:
    list of ordering-edges

Ensure: List of ordering-edges of G

 1: Initially assign the number '1' to u and the number '0' to all other vertices of G
 2: Enqueue (u, e) in a queue Q
 3: $List := ()$
 4: $C = \text{"\#"}$
 5: $i := 2$
 6: List-count:= 0
 7: $k := 1$
 8: **while** Q is not empty **do**
 9:     Dequeue a pair (v, e') of a vertex and an ordering-edge from Q
10:     Let $Length$ be the number of characters between k-th '#' and $(k + 1)$-th '#'
11:     **for** w in $\ell_G(v)$ in order starting from e' **do**
12:         **if** $List_g[\text{List-count}] \in H(g)$ **then**
13:             $Length + +$
14:             **if** $|\ell_G(v)| - Length = 0$ **then**
15:                 $C := C \circ \text{'x'}$
16:                 List-count++
17:             **end if**
18:         **else**
19:             **if** the edge labels of w and $List_g[\text{List-count}]$ are different **then**
20:                 **return** ()
21:             **end if**
22:             List-count++
23:             **if** w is an ordering-edge between v and a vertex u' assigned by '0' **then**
24:                 Append w to the end of $List$
25:                 Assign i to u'
26:                 $C := C \circ i$
27:                 $i := i + 1$
28:                 Enqueue (u', w) in Q
29:             **else**
30:                 Append w to the end of $List$
31:                 $C := C \circ j$
                    $\{w$ is an ordering-edge between v and a vertex u' assigned by $j\}$
32:             **end if**
33:         **end if**
34:     **end for**
35:     $C := C \circ \text{'\#'}$
36:     $k := k + 1$
37: **end while**
38: **if** C is equal to C_g **then**
39:     **return** $List$
40: **end if**
41: **return** ()
_____

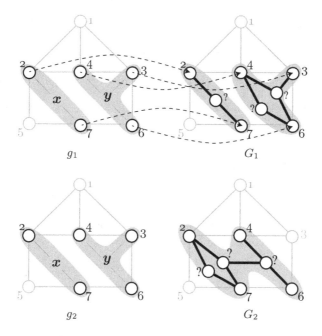

Fig. 4. Example of STRUCTUREMATCH

with dual 3.47 GHz/3.47 GHz Intel XEON W3690 processors, a main memory of 12.0 GB and a Microsoft 64-bit Windows7 SP1 operating system.

Experiment 1: Evaluating the algorithm $\mathcal{OGPL}$-MATCHING

We created two synthetic collections $OP(m)$ and $OD(m)$ of ordered graph patterns and ordered graphs, respectively, each of which had m ordering-edges and $\lfloor m/2 \rfloor$ vertices. In order to measure the execution time of $\mathcal{OGPL}$-MATCHING, we prepared pairs of an ordered graph pattern g in $OP(m_g)$ and an ordered graph G in $OD(m_G)$ as inputs so that $m_G \in \{11000, 12000, \cdots, 30000\}$ for $m_g = 10000$, $m_G \in \{21000, 22000, \cdots, 40000\}$ for $m_g = 20000$ and $m_G \in \{31000, 32000, \cdots, 50000\}$ for $m_g = 30000$, respectively, i.e., for a vertex count n of g, $m_g = 2n$ holds. The average execution times of $\mathcal{OGPL}$-MATCHING are shown in Fig. 5. All execution times of $\mathcal{OGPL}$-MATCHING varied linearly depending on the ordering-edge count of G. These results are also proportional to the ordering-edge count of g. As we discussed in Section 3, the computational complexity of our matching algorithm is $O(|V(g)| \times ||\mathcal{L}(G)||^2)$. In case that all vertices of g have the same degree and each vertex of g is a port of a variable, the execution time of $\mathcal{OGPL}$-MATCHING becomes worst. From creating method of synthetic collections of ordered graph patterns, such worst cases were not occurred in experiments. Hence, in experimental results, the execution times of $\mathcal{OGPL}$-MATCHING are proportional to $|V(g)| \times ||\mathcal{L}(G)||$.

Experiment 2: Comparing $\mathcal{OGPL}$-MATCHING with $\mathcal{PM}$-MATCHING [4]

Procedure 4. StrucureMatch

Require: g: ordered graph pattern, G: ordered graph, $List_g$: list of ordering-edges of g, $List_G$: list of ordering-edges of G;

Ensure: true or **false**

1: Let ψ be the ordering isomorphism between $g\langle\mathcal{OE}(g) - H(g)\rangle$ and $G\langle\{e \mid e \in List_G\}\rangle$ constructed by $List_g$ and $List_G$

2: Let $W = \{e \mid e \in \mathcal{OE}(G), e \notin List_G\}$

3: **for** $h = (u_0, x, u_1, \ldots, u_n) \in H(g)$ **do**

4:     **if** all vertices $\psi(u_0), \psi(u_1), \ldots, \psi(u_n)$ are in the same component of $G\langle W\rangle$ **then**

5:         **for** $(v_0, y, v_1, \ldots, v_m) \in \{H(g) - h\}$ **do**

6:             **if** $\psi(u_0)$ is connected to $\psi(v_0)$ in $G\langle W\rangle$ **then**

7:                 **return false**

8:             **end if**

9:             **if** $(x = y)$ **and** (the component in $G\langle W\rangle$ having $\psi(u_0), \psi(u_1), \ldots, \psi(u_n)$ is not ordering isomorphic to the component in $G\langle W\rangle$ having $\psi(v_0), \psi(v_1), \ldots, \psi(v_m)$) **then**

10:                 **return false**

11:             **end if**

12:         **end for**

13:         **return false**

14:     **end if**

15: **end for**

16: **return true**

Tutte [13] introduced a planar map in which a single edge in the outer frame is directed in the clockwise direction. A planar map is called a *rooted* planar map if one vertex lying on the outer frame is designated the root. Kawamoto et al. [4] defined a rooted planar map pattern as a planar map having structured variables. For a planar map M and a rooted planar map pattern p, we say that p *matches* M, denoted by $p \cong M$, if M is ordering-isomorphic to a planar map obtained from p by replacing all structured variables with arbitrary planar maps. Moreover, for a rooted planar map N and a rooted planar map pattern p, we say that p and N *match with root*, denoted by $p \cong_r N$, if there exists an ordering isomorphism φ between p' and N such that p' is a rooted planar map obtained from p by replacing all structured variables with arbitrary planar maps and for the root r of p', $\varphi(r)$ is the root of N. Kawamoto et al. [4] proposed an algorithm, denoted by $\mathcal{PM}$-MATCHING, which determines whether or not a given rooted planar map M and a given rooted planar map pattern p match with root. This algorithm first visits all vertices of p except variables. Next, for each vertex v of the outer frame of M, it visits vertices of M in the same order as p starting from v. In this experiment, we implemented $\mathcal{PM}$-MATCHING and compared its execution results with those of $\mathcal{OGPL}$-MATCHING. In order to experiment under the same conditions, we introduced the same notions of the root, the outer frame and matching as rooted planar map patterns to ordered graphs and ordered graph patterns. Such ordered graphs and ordered graph patterns call *rooted* ordered graph and *rooted* ordered graph patterns, respectively. We created four synthetic

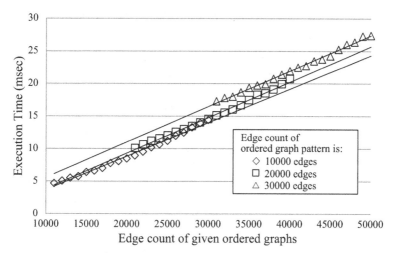

Fig. 5. Execution times of $\mathcal{OGPL}$-MATCHING

collections -$OP(m)$, $OD(m)$, $PP(m)$ and $PD(m)$- of ordered graph patterns, ordered graphs, rooted planar map patterns and planar maps, respectively, each of which had m ordering-edges including 250 ordering-edges of the outer frame.

We used the following three cases of pairs of ordered graphs G in $OD(m)$ and ordered graph patterns g in $OP(5000)$ given to $\mathcal{OGPL}$-MATCHING as inputs (resp., pairs of rooted planar map M in $PM(m)$ and planar map patterns p in $PP(5000)$ given to $\mathcal{PM}$-MATCHING as inputs).

(a) Rooted ordered graph pattern P and rooted graph pattern G such that $g \cong_r G$ (resp., rooted planar map pattern p and rooted planar map M with $p \cong_r M$)
(b) Rooted ordered graph pattern g and ordered graph pattern G with $g \cong G$ (resp., rooted planar map pattern p and planar map M with $p \cong M$)
(c) Rooted ordered graph pattern g and ordered graph pattern G without $g \cong G$ (resp., rooted planar map pattern p and planar map M without $p \cong M$)

For each cases, by varying ordering-edge counts m from 7000 to 15000, the average execution times of $\mathcal{OGPL}$-MATCHING (resp., $\mathcal{PM}$-MATCHING) were as shown in Tables 1–3, respectively. First, from Tables 1 and 2, we can see that $\mathcal{OGPL}$-MATCHING was faster than $\mathcal{PM}$-MATCHING, and in Table 2, the difference of the execution times between the two is even greater. This is probably because, for each vertex v of the outer frame of M, $\mathcal{PM}$-MATCHING visits vertices of M starting from v and the corresponding vertices of p except variables. On the other hand, $\mathcal{OGPL}$-MATCHING visits all ordering-edges of g only once, and then, for each ordering-edge e of the outer frame of G, $\mathcal{OGPL}$-MATCHING visits ordering-edges of G starting from e. Finally, Table 3 shows that our algorithm costs more than $\mathcal{PM}$-MATCHING when the G did not match g. $\mathcal{PM}$-MATCHING terminates without visiting next vertices of p and M as soon as it determines that p does not match M. On the other hand, $\mathcal{OGPL}$-MATCHING must visit all ordering-edge of g once even if g does not match G. This is why our algorithm costs more.

Table 1. Execution times of $\mathcal{OGPL}$-Matching and $\mathcal{PM}$-Matching in Case (a)

Algorithm	7000	9000	11000	13000	15000
$\mathcal{OGPL}$-Matching	**1.23**	**1.22**	**1.22**	**1.22**	**1.22**
$\mathcal{PM}$-Matching	1.45	1.47	1.45	1.44	1.45

Table 2. Execution times of $\mathcal{OGPL}$-Matching and $\mathcal{PM}$-Matching in Case (b)

Algorithm	7000	9000	11000	13000	15000
$\mathcal{OGPL}$-Matching	**1.24**	**1.24**	**1.23**	**1.24**	**1.23**
$\mathcal{PM}$-Matching	1.53	1.53	1.53	1.53	1.49

5   Conclusion and Future Work

We have formally defined a new ordered graph pattern with ordered graph structures and structured variables. We have also defined an ordered graph language $GL(g)$ for an ordered graph pattern g. Given an ordered graph G and an ordered graph pattern g, we have proposed a polynomial time pattern matching algorithm of determining whether or not $GL(g)$ contains G. We have shown that the class $\mathcal{OGPL}$ has a finite thickness. As future work, if we can present a polynomial time algorithm of solving the MINL problem for the class $\mathcal{OGPL}$, we can prove that the class $\mathcal{OGPL}$ of ordered graph languages is polynomial time inductively inferable from positive data.

For the class of *well-behaved* outerplanar graphs, which have only polynomially many simple cycles, Horváth et al. [2] and Yamasaki et al. [14] presented incremental polynomial time graph mining algorithms for enumerating frequent outerplanar graph patterns and frequent block preserving outerplanar graph patterns (bpo-graph patterns, for short) with respect to the *block and bridge preserving (BBP) subgraph isomorphism*, respectively. As practical applications, they applied their graph mining algorithms to a subset of NCI dataset[1]. NCI dataset is one of the popular graph mining datasets and 94.3% of all elements are well-behaved outerplanar graphs. NCI dataset have geometric information. Kuramochi and Karypis [6] presented an effective algorithm for discovering frequent geometric subgraphs from geometric data such as NCI dataset. For geometric data naturally modelled by ordered graphs, we plan to provide an effective graph mining algorithm for enumerating frequent ordered graph patterns with respect to the *ordering isomorphism* by employing our polynomial time pattern matching algorithm for ordered graph patterns. Moreover, in 2000, we defined a layout term graph that is a graph pattern consisting of variables and graph structures containing geometric information [12]. We previously proposed a Layout Formal Graph System (LFGS) as a new logic programming system that uses a layout term graph as a term. LFGS directly deals with graphs containing geometric information just like first order terms. By extending the concept of an ordered

---

[1] http://cactus.nci.nih.gov/

Table 3. Execution times of $\mathcal{OGPL}$-Matching and $\mathcal{PM}$-Matching in Case (c)

Algorithm	7000	9000	11000	13000	15000
$\mathcal{OGPL}$-Matching	0.65	0.61	0.61	0.60	0.61
$\mathcal{PM}$-Matching	**0.19**	**0.12**	**0.06**	**0.06**	**0.06**

graph pattern and modifying the results of this paper, we aim to design an effective knowledge discovery system using LFGS as knowledge representation.

References

1. Angluin, D.: Inductive inference of formal languages from positive data. Information and Control 45(2), 117–135 (1980)
2. Horváth, T., Ramon, J., Wrobel, S.: Frequent subgraph mining in outerplanar graphs. Data Mining and Knowledge Discovery 21(3), 472–508 (2010)
3. Jiang, X., Bunke, H.: On the coding of ordered graphs. Computing 61(1), 23–38 (1998)
4. Kawamoto, S., Suzuki, Y., Shoudai, T.: Learning characteristic structured patterns in rooted planar maps. In: IMECS 2010, IAENG, pp. 465–470 (2010)
5. Kononenko, I., Kukar, M.: Machine Learning and Data Mining: Introduction to Principles and Algorithms. Horwood Pub. (2007)
6. Kuramochi, M., Karypis, G.: Discovering frequent geometric subgraphs. Information Systems 32(8), 1101–1120 (2007)
7. Michalski, R.S., Bratko, I., Bratko, A. (eds.): Machine Learning and Data Mining; Methods and Applications. John Wiley & Sons, Inc., NY (1998)
8. Shinohara, T.: Polynomial time inference of extended regular pattern languages. In: Goto, E., Furukawa, K., Nakajima, R., Nakata, I., Yonezawa, A. (eds.) RIMS 1982. LNCS, vol. 147, pp. 115–127. Springer, Heidelberg (1983)
9. Shoudai, T., Uchida, T., Miyahara, T.: Polynomial time algorithms for finding unordered tree patterns with internal variables. In: Freivalds, R. (ed.) FCT 2001. LNCS, vol. 2138, pp. 335–346. Springer, Heidelberg (2001)
10. Suzuki, Y., Shoudai, T., Uchida, T., Miyahara, T.: Ordered term tree languages which are polynomial time inductively inferable from positive data. Theoretical Computer Science 350(1), 63–90 (2006)
11. Takami, R., Suzuki, Y., Uchida, T., Shoudai, T.: Polynomial time inductive inference of TTSP graph languages from positive data. IEICE Transactions 92-D(2), 181–190 (2009)
12. Uchida, T., Itokawa, Y., Shoudai, T., Miyahara, T., Nakamura, Y.: A new framework for discovering knowledge from two-dimensional structured data using layout formal graph system. In: Arimura, H., Sharma, A.K., Jain, S. (eds.) ALT 2000. LNCS (LNAI), vol. 1968, pp. 141–155. Springer, Heidelberg (2000)
13. Tutte, W.T.: A census of planar maps. Canadian Journal of Mathematics 15, 249–271 (1963)
14. Yamasaki, H., Sasaki, Y., Shoudai, T., Uchida, T., Suzuki, S.: Learning block-preserving graph patterns and its application to data mining. Machine Learning 76(1), 137–173 (2009)
15. Yoshimura, Y., Shoudai, T., Suzuki, Y., Uchida, T., Miyahara, T.: Polynomial time inductive inference of cograph pattern languages from positive data. In: Muggleton, S.H., Tamaddoni-Nezhad, A., Lisi, F.A. (eds.) ILP 2011. LNCS, vol. 7207, pp. 389–404. Springer, Heidelberg (2012)

Fast Parameter Learning for Markov Logic Networks Using Bayes Nets

Hassan Khosravi

School of Computing Science
Simon Fraser University
Vancouver-Burnaby, B.C., Canada
hkhosrav@cs.sfu.ca

Abstract. Markov Logic Networks (MLNs) are a prominent statistical relational model that have been proposed as a unifying framework for statistical relational learning. As part of this unification, their authors proposed methods for converting other statistical relational learners into MLNs. For converting a first order Bayes net into an MLN, it was suggested to moralize the Bayes net to obtain the structure of the MLN and then use the log of the conditional probability table entries to calculate the weight of the clauses. This conversion is exact for converting propositional Markov networks to propositional Bayes nets however, it fails to perform well for the relational case. We theoretically analyze this conversion and introduce new methods of converting a Bayes net into an MLN. An extended imperial evaluation on five datasets indicates that our conversion method outperforms previous methods.

1 Introduction

The field of statistical relational learning (SRL) has developed a number of new statistical models for the induction of probabilistic knowledge that supports accurate predictions for multi-relational structured data [1]. Markov Logic Networks (MLNs) form one of the most prominent SRL model classes; they generalize both first-order logic and Markov network models [2]. Essentially, an MLN is a set of weighted first-order formulas that compactly defines a Markov network comprising ground instances of logical predicates. The formulas are the structure or qualitative component of the Markov network; they represent associations among ground facts. The weights are the parameters or quantitative component; they assign a likelihood to a given relational database by using the log-linear formalism of Markov networks. An open-source benchmark system for MLNs is the Alchemy package [3].

MLNs were proposed as a unifying framework for SRL since they are general enough that many of the other well known SRL models can easily be converted into them. Schulte et al show that it is desirable to use models where learning is performed on First-order Bayes nets and inference is performed on MLNs [4, 5, 6]. Their model combines the scalability and efficiency of model searches in directed models with the inference power and theoretical foundations of undirected models. For converting a first order Bayes net into an MLN, the suggested method has been to moralize the Bayes net

F. Riguzzi and F. Železný (Eds.): ILP 2012, LNAI 7842, pp. 102–115, 2013.
© Springer-Verlag Berlin Heidelberg 2013

to obtain the structure of the MLN and then use the log of the conditional probability table entries to calculate the weight of the clauses [2].

In propositional data using the log of the conditional probability table entries of a Bayes net to obtain weights in a corresponding Markov random field, results in two predictively equivalent graphical models. Although there is no corresponding result for the case of relational data, Richardson and Domingos propose using the same method which is a plausible candidate [2]. In this paper, we examine this conversion theoretically and experimentally for the first time, and provide rationale for why it fails in the relational case. We propose another conversion method for converting weights from a Bayes net to an MLN that is theoretically justifiable and out-performs the current proposed method.

1.1 Related Work

Most of the work on parameter learning in MLNs is based on ideas developed for Markov networks (undirected graphical models) in the propositional case. Special issues that arise with relational data are discussed by Lowd and Domingos [7]. Most recent methods aim to maximize the regularized weighted pseudo log-likelihood [2, 8], and/or perform a scaled conjugate gradient descent using second-order derivative information [7].

We focus on parameter estimation algorithms to convert already calculated weights from directed models to MLNs, so our method can only be applied to a restricted class of MLN structures that are learned from, or can be converted into a Bayes net. The main motivation for converting the directed model into an undirected model and performing inference with an undirected model is that they do not suffer from the problem of cyclic dependencies in relational data [2, 9, 10]. Early work on this topic required ground graphs to be acyclic [11, 12]. For example, Probabilistic Relational Models allow dependencies that are cyclic at the predicate level as long as the user guarantees acyclicity at the ground level [12]. A recursive dependency of an attribute on itself is shown as a self loop in the model graph. If there is a natural ordering of the ground atoms in the domain (e.g., temporal), there may not be cycles in the ground graph; but this assumption is restrictive in general. The generalized order-search of Ramon *et al.* [13] instead resolves cycles by learning an ordering of ground atoms which complicates the learning procedure. The learn-and-join algorithm of Schulte *et al.* utilizes a pseudo-loglikelihood that measures the fit of a Bayes net to a relational database which is well-defined even in the presence of recursive dependencies [5].

1.2 Background

A **Bayes net structure** [14] is a directed acyclic graph (DAG) G, whose nodes comprise a set of random variables denoted by V. In this paper we consider only discrete finite random variables. When discussing a Bayes net, we refer interchangeably to its nodes or its variables. A Bayes net is a pair $\langle G, \theta_G \rangle$ where θ_G is a set of parameter values that specify the probability distributions of children conditional on assignments of values to their parents. We use as our basic model class **Functor Bayes Nets** (FBN)[15][1], a relatively straightforward generalization of Bayes nets, for relational data. Our methods also apply to other directed graphical formalisms.

A **functor** is a function symbol or a predicate symbol. Each functor has a set of values (constants) called the **range** of the functor. A **population** is a set of individuals, corresponding to a domain or type in logic. A functor whose range is $\{T, F\}$ is a **predicate**, usually written with uppercase letters like P, R. A **functor random variable** is of the form $f(t_1, \ldots, t_k)$ where f is a **functor** (either a function symbol or a predicate symbol) and each t_i is a first-order variable or a constant. Each functor has a set of values (constants) called the **range** of the functor. The structure of a Functor Bayes Net consists of: (1) A directed acyclic graph (DAG) whose nodes are parametrized random variables, (2) a population for each first-order variable, and (3) an assignment of a range to each functor. Figure 1 is an example of an FBN.

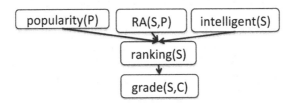

Fig. 1. A Functor Bayes Net which shows that *ranking* of a student correlates with his *intelligence*, *grades*, and *popularity* of the professors that he is a research assistant for

Moralization is a technique used to convert a directed acyclic graph (DAG) into undirected models or MLN formulas. To convert a Bayes net into an MLN using moralization, add a formula to the MLN for each assignment of values to a child and its parents [2](Sec. 12.5.3). Thus, an MLN obtained from a BN contains a formula for each CP-table entry. Figure 2 shows an arbitrary conditional probability table and its corresponding clauses for the *ranking* node in Figure 1.

While the moralization approach produces graph structures that represent the dependencies among predicates well, converting each row of each conditional probability table to an MLN clause leads to a large number of MLN clauses and hence MLN parameters. **Local** or **context-sensitive** independencies are a well-known phenomenon that can be exploited to reduce the number of parameters required in a Bayes net. A **decision tree** can compactly represent conditional probabilities [16]. The nodes in a decision tree for a functor random variable c are themselves functor random variables. An edge that originates in $f(t_1, \ldots, t_k)$ is labeled with one of the possible values in the range of f. The leaves are labeled with probabilities for the different possible values of the c variable. Khosravi et al combine decision tree learning algorithms with Bayes nets to learn a compact set of clauses for relational data [17]. In their experiments, using

---

[1] Functor Bayes nets were called Parametrized Bayes Nets by Poole. The term "Parametrized" referred to the their semantics, and does not mean that parameters have been assigned for the structure. We modified the name to overcome this confusion.

Pop, RA, Int	r =1	r=2	r=3
1, 1, True	$r_{1,1}$	$r_{1,1}$	$r_{1,1}$
--- , ---, ---	---	---	---
3, 3, False	$r_{1,1}$	$r_{1,1}$	$r_{1,1}$

$w_{1,1}$	Pop(P,1), Int(S,1), RA(P,S,True), Rank(S,1)
$w_{2,1}$	Pop(P,1), Int(S,1), RA(P,S,True), Rank(S,2)
$w_{3,1}$	Pop(P,1), Int(S,1), RA(P,S,True), Rank(S,3)
$w_{1,2}$	Pop(P,1), Int(S,1), RA(P,S,False), Rank(S,1)
....	
$w_{3,18}$	Pop(P,3), Int(S,3), RA(P,S,False), Rank(S,3)

Fig. 2. A conditional probability table for node *ranking*. Assuming that the range of *popularity*, *intelligence*, *ranking* = $\{1, 2, 3\}$ and the range of $RA = \{True, False\}$. A tabular representation requires a total of $3 \times 3 \times 2 \times 3 = 54$ conditional probability parameters. The figure on the right shows the corresponding MLN clauses for this conditional probability table.

the decision tree representation instead of "flat" conditional probability tables, reduced the number of MLN clauses by a factor of 5-25. Figure 3 shows a decision tree and its corresponding MLN clauses for the ranking node in 1.

2   Conversion of Parameters from Functor Bayes Nets to Markov Logic Networks

In this section we focus on the problem of converting the parameters of an FBN with a fixed structure and parameters to an MLN. In the following discussion, fix an FBN B and a child node v with k possible values $v_1, \ldots, v_k$ and an assignment π of values to its parents. Then the conditional probability $p(v_i|\pi)$ is defined in the CP-table or decision tree leafs of v in B. The corresponding MLN contains a formula p_j that expresses that a child node takes on the value v_i and the parents take on the values π. The weight of the formula p_j is denoted as w_j. As a running example, we are interested in predicting the ranking of a student Jack given that we have about the five courses he has taken, and the one professor that he is a research assistant for. We also know that Jack is highly intelligent.

In order to convert the conditional probabilities from a Bayes net into weights for MLNs, the use of the logarithm of the conditional probabilities was suggested by Domingos and Richardson [2]. We refer to this method as LOGPROB. The LOGPROB method sets the weights using the following formula:

$$w_j := log(p(v_i|\pi)).$$

In the propositional case, combining moralization with the log-conditional probabilities as in the LOGPROB method leads to an undirected graphical model that is predictively equivalent to the original directed graphical model [14]. Theoretical support for the LOGPROB method is provided by considering the log-likelihood function for an MLN structure obtained from a Bayes net. The standard log-likelihood for an MLN M [2] is given by

$$L_M(\mathcal{D}) = \sum_j w_j n_j(\mathcal{D}) + ln(Z),$$

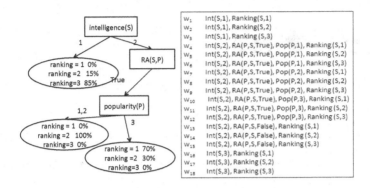

Fig. 3. A decision tree that specifies conditional probabilities for the $Ranking(S)$ node of Figure 1 and the corresponding MLN clauses generated from the decision tree. The number of the clauses has been reduced to 18.

where $n_j(\mathcal{D})$ denotes the number of instances of formula j in database $\mathcal{D}$ and Z is a normalization constant. Omitting the normalization term $ln(Z)$, this is the sum over all child-parent configurations of the corresponding Bayes net in log scale multiplied by $n_{j,\pi}(\mathcal{D})$, which is the number of instances of the child-parent configuration in the database:

$$L_M(\mathcal{D}) = \sum_{i} \sum_{\pi} \log(p(v_i|\pi)) n_{i,\pi}(\mathcal{D}).$$

This unnormalized log-likelihood is maximized by using the observed conditional frequencies in the data table. While the normalization constant is required for defining a valid probabilistic inference, it arguably does not contribute to measuring the fit of a parameter setting and hence can be ignored in model selection; the constraint that weights are derived from normalized conditional probabilities in a Bayes net already bounds their range.

Although there is no corresponding result for the case of relational data, the propositional conversion result makes the log-probabilities a plausible candidate for weights in an MLN structure obtained from a 1st-order Bayes net. We analyze the proposed method on both FBNs with and without parameter reduction, and introduce our own conversion method for converting weights from a Bayes net to an MLN. This conversion method is theoretically justifiable and out-performs the current proposed method.

2.1   Conversion in Functor Bayes Nets Without Parameter Reduction

The complication which is introduced by relational data is the existence of different objects and relations which leads to having a diverse number of groundings for different clauses. Using the moralization technique to obtain an MLN from a Bayesian network introduces two types of clauses for each functor $f(t_1, \ldots, t_k)$:

– Clauses that are produced from the conditional probability table of $f(t_1, \ldots, t_k)$.
– Clauses that are produced from the conditional probability table of other functors where $f(t_1, \ldots, t_k)$ is in the set of their parents.

Clause	#Grounds
pop(P,POP), int(S,INT), ra(S,P,RA), ranking(S,RANK)	n1
ranking(S,RANK) grade(S,C,GRADE)	n2

Fig. 4. Clauses that the $ranking$ node participates in. P, S, and C are first order variables for professors, students, and courses. POP, INT, RA, $RANK$, $GRADE$ are constant values in the range of their corresponding functors.

Figure 4 shows the two types of clauses that the $ranking$ node participates in when the FBN of Figure 1 is moralized.

The two main factors that directly influence the inference procedure of a functor are the weights and the number of groundings of the clauses where the functor node is present. This means that clauses with more groundings will have a higher impact on the final predicated value however, the number of groundings of a clause usually correlate with the number of free variables in it which does not always reflect its importance. In fact longer clauses with a diverse set of functors for different objects tend to have many groundings with limited predictive information, but a short clause with only functors related to the object may carry more predictive information but has fewer number of groundings. For example, Jack is a research assistant for only one professor but has taken 5 courses, so there are five times more groundings for the second clause compared to the first one however, it may be the case that the first clause is a better predictor for the $ranking$ node.

The other factor, as discussed, is the weight of the clauses. The LOGPROB method assigns negative weights to all the clauses. A weight distribution that punishes all clauses performs well in the propositional cases because all of the clauses have the same number of groundings however, it fails to achieve good predictive performance on the relational case.

It may seem trivial that normalizing the weight of the clauses with their number of groundings may overcome the problem (i.e. divide the weight of the clause by its number of groundings) but this is not possible as the weight of the clauses are fixed during the learning phase and the number of groundings for each model is determined during the inference phase. Schulte et al propose a new log-linear $inference$ method that uses the $geometric$ $mean$ rather than the $arithmetic$ $mean$ to use this idea.

We propose using a weight conversion, referred to as LOG-LINEAR, that has meaningful interpretation for the weights of the clauses. The LOG-LINEAR method uses the following formula to assign weight to clauses:

$$w_j := log(p(v_i|\pi)) - log(1/k).$$

Weights set using this method are measuring the information gain provided by the parent information π relative to the uniform probability baseline. These weights can be interpreted as usual in a linear model: a positive weight indicates that a predictive factor increases the baseline probability, a negative weight indicates a decreased probability relative to the baseline. A zero weight indicates a condition that is irrelevant in the sense of not changing the baseline probability. With the LOG-LINEAR transformation some of the formulas receive positive and some negative weights so a clause with many groundings can have a small impact if the weight of the clause is close to zero. Figure 5 shows the two conversion functions.

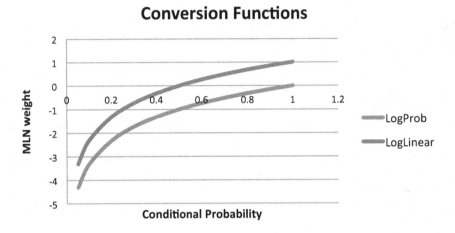

Fig. 5. Comparison of LOGPROB and LOG-LINEAR for a functor with a range of two values. Conditional probabilities smaller than 0.5 have negative weights and conditional probabilities larger than 0.5 have positive weights.

2.2   Conversion in Functor Bayes Nets with Parameter Reduction

The LOGPROB method has short-comings with conversion in FBNs without parameter reduction but still performs reasonably well. In this section we explain why this method performs very poorly when parameter reduction techniques are used on Bayes nets.

Focusing on the inference procedure in MLNs using our running example, the MLN inference model evaluates the likelihood for all possible ranking values, so likelihood of Ranking(Jack, 1), Ranking(Jack, 2), and Ranking(Jack, 3) are calculated and the most likely one will be picked as the ranking for Jack. With the tabular conditional probabilities that have no parameter reduction, all possible combinations of values are converted into MLN clauses. For instance, clauses with weights w_4 -w_{12} are all the different combinations of values assigned to the first clause in Figure 4. The same is also true for the second clause in Figure 4. This is not the case when parameter reduction techniques are used. Figure 6 shows the decision tree for *grade*, and Figure 7 shows the two types of clauses that are used for the *ranking* node when parameter reduction techniques are used.

Based on these clauses, the likelihood of $Ranking(Jack, 2)$ and $Ranking(Jack, 3)$ will be calculated using one grounding on the first clause, and the likelihood of $Ranking(Jack, 1)$ will be calculated using the grounding on the first clause plus the five groundings on the second clause. Log-probabilities are negative, so using the LOGPROB method means that frequently many negative weights will be added up in evaluating $Ranking(Jack, 1)$ compared to $Ranking(Jack, 2)$ and

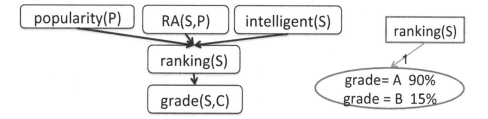

Fig. 6. The decision tree used to store the parameters of *grade*. *Ranking* node only contributes in this decision tree with value 1, so values 2 and 3 are independent of *grade*

Clause	#Grounds
Pop(P,POP), Int(S,INT), RA(S,P,RA), Rank(S,RANK)	n1
Ranking(S,RANK_1) Grade(S,C,Grade)	n2

Fig. 7. Clauses that the *ranking* node participates in. P, S, and C are first order variables for professors, students, and courses. POP, INT, RA, $RANK$, $GRADE$ are constant values in the range of their corresponding functors. $RANK_1$ in the second clause indicates that this clause can only be instantiated when Ranking=1.

$Ranking(Jack, 3)$. Thus, LOGPROB induces a bias against values that satisfy formulas with more groundings.

With the LOG-LINEAR transformation, some of the formulas receive positive and some negative weights, so there is no bias against values that are involved in formulas with more groundings. That is, the influences of the different groundings are more balanced against each other.

3   Experimental Design

The main objective of this evaluation is to show that our proposed conversion function performs better than the previously proposed method. The second objective is to show that Moralization methods are very fast and competitive with state-of-the-art MLN learners. We first introduce the datasets used, then the systems compared, and finally the comparison metrics.

3.1   Datasets

We used five benchmark real-world datasets. Table 1 lists the resulting databases and their sizes in terms of total number of tuples and number of ground atoms, which is the input format for Alchemy. Each descriptive attribute is represented as a separate function, so the number of ground atoms is larger than that of tuples.

MovieLens Database. The first dataset is the MovieLens dataset from the UC Irvine machine learning repository [10].

Mutagenesis Database. This dataset is widely used in ILP research [18]. It contains information on Atoms, Molecules, and Bonds among them.

Hepatitis Database. This data is a modified version of the PKDD02 Discovery Challenge database, following [19]. The database contains information on the laboratory examinations of hepatitis B and C infected patients.

Mondial Database. This dataset contains data from multiple geographical web data sources. We follow the modification of [20], and use a subset of the tables and features. Our dataset includes a self-relationship table *Borders* that relates two countries.

UW-CSE Database. This dataset lists facts about the Department of Computer Science and Engineering at the University of Washington (UW-CSE) (e.g., Student, Professor) and their relationships (i.e. AdvisedBy, Publication). The dataset was obtained by crawling pages on the department's Website (www.cs.washington.edu).

Table 1. Size of datasets in total number of table tuples and ground atoms

Dataset	#tuples	#Ground atoms
Movielens	82623	170143
Mutagenesis	15218	35973
Hepatitis	12447	71597
Mondial	814	3366
UW-CSE	2099	3380

3.2 Comparison Systems and Performance Metrics

Structure learning. We fix the structure for all the methods to evaluate just the parameters. We use two different structure learning methods to evaluate the parameter learning methods both with and without parameter reduction. We used the **MBN** [10] to get tabular representation without parameter reduction and **MBN-DT** [17] to get a sparse structure with decision trees. Both methods use GES search [21] and the BDeu score as implemented in version 4.3.9-0 of CMU's Tetrad package (structure prior uniform, ESS=10; [22]).

Our experiments compare the two conversion functions as well as state-of-the-art MLN learning methods.

LOGPROB. Weight learning is carried out using log-probabilities as suggested by Richardson and Domingos. Parameters are given by Tetrad's maximum likelihood estimation method and the LOGPROB conversion.

LOG-LINEAR. weight learning is the same as the LOGPROB method but we use our proposed conversion method.

MLN. Weight learning is carried out using the procedure of Lowd and Domingos [7, 23], implemented in Alchemy.

LSM. Learning Structural Motifs [8] uses random walks to identify densely connected objects in the data, and groups them and their associated relations into a motif. We input the structure of the learn-and-join algorithms to LSM. Running LSM's structure learning algorithm tries to prune the structure.

We report measurements on runtime, accuracy, and conditional log-likelihood (CLL). To define accuracy, we apply MLN inference to predict the probability of an attribute value, and score the prediction as correct if the most probable value is the true one. For example, to predict the gender of person Bob, we apply MLN inference to the atoms gender(Bob, male) and gender(Bob, female). The result is correct if the predicted probability of gender(Bob, male) is greater than that of gender(Bob, female). The conditional log-likelihood (CLL) of a ground atom in a database $\mathcal{D}$ given an MLN is its log-probability given the MLN and $\mathcal{D}$ [2]. The CLL directly measures how precise the estimated probabilities are. The values we report are averages over all attribute predicates.

Infernce. We use the MC-SAT inference algorithm [24] implemented in Alchemy to compute a probability estimate for each possible value of a descriptive attribute for a given object or tuple of objects.

We also compare our model with discriminative learning methods from Inductive Logic Programming.

4 Evaluation Results

In the following section we discuss the run time and then the accuracy of the models. We investigate the predictive performance by doing five-fold cross validation on the given datasets. All experiments were done on a QUAD CPU Q6700 with a 2.66GHz CPU and 8GB of RAM.

4.1 Run Times

Table 2 shows the time taken in seconds for learning the parameters for Markov Logic Networks using the structures generated by MBN and MBN-DT. The time for the conversion methods is basically the same, namely the time required to compute the database statistics for the entries. For the purposes of discussing runtime, we group LOGPROB

Table 2. The time taken in seconds for parameter learning. we group LOGPROB and LOG-LINEAR methods and call it Log/Lin in this table.

Structure Learning	MBN			MBN-DT		
Parameter Learning	Log/Lin	MLN	LSM	Log/Lin	MLN	LSM
UW-CSE	2	5	80	3	3	8
Mondial	3	90	260	3	15	26
MovieLens	8	10800	14300	9	1800	2100
Mutagenesis	3	9000	58000	4	600	1200
Hepatitis	3	23000	34200	5	4000	5000

and LOG-LINEAR methods and call it Log/Lin in this table. The runtime improvements are orders of magnitude faster than previous methods. Results from extending the moralization approach to parameter learning, provide strong evidence that the moralization approach leverages the scalability of relational databases for data management and Bayes nets learning to achieve scalable MLN learning on databases of realistic size for both structure and parameter learning.

4.2 Accuracy

Tables 3 and 4 show the accuracy results using the MBN and MBN-DT respectively. Higher numbers indicate better performance. For the MBN structure, LSM clearly performs worse. The LOGPROB method performs reasonably well and the results are acceptable.The LOG-LINEAR and MLN methods are competitive and out-perform the other two methods. We will call LOG-LINEAR superior to MLN since learning in LOG-LINEAR is approximately two orders of magnitude faster

For the MBN-DT structure, the LOGPROB method performs very poorly as discussed previously. Since the prediction is mainly based on the number of groundings for clauses, the performance is similar to randomly assigning values for predicates. The LOG-LINEAR and MLN methods are competitive, but MLN tends to do slightly better.

4.3 Conditional Log-Likelihood

Tables 5 and 6 show the predicted average log-likelihood of each fact in the database for the MBN and MBN-DT structure. This measure is especially sensitive to the quality of the parameter estimates. Smaller negative numbers indicate better performance.

The performance on CLL is very similar to accuracy. LOG-LINEAR method does much better than LOGPROB on both MBN and MBN-DT structure. Both MLN and LSM method minimizes CLL in their learning procedure which explains their great performance.

4.4 Comparison with Inductive Logic Programming on Mutagenesis

Table 3 and 4 show results on generative learning where averages over all predicates are reported. In this section we compare the performance of the LOGPROB and LOG-LINEAR algorithm for a classification task, discriminative learning, to predict the mutagenicity of chemical compounds. The class attribute is the mutagenicity. Compounds recorded as having positive mutagenicity are labeled active (positive examples) and compounds recoreded as having 0 or negative mutagenicity are labeled inactive (negative examples). The database contains a total of 188 compounds.

This problem has been extensively studied in Inductive Logic Programming (ILP). The purpose of this comparison is to benchmark the predictive performance of the moralization approach, using generative learning, against discriminative learning by methods that are different from Markov Logic Network learners. Table 7 presents the results of Lodhi and Muggleton [25]. For the STILL system, we followed the creators' evaluation methodology of using a randomly chosen training and test set. The other

Table 3. The 5-fold cross-validation estimate using MBN structure learning for the accuracy of predicting the true values of descriptive attributes, averaged over all descriptive attribute instances. Observed standard deviations are shown.

	LOGPROB	LOG-LINEAR	MLN	LSM
UW-CSE	0.72 ± 0.083	0.76 ± 0.022	0.75 ± 0.028	0.64 ± 0.086
Mondial	0.40 ± 0.060	0.41 ± 0.045	0.44 ± 0.050	0.32 ± 0.042
Movielens	0.64 ± 0.006	0.64 ± 0.006	0.60 ± 0.029	0.57 ± 0.016
Mutagenesis	0.55 ± 0.139	0.64 ± 0.025	0.61 ± 0.022	0.64 ± 0.029
Hepatitis	0.49 ± 0.033	0.50 ± 0.037	0.51 ± 0.025	0.30 ± 0.028

Table 4. The 5-fold cross-validation estimate using MBN-DT structure learning for the accuracy of predicting the true values of descriptive attributes, averaged over all descriptive attribute instances. Observed standard deviations are shown.

	LOGPROB	LOG-LINEAR	MLN	LSM
UW-CSE	0.06 ± 0.088	0.73 ± 0.166	0.75 ± 0.086	0.65 ± 0.076
Mondial	0.18 ± 0.036	0.43 ± 0.027	0.44 ± 0.033	0.31 ± 0.024
Movielens	0.26 ± 0.017	0.62 ± 0.026	0.62 ± 0.023	0.59 ± 0.051
Mutagenesis	0.21 ± 0.021	0.61 ± 0.023	0.60 ± 0.027	0.61 ± 0.025
Hepatitis	0.19 ± 0.024	0.48 ± 0.032	0.50 ± 0.021	0.40 ± 0.032

Table 5. The 5-fold cross-validation estimate using MBN for the conditional log-likelihood assigned to the true values of descriptive attributes, averaged over all descriptive attribute instances. Observed standard deviations are shown.

	LOGPROB	LOG-LINEAR	MLN	LSM
UW-CSE	-0.45 ± 0.122	-0.37 ± 0.090	-0.40 ± 0.151	-0.42 ± 0.058
Mondial	-2.01 ± 0.721	-1.53 ± 0.287	-1.28 ± 0.149	-1.30 ± 0.055
Movielens	-2.10 ± 0.205	-2.10 ± 0.205	-0.79 ± 0.338	-0.76 ± 0.124
Mutagenesis	-0.87 ± 0.040	-0.83 ± 0.047	-0.92 ± 0.125	-0.92 ± 0.039
Hepatitis	-2.11 ± 1.335	-2.03 ± 1.401	-1.31 ± 0.053	-1.26 ± 0.215

Table 6. The 5-fold cross-validation estimate using MBN-DT for the conditional log-likelihood assigned to the true values of descriptive attributes, averaged over all descriptive attribute instances. Observed standard deviations are shown.

	LOGPROB	LOG-LINEAR	MLN	LSM
UW-CSE	-1.24 ± 0.035	-0.76 ± 0.124	-0.46 ± 0.104	-0.54 ± 0.046
Mondial	-1.43 ± 0.071	-1.19 ± 0.101	-1.24 ± 0.083	-1.29 ± 0.071
Movielens	-2.63 ± 0.027	-1.27 ± 0.096	-0.85 ± 0.030	-0.93 ± 0.071
Mutagenesis	-1.26 ± 0.098	-0.95 ± 0.046	-0.93 ± 0.050	-0.92 ± 0.130
Hepatitis	-1.55 ± 0.547	-1.50 ± 0.039	-1.13 ± 0.052	-1.20 ± 0.048

Table 7. A comparison of the LOG-LINEAR method with standard Inductive Logic Programming systems trained to predict mutagenicity. Although Bayes net learning produces a generative model, its performance is competitive with discriminative learners.

Method	Evaluation	Accuracy	Reference
MBN-LOG-LINEAR	10-fold	0.87	
MBNDT-LOG-LINEAR	10-fold	0.87	
P-progol	10-fold	0.88	[18]
FOIL	10-fold	0.867	[26]
STILL	90%train-10%test	0.936	[27]
MBN-LOG-LINEAR	90%train-10%test	0.944	
MBNDTLOG-LINEAR	90%train-10%test	0.944	

systems are evaluated using 10-fold cross-validation. The table shows that the classification performance of the generative Moralized Bayes net model matches that of the discriminative Inductive Logic Programming models.

5 Conclusion and Future Work

The moralization approach combines Bayes net learning, one of the most successful machine learning techniques, with Markov Logic networks, one of the most successful statistical-relational formalisms. Previous work applied the moralization method to learning MLN structure; in this paper we extended it to learning MLN parameters. We motivated and empirically investigated a new method for converting Bayes net parameters to MLN weights. For converting a first order Bayes net into an MLN, it was suggested to moralize the Bayes net to obtain the structure of the MLN and then use the log of the conditional probability table entries to calculate the weight of the clauses. This conversion is exact for converting propositional Markov networks to propositional Bayes Nets; however, it fails to perform well for the relational case. We theoretically analyze this conversion and introduce new methods of converting a Bayes net into an MLN.

Our evaluation on five medium-size benchmark databases with descriptive attributes indicates that compared to previous MLN learning methods, the moralization parameter learning approach improves the scalability and run-time performance by at least two orders of magnitude. Predictive accuracy is competitive or even superior.

References

[1] Getoor, L., Tasker, B.: Introduction to statistical relational learning. MIT Press (2007)
[2] Domingos, P., Richardson, M.: Markov logic: A unifying framework for statistical relational learning. In: [1]
[3] Kok, S., Summer, M., Richardson, M., Singla, P., Poon, H., Lowd, D., Wang, J., Domingos, P.: The Alchemy system for statistical relational AI. Technical report, University of Washington, Version 30 (2009)
[4] Schulte, O., Khosravi, H.: Learning graphical models for relational data via lattice search. Machine Learning, 41 pages (2012) (to appear)

[5] Schulte, O., Khosravi, H.: Learning directed relational models with recursive dependencies. Machine Learning (2012) (Forthcoming. Extended Abstract)

[6] Khosravi, H., Schulte, O., Hu, J., Gao, T.: Learning compact markov logic networks with decision trees. Machine Learning (2012) (Forthcoming. Extended Abstract. Acceptance Rate?)

[7] Lowd, D., Domingos, P.: Efficient weight learning for Markov logic networks. In: Kok, J.N., Koronacki, J., Lopez de Mantaras, R., Matwin, S., Mladenič, D., Skowron, A. (eds.) PKDD 2007. LNCS (LNAI), vol. 4702, pp. 200–211. Springer, Heidelberg (2007)

[8] Kok, S., Domingos, P.: Learning Markov logic networks using structural motifs. In: ICML, pp. 551–558 (2010)

[9] Taskar, B., Abbeel, P., Koller, D.: Discriminative probabilistic models for relational data. In: UAI, pp. 485–492 (2002)

[10] Khosravi, H., Schulte, O., Man, T., Xu, X., Bina, B.: Structure learning for Markov logic networks with many descriptive attributes. In: AAAI, pp. 487–493 (2010)

[11] Kersting, K., de Raedt, L.: Bayesian logic programming: Theory and tool. In: [1], ch. 10, pp. 291–318

[12] Friedman, N., Getoor, L., Koller, D., Pfeffer, A.: Learning probabilistic relational models. In: IJCAI, pp. 1300–1309. Springer (1999)

[13] Ramon, J., Croonenborghs, T., Fierens, D., Blockeel, H., Bruynooghe, M.: Generalized ordering-search for learning directed probabilistic logical models. Machine Learning 70, 169–188 (2008)

[14] Pearl, J.: Probabilistic Reasoning in Intelligent Systems. Morgan Kaufmann (1988)

[15] Poole, D.: First-order probabilistic inference. In: IJCAI, pp. 985–991 (2003)

[16] Boutilier, C., Friedman, N., Goldszmidt, M., Koller, D.: Context-specific independence in bayesian networks. In: UAI, pp. 115–123 (1996)

[17] Khosravi, H., Schulte, O., Hu, J., Gao, T.: Learning compact markov logic networks with decision trees. In: Muggleton, S.H., Tamaddoni-Nezhad, A., Lisi, F.A. (eds.) ILP 2011. LNCS, vol. 7207, pp. 20–25. Springer, Heidelberg (2012)

[18] Srinivasan, A., Muggleton, S., Sternberg, M., King, R.: Theories for mutagenicity: A study in first-order and feature-based induction. Artificial Intelligence 85, 277–299 (1996)

[19] Frank, R., Moser, F., Ester, M.: A method for multi-relational classification using single and multi-feature aggregation functions. In: Kok, J.N., Koronacki, J., Lopez de Mantaras, R., Matwin, S., Mladenič, D., Skowron, A. (eds.) PKDD 2007. LNCS (LNAI), vol. 4702, pp. 430–437. Springer, Heidelberg (2007)

[20] She, R., Wang, K., Xu, Y.: Pushing feature selection ahead of join. In: SIAM SDM (2005)

[21] Chickering, D.: Optimal structure identification with greedy search. Journal of Machine Learning Research 3, 507–554 (2003)

[22] The Tetrad Group: The Tetrad project (2008), http://www.phil.cmu.edu/projects/tetrad/

[23] Domingos, P., Lowd, D.: Markov Logic: An Interface Layer for Artificial Intelligence. Morgan and Claypool Publishers (2009)

[24] Poon, H., Domingos, P.: Sound and efficient inference with probabilistic and deterministic dependencies. In: AAAI (2006)

[25] Lodhi, H., Muggleton, S.: Is mutagenesis still challenging? In: Inductive Logic Programming, pp. 35–40 (2005)

[26] Quinlan, J.: Boosting first-order learning. In: Arikawa, S., Sharma, A.K. (eds.) ALT 1996. LNCS, vol. 1160, pp. 143–155. Springer, Heidelberg (1996)

[27] Sebag, M., Rouveirol, C.: Tractable induction and classification in first order logic via stochastic matching. In: IJCAI, pp. 888–893 (1997)

Bounded Least General Generalization

Ondřej Kuželka, Andrea Szabóová, and Filip Železný

Faculty of Electrical Engineering, Czech Technical University in Prague
Technicka 2, 16627 Prague, Czech Republic
{kuzelon2,szaboand,zelezny}@fel.cvut.cz

Abstract. We study a generalization of Plotkin's *least general generalization*. We introduce a novel concept called *bounded least general generalization w.r.t. a set of clauses* and show an instance of it for which polynomial-time reduction procedures exist. We demonstrate the practical utility of our approach in experiments on several relational learning datasets.

1 Introduction

Methods for construction of hypotheses in relational learning can be broadly classified into two large groups: methods based on specialization, so-called top-down methods, and methods based on generalization, so-called bottom-up methods. Our main motivation is to be able to learn clauses in a bottom-up manner more efficiently when we assume that there exist solutions to the learning problem from some fixed potentially infinite set. In this paper we describe a novel bottom-up method based on Plotkin's least general generalization operator [1]. We start by describing generalized versions of θ-subsumption and θ-reduction. Then we define a generalized version of least general generalization and we show that its reduced form can be computed in polynomial time w.r.t. practically relevant classes of clauses such as those with bounded treewidth. Informally, if a learning problem has a solution from a given set of clauses, such as clauses with bounded treewidth, then some equally good solution not necessarily from that set can be found using bounded least general generalizations. We demonstrate practical utility of the approach in experiments on several datasets.

2 Preliminaries: Subsumption, CSP, Treewidth

A first-order-logic clause is a universally quantified disjunction of first-order-logic literals. We treat clauses as disjunctions of literals and as sets of literals interchangeably. The set of variables in a clause A is written as $vars(A)$ and the set of all terms by $terms(A)$. Terms can be variables or constants. A substitution θ is a mapping from variables of a clause A to terms of a clause B. If A and B are clauses then we say that A θ-subsumes B, if and only if there is a substitution θ such that $A\theta \subseteq B$. If $A \preceq_\theta B$ and $B \preceq_\theta A$, we call A and B θ-equivalent (written $A \approx_\theta B$). The notion of θ-subsumption was introduced by Plotkin as

F. Riguzzi and F. Železný (Eds.): ILP 2012, LNAI 7842, pp. 116–129, 2013.
© Springer-Verlag Berlin Heidelberg 2013

an incomplete approximation of implication. Let A and B be clauses. If $A \preceq_\theta B$ then $A \models B$ but the other direction of the implication does not hold in general. If A is a clause and if there is another clause R such that $A \approx_\theta R$ and $|R| < |A|$ then A is said to be θ-reducible. A minimal such R is called θ-reduction of A.

An important tool exploited in this paper, which can be used for learning clausal theories, is Plotkin's least general generalization (LGG) of clauses. A clause C is said to be a least general generalization of clauses A and B (denoted by $C = LGG(A, B)$) if and only if $C \preceq_\theta A$, $C \preceq_\theta B$ and for every clause D such that $D \preceq_\theta A$ and $D \preceq_\theta B$ it holds $D \preceq_\theta C$. A least general generalization of two clauses C, D can be computed in time $\mathcal{O}(|C| \cdot |D|)$ [2]. Least general generalization can be used as a refinement operator in searching for hypotheses [3,4]. Basically, the search can be performed by iteratively applying LGG operation on examples or on already generated LGGs. A problem of approaches based on least general generalization is that the size of a LGG of a set of examples can grow exponentially in the number of examples. In order to keep the LGGs reasonably small θ-reduction is typically applied after a new LGG of some clauses is constructed [4]. Application of θ-reduction cannot guarantee that the size of LGG would grow polynomially in the worst case, however, it is able to reduce the size of the clauses significantly in non-pathological cases.

Constraint satisfaction [5] with finite domains represents a class of problems closely related to the θ-subsumption problems. This equivalence of CSP and θ-subsumption has been exploited by Maloberti and Sebag [6] who used off-the-shelf CSP algorithms to develop a fast θ-subsumption algorithm.

Definition 1 (Constraint Satisfaction Problem). *A constraint satisfaction problem is a triple $(\mathcal{V}, \mathcal{D}, \mathcal{C})$, where $\mathcal{V}$ is a set of variables, $\mathcal{D} = \{D_1, \ldots, D_{|\mathcal{V}|}\}$ is a set of domains of values (for each variable $v \in \mathcal{V}$), and $\mathcal{C} = \{C_1, \ldots, C_{|\mathcal{C}|}\}$ is a set of constraints. Every constraint is a pair (s, R), where s (scope) is an n-tuple of variables and R is an n-ary relation. An evaluation of variables θ satisfies a constraint $C_i = (s_i, R_i)$ if $s_i\theta \in R_i$. A solution is an evaluation that maps all variables to elements in their domains and satisfies all constraints.*

The CSP representation of the problem of deciding $A \preceq_\theta B$ has the following form. There is one CSP variable X_v for every variable $v \in vars(A)$. The domain of each of these CSP variables contains all terms from $terms(B)$. The set of constraints contains one k-ary constraint $C_l = (s_l, R_l)$ for each literal $l = pred_l(t_1, \ldots, t_k) \in A$. We denote by $I_{var} = (i_1, \ldots, i_m) \subseteq (1, \ldots, k)$ the indexes of variables in arguments of l (the other arguments might contain constants). The scope s_l of the constraint C_l is $(X_{t_{i_1}}, \ldots, X_{t_{i_m}})$ (i.e. the scope contains all CSP variables corresponding to variables in the arguments of literal l). The relation R_l of the constraint C_l is then constructed in three steps.

1. A set L_l is created which contains all literals $l' \in B$ such that $l \preceq_\theta l'$ (note that checking θ-subsumption of two literals is a trivial linear-time operation).
2. Then a relation R_l^* is constructed for every literal $l \in A$ from the arguments of literals in the respective set L_l. The relation R_l^* contains a tuple of terms $(t_1', \ldots, t_k')$ if and only if there is a literal $l' \in L_l$ with arguments $(t_1', \ldots, t_k')$.

3. Finally, the relation R_l of the constraint C_l is then the projection of R_l^* on indexes I_{var} (only the elements of tuples of terms which correspond to variables in l are retained).

Next, we exemplify this transformation process.

Example 1 (Converting θ-subsumption to CSP). Let us have clauses A and B as follows

$$A = hasCar(C) \vee hasLoad(C, L) \vee shape(L, box)$$

$$B = hasCar(c) \vee hasLoad(c, l_1) \vee hasLoad(c, l_2) \vee shape(l_2, box).$$

We now show how we can convert the problem of deciding $A \preceq_\theta B$ to a CSP problem. Let $\mathcal{V} = \{C, L\}$ be a set of CSP-variables and let $\mathcal{D} = \{D_C, D_L\}$ be a set of domains of variables from $\mathcal{V}$ such that $D_C = D_L = \{c, l_1, l_2\}$. Further, let $\mathcal{C} = \{C_{hasCar(C)}, C_{hasLoad(C,L)}, C_{shape(L,box)}\}$ be a set of constraints with scopes (C), (C, L) and (L) and with relations $\{(c)\}$, $\{(c, l_1), (c, l_2)\}$ and $\{(l_2)\}$, respectively. Then the constraint satisfaction problem given by $\mathcal{V}$, $\mathcal{D}$ and $\mathcal{C}$ represents the problem of deciding $A \preceq_\theta B$ as it admits a solution if and only if $A \preceq_\theta B$ holds.

The Gaifman (or primal) graph of a clause A is the graph with one vertex for each variable $v \in vars(A)$ and an edge for every pair of variables $u, v \in vars(A)$, $u \neq v$ such that u and v appear in a literal $l \in A$. Similarly, we define Gaifman graphs for CSPs. The Gaifman graph of a CSP problem $\mathcal{P} = (\mathcal{V}, \mathcal{D}, \mathcal{C})$ is the graph with one vertex for each variable $v \in \mathcal{V}$ and an edge for every pair of variables which appear in a scope of some constraint $c \in \mathcal{C}$. Gaifman graphs can be used to define treewidth of clauses or CSPs.

Definition 2 (Tree decomposition, Treewidth). *A tree decomposition of a graph $G = (V, E)$ is a labeled tree T such that: (i) every node of T is labeled by a non-empty subset of V, (ii) for every edge $(v, w) \in E$, there is a node of T with label containing v, w, (iii) for every vertex $v \in V$, the set of nodes of T with labels containing v is a connected subgraph of T. The width of a tree decomposition T is the maximum cardinality of a label in T minus 1. The treewidth of a graph G is the smallest number k such that G has a tree decomposition of width k. The treewidth of a clause is equal to the treewidth of its Gaifman graph. Analogically, the treewidth of a CSP is equal to the treewidth of its Gaifman graph.*

For example, all trees have treewidth 1, cycles have treewidth 2, rectangular $n \times n$ grid-graphs have treewidth n. Any graph with treewidth 1 is a forest. An illustration of Gaifman graphs of two exemplar clauses and their tree-decompositions is shown in Table 1. Note that tree decompositions are not unique. That is why treewidth is defined as the *maximum* cardinality of a label minus 1. Treewidth is usually used to isolate tractable sub-classes of NP-hard problems. Constraint satisfaction problems with treewidth bounded by k can be solved in polynomial time by the k-consistency algorithm[1] [8]. For constraint satisfaction problems

---

[1] In this paper we follow the conventions of [7]. In other works, e.g. [8], what we call *k-consistency* is known as *strong k + 1-consistency*.

Table 1. An illustration of Gaifman graphs and tree-decompositions of clauses

Clause	Gaifman graph	Tree decomposition

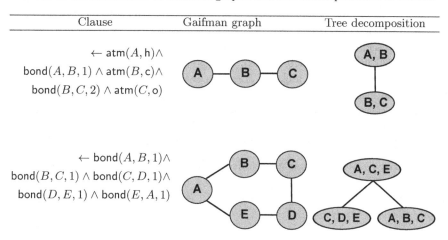

with generally unbounded treewidth, k-consistency is only a necessary but not a sufficient condition to have a solution. If the k-consistency algorithm returns *false* for a CSP problem $\mathcal{P}$ then $\mathcal{P}$ is guaranteed to have no solutions. So, equivalently, if the problem is soluble then k-consistency always returns *true*. If it returns *true* then the problem may or may not have some solutions. Finally, if the k-consistency algorithm returns *true* and $\mathcal{P}$ has treewidth bounded by k then $\mathcal{P}$ is guaranteed to have a solution.

The following description of the k-consistency algorithm is based on the presentation by Atserias et al. [7]. Let us have a CSP $\mathcal{P} = (\mathcal{V}, \mathcal{D}, \mathcal{C})$ where $\mathcal{V}$ is the set of variables, $\mathcal{D}$ is the set of domains of the variables and $\mathcal{C}$ is the set of constraints. A partial solution ϑ is an evaluation of variables from $\mathcal{V}' \subseteq \mathcal{V}$ which is a solution of the sub-problem $\mathcal{P}' = (\mathcal{V}', \mathcal{D}, \mathcal{C})$. If ϑ and φ are partial solutions, we say that φ extends ϑ (denoted by $\vartheta \subseteq \varphi$) if $Supp(\vartheta) \subseteq Supp(\varphi)$ and $V\vartheta = V\varphi$ for all $V \in Supp(\vartheta)$, where $Supp(\vartheta)$ and $Supp(\varphi)$ denote the sets of variables which are affected by the respective evaluations ϑ and φ. The k-consistency algorithm then works as follows:

1. Given a constraint satisfaction problem $\mathcal{P} = (\mathcal{V}, \mathcal{D}, \mathcal{C})$ and a positive integer k.
2. Let H be the collection of all partial solutions ϑ with $|Supp(\vartheta)| < k + 1$.
3. For every $\vartheta \in H$ with $|Supp(\vartheta)| \leq k$ and every $V \in \mathcal{V}$, if there is no $\varphi \in H$ such that $\vartheta \subseteq \varphi$ and $V \in Supp(\varphi)$, remove ϑ and all its extensions from H.
4. Repeat step 3 until H is unchanged.
5. If H is empty return *false*, else return *true*.

A basic property of k-consistency that we will also need is the following. If the k-consistency algorithm returns *true* for a CSP problem then it will also return *true* for any problem created from the original problem by removing some variables and some constraints, i.e. with a *subproblem*. This can be seen by noticing that if the k-consistency algorithm starts with a set H of partial solutions and returns *true* then it must also return *true* if it starts with a superset of this set. The set of partial solutions of the subproblem must necessarily be a superset of the set of partial solutions of the original problem projected on the variables of the subproblem (from monotonicity of constraints).

It is easy to check that if a clause A has treewidth bounded by k then also the CSP representation of the problem of deciding $A \preceq_\theta B$ has treewidth bounded by k for any clause B. It is known that due to this and due to the equivalence of CSPs and θ-subsumption, the problem of deciding θ-subsumption $A \preceq_\theta B$ can be solved in polynomial time when clause A has bounded treewidth.

Proposition 1. *We say that clause A is k-consistent w.r.t. clause B (denoted by $A \lhd_k B$) if and only if the k-consistency algorithm executed on the CSP representation of the problem of deciding $A \preceq_\theta B$ returns* true. *If A has treewidth at most k and $A \lhd_k B$ then $A \preceq_\theta B$.*

Proof. Follows directly from the solubility of CSPs with bounded treewidth by the k-consistency algorithm [7] and from the equivalence of CSPs and θ-subsumption shown earlier in this section.

3   Bounded Subsumption

In this section, we introduce *bounded* versions of θ-subsumption and develop methods for working with them. We start by defining x-subsumption and x-equivalence which are weaker versions of θ-subsumption and θ-equivalence.

Definition 3 (x-subsumption, x-equivalence). *Let X be a possibly infinite set of clauses. Let A, B be clauses not necessarily from X. We say that A x-subsumes B w.r.t. X (denoted by $A \preceq_X B$) if and only if $(C \preceq_\theta A) \Rightarrow (C \preceq_\theta B)$ for every clause $C \in X$. If $A \preceq_X B$ and $B \preceq_X A$ then A and B are called x-equivalent w.r.t. X (denoted by $A \approx_X B$). For a given set X, the relation $\preceq_X$ is called x-subsumption w.r.t. X and $\approx_X$ is called x-equivalence w.r.t. X.*

Conventional θ-subsumption is a special case of x-subsumption. It is an x-subsumption w.r.t. the set of all clauses. It is not hard to check that x-subsumption is a transitive and reflexive relation on clauses and that x-equivalence is an equivalence-relation on clauses. Definition 3 provides no efficient way to decide x-subsumption between two clauses as it demands θ-subsumption of an infinite number of clauses to be tested in some cases. However, for many practically relevant sets of clauses X, there is a relation called *x-presubsumption* which implies x-subsumption and has other useful properties as we shall see later (for example, it allows quick finding of reduced versions of clauses etc.).

Definition 4 (x-presubsumption). *Let X be a set of clauses. If $\preceq_X$ is the x-subsumption w.r.t. X and $\lhd_X$ is a relation such that: (i) if $A \lhd_X B$ and $C \subseteq A$ then $C \lhd_X B$, (ii) if $A \in X$, ϑ is a substitution and $A\vartheta \lhd_x B$ then $A \preceq_\theta B$, (iii) if $A \preceq_\theta B$ then $A \lhd_X B$. Then we say that $\lhd_X$ is an x-presubsumption w.r.t. the set X.*

The next proposition shows that if X is a set of clauses and $\lhd_X$ is an x-presubsumption w.r.t. X then $\lhd_X$ provides a sufficient condition for x-subsumption w.r.t. X.

Proposition 2. *Let X be a set of clauses. If $\preceq_X$ is x-subsumption on X and $\lhd_X$ is an x-presubsumption w.r.t. X then $(A \lhd_X B) \Rightarrow (A \preceq_X B)$ for any two clauses A, B (not necessarily from X).*

Proof. We need to show that if $A \lhd_x B$ then $(C \preceq_\theta A) \Rightarrow (C \preceq_\theta B)$ for all clauses $C \in X$. First, if $A \lhd_x B$ and $C \npreceq_\theta A$ then the proposition holds trivially. Second, $C \preceq_\theta A$ means that there is a substitution ϑ such that $C\vartheta \subseteq A$. This implies $C\vartheta \lhd_X B$ using the condition 1 from definition of x-presubsumption. Now, we can use the second condition which gives us $C \preceq_\theta B$ (note that $C \in X$ and $C\vartheta \lhd_X B$).

We will use this proposition in the next section where we deal with bounded reduction of clauses. We will use it for showing that certain procedures which transform clauses always produce clauses which are x-equivalent w.r.t. a given set X.

In the experiments presented in this paper, we use x-presubsumption w.r.t. bounded-treewidth clauses based on k-consistency algorithm.

4   Bounded Reduction

Proposition 2 can be used to check whether two clauses are x-equivalent w.r.t. a given set of clauses X. It can be therefore used to search for clauses which are smaller than the original clause but are still x-equivalent to it. This process, which we term *x-reduction*, will be an essential tool in the bottom-up relational learning algorithm presented in this paper.

Definition 5 (x-reduction). *Let X be a set of clauses. We say that a clause $\widehat{A}$ is an x-reduction of clause A if and only if $\widehat{A} \preceq_\theta A$ and $A \preceq_X \widehat{A}$ (where $\preceq_X$ denotes x-subsumption w.r.t. X) and if this does not hold for any clause $B \subsetneq \widehat{A}$ (i.e. if there is no $B \subsetneq \widehat{A}$ such that $B \preceq_\theta A$ and $A \preceq_X B$).*

For a given clause, there may be even smaller x-equivalent clauses than its x-reductions but those might be *more specific* than the original clause which would be a problem for computing bounded least general generalizations (introduced later in this paper). There may also be multiple x-reductions differing by their lengths for a single clause.

Example 2. Let $X = \{C\}$ be the set containing just the clause $C = e(V, W) \lor e(W, X) \lor e(X, Y) \lor e(Y, Z)$. Let us have another clause $A = e(A, B) \lor e(B, C) \lor e(C, A)$ for which we want to compute its x-reduction w.r.t. the set X. We can check relatively easily (e.g. by enumerating all subsets of literals from A) that the only x-reduction of A is A itself (up to renaming of variables). However, there is also a smaller clause x-equivalent to A and that is $A' = e(X, X)$. The

x-equivalence of A and A' follows from the fact that $C \preceq_\theta A$ and $C \preceq_\theta A'$ and there is no other clause other than C in the set X. It might seem that the clauses A and A' are x-equivalent only because the set X used in this example is rather pathological but, in fact, the two clauses are also x-equivalent w.r.t. the set of all clauses with treewidth at most 1.

In order to be able to compute x-reductions, we would need to be able to decide x-subsumption. However, we very often have efficient decision procedures for x-presubsumption, but not for x-subsumption. Importantly, if there is an x-presubsumption $\lhd_X$ w.r.t. a set X decidable in polynomial time then there is a polynomial-time algorithm for computing good approximations of x-reductions. We call this algorithm *literal-elimination algorithm*.

Literal-Elimination Algorithm:

1. Given a clause A for which the x-reduction should be computed.
2. Set $A' := A$, $CheckedLiterals := \{\}$.
3. Select a literal L from $A' \setminus CheckedLiterals$. If there is no such literal, return A' and finish.
4. If $A \lhd_X A' \setminus \{L\}$ then set $A' := A' \setminus \{L\}$, else add L to $CheckedLiterals$.
5. Go to step 3.

The next proposition states formally the properties of the literal-elimination algorithm. It also gives a bound on the size of the reduced clause which is output of the literal-elimination algorithm.

Proposition 3. *Let us have a set X and a polynomial-time decision procedure for checking $\lhd_X$ which is an x-presubsumption w.r.t. the set X. Then, given a clause A on input, the literal-elimination algorithm finishes in polynomial time and outputs a clause $\widehat{A}$ satisfying the following conditions:*

1. *$\widehat{A} \preceq_\theta A$ and $A \preceq_X \widehat{A}$ where $\preceq_X$ is an x-subsumption w.r.t. the set X.*
2. *$|\widehat{A}| \leq |\widehat{A}_\theta|$ where $\widehat{A}_\theta$ is a θ-reduction of a subset of A's literals with maximum length.*

Proof. We start by proving $\widehat{A} \preceq_\theta A$ and $A \preceq_X \widehat{A}$. This can be shown as follows. First, $A \lhd_X A'$ holds in any step of the algorithm which also implies $A \preceq_X A'$ using Proposition 2 and consequently also $A \preceq_X \widehat{A}$ because $\widehat{A} = A'$ in the last step of the algorithm. Second, $\widehat{A} \preceq_\theta A$ because $\widehat{A} \subseteq A$. Now, we prove the second part of the proposition. What remains to be shown is that the resulting clause $\widehat{A}$ will not be bigger than $\widehat{A}_\theta$. Since $\widehat{A} \subseteq A$ and $A \preceq_\theta \widehat{A}_\theta$, it suffices to show that $\widehat{A}$ cannot be θ-reducible. Let us assume, for contradiction, that it is θ-reducible. If $\widehat{A}$ was θ-reducible, there would have to be a literal $L \in \widehat{A}$ such that $\widehat{A} \preceq_\theta \widehat{A} \setminus \{L\}$. The relation $\lhd_X$ satisfies $(A \preceq_\theta B) \Rightarrow (A \lhd_X B)$ therefore it would also have to hold $\widehat{A} \lhd_X \widehat{A} \setminus \{L\}$. However, then L should have been removed by the literal-elimination algorithm which is a contradiction with $\widehat{A}$ being output of it. The fact that the literal-elimination algorithm finishes in polynomial time follows from the fact that, for a given clause A, it calls the polynomial-time procedure for checking the relation $\lhd_X$ at most $|A|$ times (the other operations of the literal-elimination algorithm can be performed in polynomial time as well).

So, the output $\widehat{A}$ of the literal elimination algorithm has the same properties as x-reduction ($\widehat{A} \preceq_\theta A$ and $A \preceq_X \widehat{A}$) with one difference and that is that it may not be minimal in some cases, i.e. that there may be some other clause $B \subsetneq \widehat{A}$ and having these properties.

5   Bounded Least General Generalization

In this section, we show how x-reductions in general, and the literal-elimination algorithm in particular, can be used in bottom-up approaches to relational learning. We introduce a novel concept which we term *bounded least general generalization*. This new concept generalizes Plotkin's *least general generalization* of clauses.

Definition 6 (Bounded Least General Generalization). *Let X be a set of clauses. A clause B is said to be a bounded least general generalization of clauses $A_1, A_2, \ldots, A_n$ w.r.t. the set X (denoted by $B = LGG_X(A_1, A_2, \ldots, A_n)$) if and only if $B \preceq_\theta A_i$ for all $i \in \{1, 2, \ldots, n\}$ and if for every other clause $C \in X$ such that $C \preceq_\theta A_i$ for all $i \in \{1, 2, \ldots, n\}$, it holds $C \preceq_\theta B$.*

The set of all bounded least general generalizations of clauses $A_1, A_2, \ldots A_n$ w.r.t. a set X is a superset of the set of conventional least general generalizations of these clauses. This set of all bounded least general generalizations of clauses $A_1, A_2, \ldots A_n$ is also a subset of the set of all clauses which θ-subsume all $A_1, A_2, \ldots A_n$. There are two main advantages of bounded least general generalization over the conventional least general generalization. The first main advantage is that the reduced form of bounded least general generalization can be computed in polynomial time for many practically interesting sets X. The second main advantage is that this reduced form can actually be smaller than the reduced form of conventional least general generalization in some cases.

It is instructive to see why we defined bounded least general generalization in this particular way and not in some other, seemingly more meaningful, way. Recall that our main motivation in this paper is to be able to learn clauses more efficiently when we know (or assume) that there exist solutions to the learning problem (clauses) from some fixed potentially infinite set. Having this motivation in mind, one could argue that, for example, a more meaningful definition of bounded least general generalization should require the resulting clause to be from the set X. However, least general generalization would not exist in many cases if defined in this way, which is demonstrated in the next example.

Example 3. Let $X = \{C_1, C_2, \ldots\}$ be a set of clauses of the following form: $C_1 = e(A_1, A_2), C_2 = e(A_1, A_2) \vee e(A_2, A_3), C_3 = \ldots$. Let us also have the following two clauses: $A = e(K, L) \vee e(L, K), B = e(K, L) \vee e(L, M) \vee e(M, K)$. We would like to find a clause from X which would be their least general generalization but this is impossible for the following reason. Any clause from X θ-subsumes both A and B but none of them is least general because for any $C_i \in X$ we have $C_{i+1} \not\preceq_\theta C_i$, $C_{i+1} \preceq_\theta A$ and $C_{i+1} \preceq_\theta B$. On the other hand, bounded least general generalization, as actually defined, always exists which follows trivially from the fact that the conventional least general generalization as computed by Plotkin's algorithm is also a bounded least general generalization.

Note that we would have to face the same problems even if X consisted of more general clauses, for example, if X consisted of clauses of treewidth bounded by k or of acyclic clauses etc, so they are not caused by some specificities of the rather artificial set X.

The reduced forms of bounded least general generalizations can often be computed in polynomial time using the literal-elimination algorithm.

Proposition 4. *Let X be a set of clauses and let $\lhd_X$ be an x-presubsumption w.r.t. the set X then the clause*

$$B_n = litelim_X(LGG(A_n, litelim_X(LGG(A_{n-1}, litelim_X(LGG(A_{n-2}, \ldots))))))$$

is a bounded least general generalization of clauses A_1, A_2, ..., A_n w.r.t. X (here, $litelim_X(\ldots)$ denotes calls of the literal-elimination algorithm using $\lhd_X$).

Proof. First, we show that $B \preceq_\theta A_i$ for all $i \in \{1, 2, \ldots, n\}$ using induction on n. The base case $n = 1$ is obvious since then $B_1 = A_1$ and therefore $B_1 \preceq_\theta A_1$. Now, we assume that the claim holds for $n - 1$ and we will show that then it must also hold for n. First, $B_n = LGG(A_n, B_{n-1})$ θ-subsumes the clauses $A_1, \ldots, A_n$ which can be checked by recalling the induction hypothesis and definition of LGG. Second, $litelim_k(LGG(A_n, B_{n-1}))$ must also θ-subsume the clauses $A_1, \ldots, A_n$ because $litelim_k(LGG(A_n, B_{n-1})) \subseteq LGG(A_n, B_{n-1})$.

Again using induction, we now show that $C \preceq_\theta B_n$ for any $C \in X$ which θ-subsumes all A_i where $i \in \{1, \ldots, n\}$. The base case $n = 1$ is obvious since then $B_1 = A_1$ and therefore every C which θ-subsumes A_1 must also θ-subsume B_1. Now, we assume that the claim holds for $n - 1$ and we prove that it must also hold for n. That is we assume that

$$C' \preceq_\theta B_{n-1} = litelim_X(LGG(A_{n-1}, litelim_X(LGG(A_{n-2}, litelim_X(\ldots)))))$$

for any $C' \in X$ which θ-subsumes the clauses $A_1, A_2, \ldots, A_{n-1}$. We show that then it must also hold $C \preceq_E B_n = litelim_X(LGG(A_n, B_{n-1}))$ for any $C \in X$ which θ-subsumes the clauses $A_1, A_2, \ldots, A_n$. We have $C \preceq_\theta LGG(A_n, B_{n-1})$ because $C \preceq_\theta B_{n-1}$ which follows from the induction hypothesis and because any clause which θ-subsumes both A_n and B_{n-1} must also θ-subsume $LGG(A_n, B_{n-1})$ (from the definition of LGG). It remains to show that C also θ-subsumes $litelim_X(LGG(A_n, B_{n-1}))$. This follows from

$$LGG(A_n, B_{n-1}) \preceq_X litelim_X(LGG(A_n, B_{n-1}))$$

(which is a consequence of Proposition 3) because if

$$LGG(A_n, B_{n-1}) \preceq_X litelim_X(LGG(A_n, B_{n-1}))$$

then

$$(C \preceq_\theta LGG(A_n, B_{n-1})) \Rightarrow (C \preceq_\theta litelim_X(LGG(A_n, B_{n-1})))$$

for any clause $C \in X$ (this is essentially the definition of x-subsumption).

One of the classes of clauses w.r.t. which the reduced forms of bounded LGGs can be computed efficiently is the class of clauses with bounded treewidth. What we need to show is that there is a polynomial-time decidable x-presubsumption relation. In the next proposition, we show that k-consistency algorithm [7] can be used to obtain such an x-presubsumption.

Proposition 5. *Let $k \in N$ and let $\lhd_k$ be a relation on clauses defined as follows: $A \lhd_k B$ if and only if the k-consistency algorithm run on the CSP-encoding (described in Section 2) of the θ-subsumption problem $A \preceq_\theta B$ returns true. The relation $\lhd_k$ is an x-presubsumption w.r.t. the set X_k of all clauses with treewidth at most k.*

Proof. We need to verify that $\lhd_k$ satisfies the conditions stated in Definition 4.

1. *If $A \lhd_k B$ and $C \subseteq A$ then $C \lhd_k B$.* This holds because if the k-consistency algorithm returns *true* for a problem then it must also return *true* for any of its subproblems (recall the discussion in Section 2). It is easy to check that if $C \subseteq A$ are clauses then the CSP problem encoding the θ-subsumption problem $C \preceq_\theta B$ is a subproblem of the CSP encoding of the θ-subsumption problem $A \preceq_\theta B$. Therefore this condition holds.
2. *If $A \in X$, ϑ is a substitution and $A\vartheta \lhd_k B$ then $A \preceq_\theta B$.* The CSP encoding of the problem $A \preceq_\theta B$ is a subproblem of the problem encoding $A\vartheta \preceq_\theta B$, in which there are additional constraints enforcing consistency with the substitution ϑ (because the set of constraints of the former is a subset of the constraints of the latter). Therefore if $A\vartheta \lhd_k B$ then also $A \lhd_k B$ and, since $A \in X$, it also holds $A \preceq_\theta B$.
3. *If $A \preceq_\theta B$ then $A \lhd_k B$.* This is a property of k-consistency (recall the discussion in Section 2).

6   Experiments

In this section, we describe experiments with a simple learning method based on three main ingredients: (i) bounded least general generalization w.r.t. clauses of treewidth 1, (ii) sampling and (iii) propositionalization. The method is very simple once we use the machinery introduced in the previous sections. It repeats the following process for each different class label K_i until a specified number of features covering different subsets of examples is reached: It samples a random set of k examples $\{e_1, \ldots, e_k\}$ with class label K_i. Then it computes clauses $C_1 = litelim(LGG_X(e_1, e_2))$, $C_2 = litelim(LGG_X(C_1, e_3))$, $\ldots$, $litelim(LGG_X(C_{k-2}, e_k))$ and stores them. The LGGs are taken w.r.t. the set X of all clauses with treewidth 1 having constants in arguments specified by a simple language bias. In the end, it computes extensions of the stored clauses (i.e. computes the sets of examples covered by particular clauses). The constructed clauses and their extensions give rise to an attribute-value table in which attributes are clauses and the values of these attributes are Boolean values indicating whether a clause covers an example. This attribute-value table can be then used to learn an attribute-value classifier such as decision tree or random forest.

Computation of attribute-value tables is the only place where the exponential-time θ-subsumption is used. Using the rather costly θ-subsumption as a covering relation only for construction of the attribute-value table is a good compromise. From the positive side, θ-subsumption is intuitive (certainly more so than the polynomial-time x-subsumption) and also quite expressive (again, more expressive than x-subsumption) so it makes sense to use it for computing extensions. However, using θ-subsumption for reduction of clauses, for which we use the polynomial-time k-subsumption, would be much more costly.

We evaluated the novel method called Bottom in experiments with four relational datasets: Mutagenesis [9], Predictive Toxicology Challenge [10], Antimicrobial Peptides [11] and CAD [12]. The first two datasets are classical datasets used in ILP. The first dataset contains descriptions of 188 molecules labelled according to their mutagenicity. The second dataset consists of four datasets of molecules labelled according to their toxicity to female mice, male mice, female rats and male rats. The third dataset contains spatial structures of 101 peptides, which are short sequences of amino acids, labelled according to their antimicrobial activity. The last dataset contains description of class-labeled CAD documents (product structures).

Table 2. Accuracies estimated by 10-fold cross-validation using transformation-based learning with random forests and our propositionalization method (Bottom)

	Mutagenesis	PTC(FM)	PTC(MM)	PTC(FR)	PTC(MR)	Peptides	CAD
nFOIL	76.6	60.2	63.1	**67.0**	57.3	77.2	92.7
Bottom	**78.9**	**62.4**	**65.2**	59.5	**62.2**	**82.2**	**95.8**

In the experiments, we use random forest classifiers learned on attribute-value representations of the datasets constructed using our novel method. We compare the results obtained by our method with nFOIL [13] which is a state-of-the-art relational learning algorithm combining FOIL's hypothesis search and Naive Bayes classifier. Cross-validated predictive accuracies are shown in Table 2. Our method achieved higher predictive accuracies than nFOIL in all but one case. This could be attributed partly to the use of random forests instead of Naive Bayes. However, it indicates the ability of our method to construct meaningful features in reasonable time. We used sample size equal to five for all experiments. The number of features constructed for every class was set to 100. Features were always constructed only using training examples from the given fold. Numbers of trees of random forests were selected automatically using internal cross-validation. The average runtimes of our method were: 0.3 min for Peptides, 0.6 min for CAD, 2.4 min for Mutagenesis, 4.4 min for PTC(MR), 5.1 min for PTC(FM), 9.8 min for PTC(FR) and 9.9 min for PTC(MM).

7   Related Work

The first method that used least general generalization for clause learning was Golem [3]. Golem was restricted to ij-determinate clauses in order to cope with the possibly exponential growth of LGGs. However, most practical relational learning problems are highly non-determinate. A different approach was taken in [14] where an algorithm called ProGolem was introduced. ProGolem is based on so-called asymmetric relative minimal generalizations (ARMGs) of clauses relative to a bottom clause. Size of ARMGs is bounded by the size of bottom clause so there is no exponential growth of the sizes of clauses. However, ARMGs are not unique and are not *least*-general.

Recently, an approach related to ours has been introduced [15] in which arc-consistency was used for structuring the space of graphs. There, arc-consistency was used as a covering operator called AC-projection. In contrast, we do not use the weaker versions of θ-subsumption (x-subsumptions) as covering operators in this paper but we use them only for reduction of clauses which allows us to guarantee that if a solution of standard learning problems with bounded-tree-width exists, our method is able to solve the learning problem. Thus, our approach provides theoretical guarantees which relate directly to learning problems with standard notions of covering (i.e. θ-subsumption), whereas the other approach can provide guarantees only w.r.t to the weaker (and less intuitive) AC-projection covering relation. Our framework is also more general in that it allows various different classes of clauses w.r.t. which it can compute bounded LGGs.

Another approach related to ours is the work of Horváth et al. [4] which is also based on application of least general generalization. Their approach relies on the fact that least general generalization of clauses with treewidth 1 is again a clause with treewidth 1. Since clauses with treewidth 1 can be reduced in polynomial time and since θ-subsumption problems $A \preceq_\theta B$ where A has treewidth 1 can be decided in polynomial time as well, it is possible to construct features in a manner similar to ours using only polynomial-time reduction and θ-subsumption. As in our approach, the size of the clauses constructed as least general generalizations may grow exponentially with the number of learning examples. However, unlike our approach which does not put any restriction on learning examples, the approach of Horváth et al. requires learning examples to have treewidth 1. Our approach is therefore more general even if we consider just bounded least general generalization w.r.t. the set of clauses with treewidth 1.

In a similar spirit, Schietgat et al. [16] introduced a new method based on computing maximum common subgraphs of outerplanar graphs under so-called *block-and-bridge-preserving isomorphism* which can be done in polynomial time.

This method was demonstrated to be highly competitive to *all-inclusive* strategies based on enumeration of all frequent graphs while using much lower number of maximum common subgraphs. Since it requires learning examples to be outerplanar graphs, it could not be applied to some of our datasets (e.g. the CAD dataset or the dataset of antimicrobial peptides). Aside this, another difference to our method is that it is based on a restricted form of subgraph isomorphism whereas our method is based on θ-subsumption, i.e. on homomorphism.

8   Conclusions

We have introduced a new weakened version of least general generalization of clauses which has the convenient property that its reduced form can be computed in polynomial time for practically relevant classes of clauses. Although this paper is mostly theoretical, we have also shown the practical utility of the weakened LGG in experiments where it was able to quickly find good sets of features. In our ongoing work we are developing a learning system based on the concepts presented in this paper. The system uses the bounded operations for hypothesis search and it avoids using the exponential time procedures in the learning phase altogether. Its description would not fit in the limited space available. It will be described in a longer version of this paper.

Acknowledgements. This work was supported by the Czech Grant Agency through project 103/10/1875 *Learning from Theories.*

References

1. Plotkin, G.: A note on inductive generalization. Edinburgh University Press (1970)
2. Nienhuys-Cheng, S.-H., de Wolf, R. (eds.): Foundations of Inductive Logic Programming. LNCS, vol. 1228. Springer, Heidelberg (1997)
3. Muggleton, S., Feng, C.: Efficient induction of logic programs. In: ALT, pp. 368–381 (1990)
4. Horváth, T., Paass, G., Reichartz, F., Wrobel, S.: A logic-based approach to relation extraction from texts. In: De Raedt, L. (ed.) ILP 2009. LNCS, vol. 5989, pp. 34–48. Springer, Heidelberg (2010)
5. Dechter, R.: Constraint Processing. Morgan Kaufmann Publishers (2003)
6. Maloberti, J., Sebag, M.: Fast theta-subsumption with constraint satisfaction algorithms. Machine Learning 55(2), 137–174 (2004)
7. Atserias, A., Bulatov, A., Dalmau, V.: On the power of k-consistency. In: Arge, L., Cachin, C., Jurdziński, T., Tarlecki, A. (eds.) ICALP 2007. LNCS, vol. 4596, pp. 279–290. Springer, Heidelberg (2007)
8. Rossi, F., van Beek, P., Walsh, T. (eds.): Handbook of Constraint Programming. Elsevier (2006)
9. Srinivasan, A., Muggleton, S.H.: Mutagenesis: ILP experiments in a nondeterminate biological domain. In: ILP, pp. 217–232 (1994)
10. Helma, C., King, R.D., Kramer, S., Srinivasan, A.: The predictive toxicology challenge 2000-2001. Bioinformatics 17(1), 107–108 (2001)

11. Cherkasov, A., Jankovic, B.: Application of 'inductive' qsar descriptors for quantification of antibacterial activity of cationic polypeptides. Molecules 9(12), 1034–1052 (2004)
12. Žáková, M., Železný, F., Garcia-Sedano, J.A., Masia Tissot, C., Lavrač, N., Křemen, P., Molina, J.: Relational data mining applied to virtual engineering of product designs. In: Muggleton, S.H., Otero, R., Tamaddoni-Nezhad, A. (eds.) ILP 2006. LNCS (LNAI), vol. 4455, pp. 439–453. Springer, Heidelberg (2007)
13. Landwehr, N., Kersting, K., De Raedt, L.: Integrating naïve bayes and FOIL. Journal of Machine Learning Research 8, 481–507 (2007)
14. Muggleton, S., Santos, J., Tamaddoni-Nezhad, A.: Progolem: A system based on relative minimal generalisation. In: De Raedt, L. (ed.) ILP 2009. LNCS, vol. 5989, pp. 131–148. Springer, Heidelberg (2010)
15. Liquiere, M.: Arc consistency projection: A new generalization relation for graphs. In: Priss, U., Polovina, S., Hill, R. (eds.) ICCS 2007. LNCS (LNAI), vol. 4604, pp. 333–346. Springer, Heidelberg (2007)
16. Schietgat, L., Costa, F., Ramon, J., De Raedt, L.: Effective feature construction by maximum common subgraph sampling. Machine Learning 83(2), 137–161 (2011)

Itemset-Based Variable Construction in Multi-relational Supervised Learning

Dhafer Lahbib[1], Marc Boullé[1], and Dominique Laurent[2]

[1] Orange Labs - 2, avenue Pierre Marzin, 23300 Lannion
{dhafer.lahbib,marc.boulle}@orange.com
[2] ETIS-CNRS-Université de Cergy Pontoise-ENSEA, 95000 Cergy Pontoise
dominique.laurent@u-cergy.fr

Abstract. In multi-relational data mining, data are represented in a relational form where the individuals of the target table are potentially related to several records in secondary tables in one-to-many relationship. In this paper, we introduce an itemset based framework for constructing variables in secondary tables and evaluating their conditional information for the supervised classification task. We introduce a space of itemset based models in the secondary table and conditional density estimation of the related constructed variables. A prior distribution is defined on this model space, resulting in a parameter-free criterion to assess the relevance of the constructed variables. A greedy algorithm is then proposed in order to explore the space of the considered itemsets. Experiments on multi-relationalal datasets confirm the advantage of the approach.

Keywords: Supervised Learning, Multi-Relational Data Mining, one-to-many relationship, variable selection, variable construction.

1 Introduction

Most of existing data mining algorithms are based on an attribute-value representation. In this flat format, each record represents an individual and the columns represent variables describing these individuals. In real life applications, data usually present an intrinsic structure which is hard to express in a tabular format. This structure may be naturally described using the relational formalism where each object (target table record) refers to one or more records in other tables (secondary tables) through a foreign key.

Example 1. In the context of a Customer Relationship Management (CRM) problem, Figure 1 shows an extract of a virtual CRM relational database schema. The table *Customer* is the target table, whereas *Call detail record (CDR)* is a secondary table related to *Customer* through the foreign key CID. The problem may be, for instance, to identify the customers likely to be interested in a certain product. This problem turns into a classification problem where the target variable is the variable *Appetency*, which denotes whether the customer is likely to order a particular product.

F. Riguzzi and F. Železný (Eds.): ILP 2012, LNAI 7842, pp. 130–150, 2013.
© Springer-Verlag Berlin Heidelberg 2013

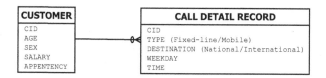

Fig. 1. Relational schema of a CRM database

Learning from relational data has recently received increasing attention in the literature. The term Multi-Relational Data Mining (MRDM) was initially introduced by [1] to address novel knowledge discovery techniques from multiple relational tables. The common point between these techniques is that they need to transform the relational representation. In Inductive Logic Programming (ILP) [2], data is recoded as logic formulas. Other methods known as propositionalisation [3] flatten the relational data by creating new variables.

Our goal in this article is to directly exploit the informativeness of secondary variables w.r.t. the target variable. We propose a multivariate pre-processing of secondary variables in order to construct multivariate itemsets from secondary tables. This pre-processing consists in a discretization in the numerical case and a value grouping in the categorical case. To the best of our knowledge, only few studies have considered the variable pre-processing problem within the multi-relational setting with one-to-many relationship (in particular, discretization and variables selection) [4,5,6]. In these approaches, secondary variables are considered independently from each other, which does not make it possible to take into account their correlation to predict the class label.

In our approach, the considered itemsets are in secondary tables while the class labels are in the target one. In order to evaluate these itemsets and exploit their information for classification, we construct new binary variables in the secondary tables. We propose a conditional density estimation of the constructed variables in order to extend the Naive Bayes classifier to multi-relational data, and we define a prior distribution on the itemsets model space. As a result, we obtain a parameter-free relevance criterion for the constructed variables.

Example 2. Given our CRM example, let us consider the following itemset π in the CDR secondary table: $(WeekDay \in \{Saturday\}) \wedge (10 : 00 : 00 \leq Time < 11 : 30 : 00) \wedge (Destination \in \{International\})$ where $WeekDay$ and $Destination$ are categorical variables and $Time$ is a numerical variable. This itemset allows constructing a new binary variable in the secondary table, according to whether the secondary records are covered or not by the itemset. For example, the secondary table record ("*C901*", "*Mobile*", "*International*", "*Saturday*", "10 : 30 : 00") is covered by π and therefore the value of A_π for that record is "1".

Computing features as a pre-processing step is a classical solution in MRDM in order to be able to use a propositional classifier. In the 1BC system, [7] compute a set of conjunctive patterns consisting of first-order conditions which are used as features in a classical Naive Bayes classifier. 1BC2 [8] and Mr-SBC [9] extend this approach with more accurate estimation of conditional probabilities and with improved results. It is worthy of mention that our method is not a propositionalisation approach [10]. The new binary variables are created in the secondary table, not in the target one. Their conditional probability is estimated directly by using the multi-relational approach introduced in [11].

The remaininder of this paper is organized as follows. In the next section, the present work is motivated and related to alternative approaches. Section 3 recalls the method [11] exploited to estimate the conditional probability of a binary secondary variable. Section 4 introduces the space of constructed itemset-based secondary variables and presents their evaluation criterion. This section also gives a heuristic algorithm in order to explore the itemset space. In Section 5, an experimental evaluation of the proposed approach on real-world multi-relational datasets is reported. Finally, Section 6 gives a summary and discusses future work.

2   Motivation and Related Work

Classifying data scattered over the multiple tables of a relational database has recently received a growing attention within the data mining community. In this paper we are interested in classifying individuals contained in a target table with a one-to-many relationship with secondary tables.

The novelty of this multi-relational setting, compared to classical attribute-value methods, consists in exploiting the predictive power of secondary variables belonging to secondary tables. The difficulty when dealing with these variables arises from the presence of one-to-many associations. In the attribute-value single table case, each individual has a single value per variable, while in multiple table setting, for a secondary variable, an individual may have a set of values (possibly empty) of varying size.

2.1   Motivation

The idea behind using itemsets on non target tables is to discover multivariate patterns between secondary variables in order to detect significant differences between individuals of distinct classes. In this paper we propose to use a multivariate approach. Instead of considering only one variable at a time, we introduce itemsets of secondary variables in order to take into account correlations between these variables w.r.t. to the target variable. In some problems, a single secondary variable may not be relevant to predict the class label, whereas several secondary variables considered jointly may help predicting the target variable.

Example 3. Let us consider the Digits[1] dataset where the task is to recognize handwritten digits (classes are digits from 0 to 9) [12]. In its original version, this dataset has an attribute-value tabular format where each line represents an image with 28×28 pixels. This database can be represented in a relational format

---

[1] Available at http://yann.lecun.com/exdb/mnist/

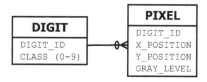

Fig. 2. Relational schema of the Digits database

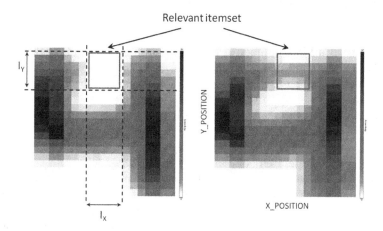

Fig. 3. Relevant itemset in the table Pixel of the Digits database

composed of two tables: a target table Digit as well as a secondary table Pixel which describes pixels composing each image. Figure 2 shows the corresponding relational schema. A handwritten digit is then associated with 784 lines in the table Pixel. A pixel is described by three secondary variables: *Gray_level* as well as *X_position* and *Y_position* which represent the position of the pixel in the original image.

The problem turns into a multi-relational classification problem. It is clear from Figure 3 that identifying the digit represented by the handwritten character (for example whether it is a 9 or a 4) requires taking into account simultaneously the values of the three secondary variables: *Gray_level* , *X_position* and *Y_position*. The itemset

$$\pi : (X\_position \in I_X) \wedge (Y\_position \in I_Y) \wedge (Gray\_level > 0)$$

gives a discriminant pattern characterizing a 9 or a 4 digit. The binary secondary variable which denotes whether a pixel is covered or not by this itemset can accurately predict the class.

2.2 Related Work

Mining itemsets of secondary variables can be seen as a descriptive task which joins many studies on association rules and frequent patterns. Classical solutions assume that data are stored in a single attribute-value data table. But many attempts have been proposed recently to deal with relational data.

First multi-relational association rules and frequent patterns are based on ILP in order to discover frequent Prolog queries ([13,14]) or frequent predicates ([15]). These methods need to transform the initial relational database into a deductive one. Furthermore, they have a high complexity and the mined patterns may be difficult to understand [16]. Some other approaches use a classical attribute-value association rule algorithm on a single table obtained by joining all the tables in order to generate a cross table [16] or by propagating the target variable to the secondary tables [17]. Such transformations may lead to statistical skews since individuals with a large number of related records in a secondary table will be overestimated thereby causing overfitting.

Beside descriptive tasks [18,19], association rules have been proposed for a classification purpose. [20] investigated the use of logical association rules in order to classify spatial data. Discrimininant features are generated based on these rules and are exploited for propositionalization and to propose an extension of the Naive Bayes classifier. The proposed approach is limited to spatial relational data described by a hierarchy of concepts.

[21] used emergent patterns over multiple tables in order to extend the Naive Bayes classifier to relational data. The considered emergent patterns are conjunctions of logical predicates modeling properties of the relational objects and associations between them. Using these patterns, the authors propose a decomposition of the posterior probability based on the naive Bayes assumption to simplify the probability estimation problem. The problem of this approach is that it suffers from a high number of considered emerging patterns and scalability limits since it is based on logical inference.

We notice that in this article we restrict ourselves to secondary variables located in tables with direct one-to-many relationships with the target table. More generally, we consider a star schema with a central target table related to secondary tables. The second level of one-to-many relations is intrinsically challenging and is left for future work.

3   Evaluation of Binary Secondary Variables

In this section, we summarize the method introduced in [11] to evaluate the relevance of a binary secondary variable A with values a and b.

3.1   Binary Secondary Variable Evaluation

In this case, each individual of the target table is described by a bag of secondary values among a and b, and summarized without loss of information by the numbers n_a of a and n_b of b. Thus, the whole information about the initial secondary variable can be captured by considering jointly the pair (n_a, n_b) of primary variables. We emphasize that the two variables are considered jointly so that information is preserved, as illustrated in Figure 4.

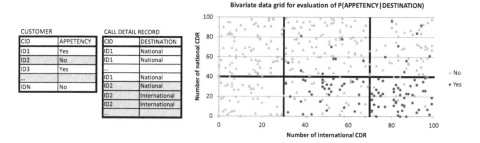

Fig. 4. Evaluation of the secondary variable *Destination* for the prediction of the target variable *Appetency*. Here, customers with a small number of national CDR and a large number of international CDR are likely to have the value *Yes* of *Appetency*.

By doing so, the conditional probability $P(Y \mid A)$ is equivalent to the probability $P(Y \mid n_a, n_b)$. Bivariate data grid models [22] are used to qualify the information contained in the pair (n_a, n_b). The couple of numeric variables are partitioned jointly into intervals. Individuals are then partitioned into a data grid whose cells are defined by interval pairs, and the target variable distribution is defined locally in each cell. Therefore, the purpose is to find the optimal bivariate discretization which maximizes the class distribution, in other words, the purpose is to obtain the optimal grid with homogeneous cells according to the class values. Applying a Bayesian model selection approach, a criterion $c_e(A)$ is obtained to assess the relevance of a secondary binary variable.

Notation 1.

- N : number of individuals (number of target table records)
- J : number of target values (classes),
- I_a, I_b : number of discretization intervals respectively for n_a and n_b
- $N_{i_a..}$: number of individuals in the interval i_a $(1 \le i_a \le I_a)$ for variable n_a
- $N_{.i_b.}$: number of individuals in the interval i_b $(1 \le i_b \le I_b)$ for variable n_b
- $N_{i_a i_b.}$: number of individuals in the cell (i_a, i_b)
- $N_{i_a i_b j}$: number of individuals in the cell (i_a, i_b) for the target value j

$$c_e(A) = \log N + \log N + \log \binom{N + I_a - 1}{I_a - 1} + \log \binom{N + I_b - 1}{I_b - 1} \tag{1}$$

$$+ \sum_{i_a=1}^{I_a} \sum_{i_b=1}^{I_b} \log \binom{N_{i_a i_b.} + J - 1}{J - 1} + \sum_{i_a=1}^{I_a} \sum_{i_b=1}^{I_b} \log \frac{N_{i_a i_b.}!}{N_{i_a i_b 1}! N_{i_a i_b 2}! \dots N_{i_a i_b J}!}$$

The first five terms of Formula 1 stand for the prior probability: choosing the numbers of intervals, their frequencies and the distribution parameters for the target values in each grid cell[2]. The last term represents the conditional likelihood of the data given the model. The bivariate discretization criterion is

---

[2] Notation $\binom{n}{k}$ represents the binomial coefficient: number of k-combinations of n elements.

optimized starting from an initial random solution, using a bottom-up greedy heuristic. Pre and post-optimization steps are employed based on alternating partial optimizations per variable. The overall complexity of the algorithm is $O(JN \log(N))$ [22]. Details regarding this criterion and the optimization algorithm can be found in [22]. Beyond the evaluation of binary secondary variables, our goal is to extend the method to numerical and categorical secondary variables while capturing the potential correlations that may exist between secondary variables. In the next section, we introduce a similar criterion for itemset-based models defined over sets of secondary variables, in order to take into account this multivariate correlation.

3.2   Naive Bayes Extension

The constructed secondary variable is used to build a Naive Bayes classifier which aims to classify an object o by maximizing the posterior probability $P(Y_j \mid o)$ that o is of class Y_j. This probability can be reformulated by applying the Bayes rule:

$$P(Y_j \mid o) = \frac{P(Y_j) P(o \mid Y_j)}{P(o)} \tag{2}$$

$P(Y_j)$ is the prior probability of the class Y_j and the term $P(o \mid Y_j)$ is estimated by using the Naive Bayes assumption: if X_k are descriptive variables, then $P(o \mid Y_j) = P(X_1, X_2, \cdots, X_K \mid Y = j) = \prod_{k=1}^{K} P(X_k \mid Y_j)$.

The above formulation of the Naive Bayesian classifier is clearly limited to the attribute-value representation. In order to take into account the multi-table data, in particular the secondary variables, we need to assume that the secondary variables are independent given the target variable, and then estimate the conditional probabilities $P(X_k \mid Y_j)$ where X_k are secondary variables.

Estimating $P(X_k \mid Y_j)$ is equivalent to evaluating $P(n_a, n_b \mid Y_j)$ which is performed by simple counting locally in each cell of the optimal bivariate data grid. More explicitly, if for an object o in test, the corresponding cell is (i_a, i_b), and if N_j denotes the number of objects of class j, $P(X_k \mid Y_j)$ can be estimated as follows :

$$P(X_k \mid Y_j) = \frac{N_{i_a i_b j}}{N_j}, \tag{3}$$

where $N_{i_a i_b j}$ stands for the number of objects in the cell (i_a, i_b) for the target value j.

4   Itemset Based Variable Construction

In this section, we introduce a method to construct new binary variables in secondary tables, based on itemset models. We first introduce a model of itemset based secondary variables, then present a criterion to evaluate these constructed variables, and finally propose an algorithm to construct itemset based variables and evaluate them.

4.1   Variable Construction Model

Let us first introduce a model of itemset based variable construction in a secondary table. We exploit the model of [23] to define itemsets with numerical or categorical variables, where the itemset is defined by a conjunction of intervals in the numerical case and sets of values in the categorical one, for a secondary table with a one-to-many path from the target table.

Definition 1 (Itemset Based Construction Model). *An (IBCM) itemset π, at the basis of a secondary Boolean constructed variable A_π, is defined by:*

- *the secondary table with a one-to-many path from the target table,*
- *the constituent variables of the itemset,*
- *the group of values involved in the itemset, for each categorical variable of the itemset,*
- *the interval involved in the itemset, for each numerical variable of the itemset,*

where the value of A_π is true for secondary records that are covered by the itemset, false otherwise.

An example of itemset is provided in Example 2.

In [24] we considered any interval and groups of values for the constituents of itemsets. In the numerical case, a number of discretization intervals is chosen, then the bounds of the interval, and finally, the index of the interval belonging to the itemset. In the categorical case, a number of groups of values is chosen, then the partition of the values into groups and finally the index of the group belonging to the itemset. In this paper, we focus on quantile partitions for each secondary variable. Given a partition, a quantile is defined solely by an index, whereas an interval requires two bounds. This allows us to consider itemset models which are both more interpretable and parsimonious, and enables efficient optimization heuristics. An itemset is then defined by a choice of a quantile part for each constituent variable, where the quantile parts are themselves defined solely by a size of partition and a part index. In definitions 2 and 3, we precisely define quantile partitions both for numerical and categorical variables.

Definition 2 (Numerical quantile partition). *Let D be a dataset of N instances and X a numerical variable. Let $x_1, x_2, \ldots, x_N$ be the N sorted values of X in dataset D. For a given number of parts P, the dataset is divided into P equal frequency intervals $] - \infty, x_{\lfloor 1 + \frac{N}{P} \rfloor}[, [x_{\lfloor 1 + \frac{N}{P} \rfloor}, x_{\lfloor 1 + 2\frac{N}{P} \rfloor}[, \ldots, [x_{\lfloor 1 + i\frac{N}{P} \rfloor}, x_{\lfloor 1 + (i+1)\frac{N}{P} \rfloor}[, \ldots, [x_{\lfloor 1 + (P-1)\frac{N}{P} \rfloor}, +\infty[.$*

Definition 3 (Categorical quantile partition). *Let D be a dataset of N instances and X a categorical variable with V values. For a given number of parts P, let $N_P = \lceil \frac{N}{P} \rceil$ be the expected minimum frequency per part. The categorical quantile partition into (at most) P parts is defined by singleton parts for each value of X with frequency beyond the threshold frequency N_P and a "garbage" part consisting of all values of X below the threshold frequency.*

Now, we can give a formal definition of IBCM itemset models, using the following notations.

Notation 2.

- $\mathcal{T} = \{T_1, T_2, \ldots\}$: *set of secondary tables having a one-to-many relation with the target table*
- $|\mathcal{T}|$: *number of secondary tables*
- $T \in \mathcal{T}$: *secondary table containing the variables which compose the itemset*
- N_s : *number of records in the secondary table T*
- m: *number of variables in the secondary table T*
- $X = \{x_1, \ldots, x_k\}$: *set of k variables of T which compose the itemset*
- I_x: *size of the quantile partition of variable x*
- i_x: *index of the quantile part of variable x involved in the itemset*

An IBCM itemset model π is then defined by the secondary table T, the set X of variables of T which compose the itemset, and for each constituent variable by the size I_x of the quantile partition and the index i_x of the quantile part involved in the itemset.

4.2   Evaluation of Constructed Variables

The new variable A_π that we have built is seen as a binary secondary variable which can be evaluated using the cost $c_e(A_\pi)$ of Formula 1. Let $c_e(\emptyset)$ be the null cost when no input variable is used estimate the target variable. This corresponds to a bivariate data grid with one single cell, whose cost is

$$c_e(\emptyset) = 2 \log N + \log \frac{N!}{N_1! N_2! \ldots N_J!}, \tag{4}$$

$$= N Ent(Y) + O(\log N), \tag{5}$$

where $Ent(Y)$ is the Shannon entropy of the target variable Y (cf. Formula 1 and using [25]). Therefore, any constructed variable with an evaluation cost beyond the null cost can be discarded, as being less informative than the target variable alone. When the number of constructed variables increases, the risk of wrong detection of informative variables grows. In order to prevent this risk of overfitting, we suggest to introduce a prior distribution over itemset based constructed variables, so as to get a construction cost $c_c(A_\pi)$ derived from a Bayesian approach. We then evaluate the overall relevance $c_r(A_\pi)$ of A_π by taking into account the construction cost $c_c(A_\pi)$ as well as the evaluation cost $c_e(A_\pi)$:

$$c_r(A_\pi) = c_c(A_\pi) + c_e(A_\pi). \tag{6}$$

4.3   Prior Distribution on Itemset Models

To apply the Bayesian approach, we need to define a prior distribution on the itemset based construction model space. We apply the following principles in order to guide the choice of the prior:

1. the prior is as flat as possible, in order to minimize the bias,
2. the prior exploits the hierarchy of the itemset models.

MODL hierarchical prior. We use the following distribution prior on IBCM item-sets, called the MODL hierarchical prior. Notice that a uniform distribution is used at each stage[3] of the parameters hierarchy of the IBCM models:

1. the itemset table T is uniformly distributed among the tables of $\mathcal{T}$

$$p(T) = \frac{1}{|\mathcal{T}|}. \tag{7}$$

2. the number of variables k in the itemset ($k \geq 0$) is distributed according to the universal prior for integers [4] [26]

$$p(k) = 2^{-L(k+1)}. \tag{8}$$

3. for a given number k of variables, every set of k constituent variables of the itemset is equiprobable, given a drawing with replacement. The number of such sets is given by $\binom{m+k-1}{k}$. We obtain

$$p(X|k) = \frac{1}{\binom{m+k-1}{k}}. \tag{9}$$

4. for each constituent variable x, the size I_x of the quantile partition is necessarily greater or equal to 2, and distributed according to the universal prior for integers

$$p(I_x) = 2^{-L(I_x-1)}. \tag{10}$$

5. for each constituent variable x_k, the index i_x of the quantile part is uniformly distributed between 1 and I_x

$$p(I_{i_x|I_x}) = \frac{1}{I_x}. \tag{11}$$

Given the definition of the model space and its prior distribution, we can now express the prior probabilities of an IBCM model.

---

[3] It does not mean that the hierarchical prior is a uniform prior over the itemset space, which would be equivalent to a maximum likelihood approach.

[4] This universal prior is defined so that the small integers are more probable than the large integers, and the rate of decay is taken to be as small as possible. The code length of the universal prior for integers is given by

$$L(n) = \log_2(c_0) + \log_2^*(n) = \log_2(c_0) + \sum_{j>1} max(\log_2^{(j)}(n), 0),$$

where $\log_2^{(j)}(n)$ is the jth composition of $\log_2$ ($\log_2^{(1)}(n) = \log_2(n)$, $\log_2^{(2)}(n) = \log_2(\log_2(n))$, ...) and $c_0 = \sum_{n>1} 2^{-\log_2^*(n)} = 2.865\ldots$ The universal prior for integers is then $p(n) = 2^{-L(n)}$.

Construction cost of an IBCM *variable.* We now have an analytical formula for the construction cost $c_c(A_\pi)$ of a secondary variable A_π constructed from an IBCM itemset π:

$$c_c(A_\pi) = \log |\mathcal{T}| + L(k+1)\log 2 + \log \binom{m+k-1}{k} \tag{12}$$

$$+ \sum_{x \in X} (L(I_x - 1)\log 2 + \log I_x).$$

The cost of an IBCM variable is the negative logarithm of probabilities which is no other than a coding length according to Shannon [27]. Here, $c_c(A_\pi)$ may be interpreted as a variable construction cost, that is the encoding cost of the itemset π. The first line in Formula 12 stands for the choice of the itemset table, the number of variables and the variables involved in the itemset. The second line is related to the choice of the size of the quantile partition and the quantile part for each variable involved in the itemset.

In Formula 6, the construction cost $c_c(A_\pi)$ acts as a regularization term. Constructed variables based on complex itemsets, with multiple constituent variables or with fine-grained constituent parts in the itemset, are penalized compared to simple constructed variables.

4.4   Variable Construction Algorithm

The objective is to construct a set of itemset based secondary variables in order to obtain a data representation suitable for supervised classification. The space of IBCM variables is so large that exhaustive search is not possible. We propose a greedy Algorithm 1 that constructs all potential variables based on quantile partitions of power of 2 sizes, given a maximum number Max_k of constituent variables in a secondary table T, a maximum size Max_s of partitions and a maximum number Max_c of constructed variables. For clarity purpose, this algorithm is described for one single secondary table T. It just has to be applied in a loop over the tables of $\mathcal{T}$ to construct itemset-based variables for all secondary tables of $\mathcal{T}$.

Filtering Actual Quantile Partitions. Let us first notice that definitions 2 and 3 relate to formal descriptions of quantile partitions. Actual partitions may contain empty parts or fine grained parts that are redundant with coarse grained parts. This is the case when the partition size is greater than the number of values, especially when the number of values is below the number of instances in the dataset. To illustrate this, let us consider a variable with only three values 1, 2 and 3 and a dataset of 10 instances, with the following sorted instances values: $1, 1, 1, 2, 2, 2, 3, 3, 3, 3$. According to Definition 2, the 2-quantile partition is $\{\,]-\infty, 2[,\ [2, +\infty[\,\}$, the 3-quantile partition is $\{\,]-\infty, 2[,\ [2, 3[,\ [3, +\infty[\,\}$ and the 4-quantile partition is $\{\,]-\infty, 1[,\ [1, 2[,\ [2, 3[,\ [3, +\infty[\,\}$. In the 4-quantile partition, the first part $]-\infty, 1[$ is empty while the last two ones are redundant with those of the 3-quantile partition. Overall, we can filter the quantile partitions

---

Algorithm 1. Greedy construction of itemset based variables

Require: T {Input secondary table}
Require: Max_k {Maximum number of constituent variables}
Require: Max_s {Maximum size of quantile distribution of constituent variables}
Require: Max_c {Maximum number of constructed variables}
Ensure: $\mathcal{A}_\pi = \{A_\pi, c_c(A_\pi) + c_e(A_\pi) < c_e(\emptyset)\}$ {Set of relevant constructed variables}
 1: {**Step 1: Compute quantile partitions for power of 2 sizes**}
 2: Read secondary table T
 3: **for** $x \in T$ **do**
 4:    **for all** $s = 2^i, 1 \le i \le \log_2 Max_s$ **do**
 5:       Compute quantile partition of size s for variable x {cf. definitions 2 and 3}
 6:    **end for**
 7: **end for**
 8:
 9: {**Step 2: Construct itemset based variables**}
10: {Exhaustive construction by increasing number of variables and partition size}
11: $\mathcal{A}_\pi \leftarrow \emptyset, varNb \leftarrow 0$
12: **for** $k = 0$ to $\max(Max_k, m)$ **do**
13:    $subsetNb \leftarrow \frac{m!}{k!(m-k)!}$ {Number of subset of k variables among m}
14:    **if** $varNb + subsetNb * 2^k \le Max_c$ **then**
15:       $maxSize_k \leftarrow \arg\max_{s \in \{2,4,8,...\}} (varNb + subsetNb * (2s - 2)^k \le Max_c)$
16:       $varNb \leftarrow varNb + subsetNb * (2 * maxSize_k - 2)^k$
17:       **for all** A_π with k variables and parts in partition of size $2, 4, 8, \ldots, maxSize_k$
          do
18:          {Construct only new variables by avoiding missing or redundant parts}
19:          **if** A_π contains only non empty and non redundant parts **then**
20:             $\mathcal{A}_\pi \leftarrow \mathcal{A}_\pi \cup \{A_\pi\}$
21:          **end if**
22:       **end for**
23:    **end if**
24: **end for**
25:
26: {**Step 3: Evaluate constructed variables and keep relevant variables
    only**}
27: Read secondary table T
28: **for all** $A_\pi \in \mathcal{A}_\pi$ **do**
29:    Compute values of constructed variable A_π
30:    Evaluate A_π according to $c_r(A_\pi) = c_c(A_\pi) + c_e(A_\pi)$
31:    **if** $c_r(A_\pi) > c_e(\emptyset)$ **then**
32:       $\mathcal{A}_\pi \leftarrow \mathcal{A}_\pi - \{A_\pi\}$
33:    **end if**
34: **end for**

---

by keeping $\{[2,3[, \; [3,+\infty[\}$ for the 3-quantile partition and $\{[1,2[\}$ for the 4-quantile partition.

Now, we can detail the variable construction Algorithm 1, that consists in three steps.

1. In the first step (line 1), Algorithm 1 reads the N_s records of table T and computes all quantile partitions for power of 2 sizes up to Max_s, according to definitions 2 and 3. This step requires sorting the records for each secondary variable (among m), then processing them for partition sizes $2, 4, 8, \ldots, \min(Max_s, N_s)$, that is at most $\log_2 N_s$ times. Empty or redundant parts in actual quantile partitions are removed at this step, and the overall number of parts per variable that need to be stored is less than or equal to the number N_s of records.

   Overall, the first step of the algorithm requires $O(m\, N_s \log N_s)$ time and $O(N_s\, m)$ space.

2. In the second step (line 9), Algorithm 1 iterates on itemsets by increasing number of constituent variables, and for each number of variables, by increasing the size of partitions, considering only power of 2 sizes. For a given number k of constituent variables, the number of subsets of variables is $subsetNb = \frac{m!}{k!(m-k)!}$. For a given maximum size of partition $maxSize = 2^i, i \geq 1$, the total number of usable parts is $2 * maxSize - 2 = 2 + 4 + 8 + \ldots + maxSize$. The total number of potential itemsets with k constituent variables and part from quantile partitions with power of 2 sizes less than or equal to $maxSize$ is $subsetNb * (2 * maxSize)^k$. In line 15, Algorithm 1 computes the maximum size $maxSize_k$ of quantile partitions that can be considered to build all related itemset based variables, while not exceeding the maximum number of requested constructed variables Max_c. In line 19, Algorithm 1 exploits the actual quantile partitions obtained after the first step, so as to filter the itemset based constructed variables. Any constructed variable involving an empty or a redundant part is removed, since the same records will be covered by a simpler itemset, with a lower construction cost. The variable construction step is similar to a breadth first tree search of the space of itemsets, constrained by a maximum size of quantile partitions, of number of variable s and of total constructed variables. Overall, this step requires $O(Max_c\, Max_k)$ time and $O(Max_c\, Max_k)$ space, since at most Max_c variables are constructed, each involving at most Max_k constituent variables in the itemsets.

3. In the third step (line 26), Algorithm 1 reads all the dataset (N instances of the target table and N_s records of the secondary table) to compute the values of all constructed variables, that is new binary values in secondary tables. To evaluate these binary secondary variables, the method described in Section 3 needs two count variables per itemset-based constructed binary variable in the target table: the number of secondary records covered or not by the itemset. The evaluation algorithm (see Section 3) requires $O(N \log N)$ time to evaluate a binary secondary variable A_π. In line 31 of Algorithm 1,

a comparison between the relevance criterion $c_r(A\pi)$ and the null cost $c_e(\emptyset)$ allows to filter the constructed variables and to keep only the relevant ones. Overall, this step requires $O(Max_c\, Max_k\, Ns)$ time to compute the values of the binary itemset based constructed variables and $O(Max_c\, N)$ space to keep the count values in the target table. Evaluating the relevance of all variables requires $O(Max_c\, N \log N)$ time and $O(Max_c\, N)$ space.

Overall, Algorithm 1 needs to read the whole dataset twice, one in the first step to build the actual quantile partitions and one in the third step to compute the values of all constructed variables. The time complexity is $O(m\, N_s \log N_s + Max_c\, (Max_k\, Ns + N \log N))$ and the space complexity is $O(N_s\, m + Max_c\, N)$. For itemsets involving few constituent variables, it is approximatively super-linear with the number of instances in the target table, of records in the secondary table, of secondary variables and of constructed variables.

5   Evaluation

We evaluate the proposed method by focusing on the following aspects: ability to generate large numbers of variables without combinatorial explosion, resistance to overfitting and contribution for the prediction task.

5.1   Evaluation on 20 Benchmark Datasets

In this first evaluation, we use 20 datasets from the multi-relational data mining community. Since we are interested in secondary variables, we ignore those of the target table. We focus on the itemsets-based variable construction method presented in Section 4.

After the variable construction step, we exploit the extension of the Naive Bayes classifier to secondary variables described in section 3.2. In this article, we use the Selective Naive Bayes (SNB) classifier [28]. It is a variant of the Naive Bayes with variable selection and model averaging, which is robust and efficient in the case of very large numbers of variables.

In order to have a baseline of comparison, we consider the method Relaggs [10], based on the following propositionalisation rules:

- for each secondary numerical variable: Mean, Median, Min, Max, StdDev, Sum
- for each categorical secondary variable: Mode, CountDistinct (number of distinct values) and the number of occurrences per value.
- the number of records in the secondary table.

The classifier used after propositionalisation is also the SNB classifier.

The used multi-relational datasets[5] belong to different domains : image processing domain (datasets Elephant, Fox, Tiger [29], and the Miml dataset [30],

---

[5] Miml: http://lamda.nju.edu.cn/data_MIMLimage.ashx, Fox, Elephant, Tiger, Mutagenesis, Musk1, Musk2: http://www.uco.es/grupos/kdis/mil/dataset.html, Diterpenses: http://cui.unige.ch/~woznica/rel_weka/, Stulong: http://euromise.vse.cz/challenge2003

with target variables Desert, Mountains, Sea, Sunset, Trees), molecular chemistry domain (Diterpenses [31], Musk1, Musk2 [32], and Mutagenesis [33] with three representations), health domain (Stulong[6] [34], with target variables Cholrisk, Htrisk, Kourisk, Obezrisk and Rarisk), game domain (TicTacToe [35], considered as multi-relational with the nine cells of the game in the secondary table). These datasets have a small size, containing from 100 to 2000 individuals. A description of the datasets is provided in Table 1

In all experiments, Algorithm 1 is used with at most five variables per itemset, and quantiles partition of at most 100 parts. By using Algorithm 1, we are able to control the size of the representation by generating 1, 10, 100, 1,000, 10,000 and 100,000 variables per dataset in the training samples of the 10-folds stratified cross-validation process, which leads to almost 20 million constructed variables.

Interpretability. To see an example of how we can interpret an itemset on real world data, let us consider the Stulong dataset [34]. It is medical database composed of two tables in a one-to-many relationship: (i) the target table Entry contains patients, and (ii) the table Control describes results of clinical examinations for each patient. We consider the target variable CHOLRISK (with two values: Normal and Risky) which denotes whether the patient presents a cholesterol risk. The task is to predict the value of this class by considering the secondary variables in table Control. We give here an example of a relevant itemset proposed by our approach: $\pi : HYPCHL \in \{1\}$. π contains only one secondary variable HYPCHL. The corresponding binary constructed variable A_π means whether the control performed by the patient presents or not a hypercholesterolemia. Figure 5 depicts the optimal bivariate data grid related to A_π. It can be seen that we obtain a contrast of the target values (Normal and Risky) in each cell of this grid. For example the top-left cell gives an interpretable rule: if the patient has at most one control with hypercholesterolemia and at least one control without hypercholesterolemia then he is not likely to have a cholesterol risk (i.e. CHOLRISK=normal) in 90% of cases.

Performance evaluation. In a first analysis, we collect the average test accuracy for each number of generated variables. The method Relaggs, which relies on variable construction by applying systematic aggregation rules, cannot control the combinatorial number of generated variables which varies from a dataset to

---

[6] The study (STULONG) was realized at the 2nd Department of Medicine, 1st Faculty of Medicine of Charles University and Charles University Hospital, U nemocnice 2, Prague 2 (head. Prof. M. Aschermann, MD, SDr, FESC), under the supervision of Prof. F. Boudík, MD, ScD, with collaboration of M. Tomečková, MD, PhD and Ass. Prof. J. Bultas, MD, PhD. The data were transferred to the electronic form by the European Centre of Medical Informatics, Statistics and Epidemiology of Charles University and Academy of Sciences (head. Prof. RNDr. J. Zvárová, DrSc). The data resource is on the web pages http://euromise.vse.cz/challenge2004. At present time the data analysis is supported by the grant of the Ministry of Education CR Nr LN 00B 107.

Table 1. Description of the datasets

Dataset	Target Table				Secondary table		
	Name	(# classes)	% Majority class	# records	Name	# variables	# records
Diterpenses	Compound	23	29.80	1503	Spectrum	3	8622
Elephant	Elephant-Image	2	50	200	Regions	230	1391
Fox	Fox-Image	2	50	200	Regions	230	1320
Miml_Desert	Desert-Image	2	79.55	2000	Regions	15	18000
Miml_Mountains	Mountains-Image	2	77.10	2000	Regions	15	18000
Miml_Sea	Sea-Image	2	71	2000	Regions	15	18000
Miml_Sunset	Sunset-Image	2	76.75	2000	Regions	15	18000
Miml_Trees	Trees-Image	2	72	2000	Regions	15	18000
TicTacToe	TicTacToe	2	65.34	958	TicTacToeCell	3	8622
Musk1	Molecule	2	51.08	92	Conformations	166	476
Musk2	Molecule	2	61.76	102	Conformations	166	6575
MutagenesisAtoms	Molecule	2	66.48	188	Mutagenesis Atoms	3	1618
MutagenesisBonds	Molecule	2	66.48	188	Mutagenesis Bonds	6	3995
MutagenesisChains	Molecule	2	66.48	188	Mutagenesis Chains	11	5349
Stulong_Cholrisk	Entry	3	72.05	1417	Control	65	10572
Stulong_Htrisk	Entry	3	72.61	1417	Control	65	10572
Stulong_Kourrisk	Entry	3	56.17	1417	Control	65	10572
Stulong_Obezrisk	Entry	3	77.91	1417	Control	65	10572
Stulong_Rarisk	Entry	3	81.72	1417	Control	65	10572
ImagesTiger	Tiger-Images	2	50	200	Regions	230	1220

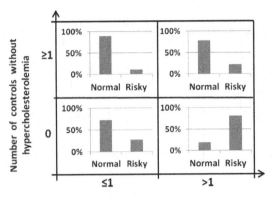

Fig. 5. Example of itemset interpretation on Stulong dataset

another (from about ten variables to 1,400 variables). It is still applicable in the case of small bases and provides here a competitive baseline performance.

The results depicted in Figure 6 show that the performance of our approach systematically increases with the number of constructed variables, and reaches or exceeds the performance of Relaggs over 15 of the 20 datasets when the number of the constructed variables is sufficient. For five datasets (Fox, Musk1, Musk2, MutagenesisAtoms and Tiger), the performance of our approach is significantly worse than Relaggs. These five datasets are either very noisy (Fox with $Acc = 0.63 \pm 0.10$ for Relaggs), very small (Musk1, Musk2 and MutagenesisAtoms with less than 200 instances) or with large variance in the results for Relaggs (from 0.07 to 0.14 of standard deviation in accuracy, eg. 0.9 for Tiger). For these small datasets, there is not sufficient number of instances to reliably recognize a potential pattern. In this case, the regularization (criterion 6) eliminates most of the constructed variables (cf. Figure 7), which brings down the performance. Another explanation that can be provided is that for certain datasets, the pattern in the secondary variables may be easily expressed with an aggregate function. In the case of such a favorable bias, Relaggs is likely to perform better than any other approach.

In order to see the ability of the regularization cost of criterion 6 to eliminate non relevant secondary variables, we report in Figure 7 the number of selected variables with respect to the number of the constructed ones. The results show that criterion 6 significantly prunes the space of the constructed variables. Only a very small number among these variables are considered to be relevant. For example, for 100,000 constructed variables, the proportion of relevant variables is inferior to about 1% for most of the datasets. For some datasets (Elephant, Miml-Sea, TicTacToe, Stulong-Obezrisk and Tiger), only about 10 variables are selected among the 100,000.

Robustness Evaluation. In a second analysis, the experiment is performed after a random reassignment of classes for each dataset in order to assess the robustness

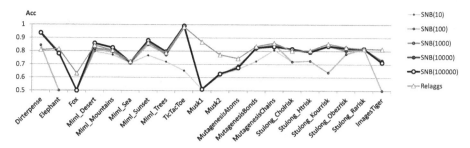

Fig. 6. Test accuracy with respect to the number of constructed variables

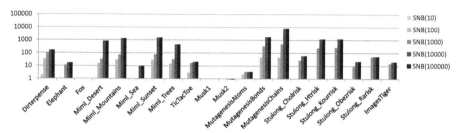

Fig. 7. Number of selected variables with respect to the number of constructed variables

of the approach. We collect the number of selected variables, taking into account the construction cost of the variables according to criterion 6. The method is extremely robust since all 20 million generated variables are identified as non-informative, without exception.

5.2   Handwritten Digits Dataset

In a second evaluation, we use the Digits dataset mentioned earlier in Example 3. In Figure 8, results on the Digits dataset are reported. We used a train and test evaluation with 60,000 instances in training and 10,000 in testing. It is a relatively large dataset with about 50 million records in the secondary table. We report the test accuracy results of the SNB classifier with 1, 10, 100, 1,000, 10,000 and 100,000 constructed variables. These results are compared to those obtained with Relaggs as well as with an SNB on the initial tabular attribute-value representation. We report in the same graphic the number of relevant and filtered variables according to criterion 6 for the different numbers of constructed variables.

The first observation is that the problem is difficult in its initial tabular representation for the SNB classifier which obtains 87.5% of test accuracy. Relaggs only gets 22.4%. In a second observation, Figure 8 shows that the performance of our approach increases with the number of constructed variables, and significantly exceeds that of the SNB classifier obtained on the initial tabular representation. Our rapproach reaches 89.8% with 1,144 relevant variables among

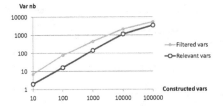 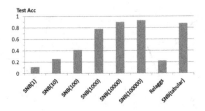

Fig. 8. On the left, number of filtered variables and of relevant variables per number of constructed variables; on the right, test accuracy obtained with Relaggs, SNB with the initial tabular format, and SNB with increasing numbers of constructed variables

10,000 constructed variables, and 92.6% with only 3,476 relevant variables among 100,000 constructed variables.

6   Conclusion

In this paper, we have proposed an approach for constructing new variables and assessing their relevance in the context of multi-relational supervised learning. The method consists in defining an itemset in a secondary table, leading to a new secondary variable that collects whether secondary records are covered or not by the itemset. The relevance of this new variable is evaluated using a bivariate supervised data grid model [11], which provides a regularized estimator of the conditional probability of the target variable. To take into account the risk of overfitting that increases with the number of constructed variables, we have applied a Bayesian model selection approach for both the itemset-based construction model and the conditional density evaluation model, and obtained an exact analytical criterion for the posterior probability of any constructed variable.

A greedy algorithm has been proposed in order to explore the itemset space. We evaluated our approach on several real world multi-relational datasets. Obtained classification performance are very promising. The experiments showed also that our approach is able to deal with relatively large datasets and generate an important number of itemsets while controlling the combinatorial explosion. Furthermore, it is a robust approach. Even with a great number of constructed variables, it remains resistant to overfitting. Future works are envisaged to provide improved search heuristics to better explore the space of constructed variables.

References

1. Knobbe, A.J., Blockeel, H., Siebes, A., Van Der Wallen, D.: Multi-Relational Data Mining. In: Proceedings of Benelearn 1999 (1999)
2. Džeroski, S., Lavrač, N.: Relational Data Mining. Springer-Verlag New York, Inc. (2001)

3. Kramer, S., Flach, P.A., Lavrač, N.: Propositionalization approaches to relational data mining. In: Džeroski, S., Lavrač, N. (eds.) Relational Data Mining, pp. 262–286. Springer, New York (2001)
4. Van Laer, W., De Raedt, L., Džeroski, S.: On multi-class problems and discretization in inductive logic programming. In: Raś, Z.W., Skowron, A. (eds.) ISMIS 1997. LNCS, vol. 1325, pp. 277–286. Springer, Heidelberg (1997)
5. Knobbe, A.J., Ho, E.K.Y.: Numbers in multi-relational data mining. In: Jorge, A.M., Torgo, L., Brazdil, P.B., Camacho, R., Gama, J. (eds.) PKDD 2005. LNCS (LNAI), vol. 3721, pp. 544–551. Springer, Heidelberg (2005)
6. Alfred, R.: Discretization Numerical Data for Relational Data with One-to-Many Relations. Journal of Computer Science 5(7), 519–528 (2009)
7. Lachiche, N., Flach, P.A.: A first-order representation for knowledge discovery and Bayesian classification on relational data. In: PKDD 2000 Workshop on Data Mining, Decision Support, Meta-learning and ILP, pp. 49–60 (2000)
8. Flach, P.A., Lachiche, N.: Naive Bayesian Classification of Structured Data. Machine Learning 57(3), 233–269 (2004)
9. Ceci, M., Appice, A., Malerba, D.: Mr-SBC: A Multi-relational Naïve Bayes Classifier. In: Lavrač, N., Gamberger, D., Todorovski, L., Blockeel, H. (eds.) PKDD 2003. LNCS (LNAI), vol. 2838, pp. 95–106. Springer, Heidelberg (2003)
10. Krogel, M.-A., Wrobel, S.: Transformation-based learning using multirelational aggregation. In: Rouveirol, C., Sebag, M. (eds.) ILP 2001. LNCS (LNAI), vol. 2157, pp. 142–155. Springer, Heidelberg (2001)
11. Lahbib, D., Boullé, M., Laurent, D.: Informative variables selection for multi-relational supervised learning. In: Perner, P. (ed.) MLDM 2011. LNCS, vol. 6871, pp. 75–87. Springer, Heidelberg (2011)
12. Lecun, Y., Bottou, L., Bengio, Y., Haffner, P.: Gradient-based learning applied to document recognition. Proceedings of the IEEE (11), 2278–2324 (1998)
13. De Raedt, L., Dehaspe, L.: Mining Association Rules in Multiple Relations. In: Džeroski, S., Lavrač, N. (eds.) ILP 1997. LNCS, vol. 1297, pp. 125–132. Springer, Heidelberg (1997)
14. Nijssen, S., Kok, J.N.: Faster association rules for multiple relations. In: Proceedings of the 17th International Joint Conference on Artificial Intelligence, vol. (1) (2001)
15. Guo, J., Bian, W., Li, J.: Multi-relational Association Rule Mining with Guidance of User. In: Fourth International Conference on Fuzzy Systems and Knowledge Discovery (FSKD 2007), pp. 704–709 (2007)
16. Gu, Y., Liu, H., He, J., Hu, B., Du, X.: MrCAR: A Multi-relational Classification Algorithm Based on Association Rules. In: 2009 International Conference on Web Information Systems and Mining, pp. 256–260 (2009)
17. Crestana-Jensen, V., Soparkar, N.: Frequent itemset counting across multiple tables. In: Terano, T., Liu, H., Chen, A.L.P. (eds.) PAKDD 2000. LNCS, vol. 1805, pp. 49–61. Springer, Heidelberg (2000)
18. Goethals, B., Le Page, W., Mampaey, M.: Mining interesting sets and rules in relational databases. In: Proceedings of the 2010 ACM Symposium on Applied Computing, p. 997 (2010)
19. Goethals, B., Laurent, D., Le Page, W., Dieng, C.T.: Mining frequent conjunctive queries in relational databases through dependency discovery. Knowledge and Information Systems 33(3), 655–684 (2012)
20. Ceci, M., Appice, A.: Spatial associative classification: propositional vs structural approach. Journal of Intelligent Information Systems 27(3), 191–213 (2006)

21. Ceci, M., Appice, A., Malerba, D.: Emerging pattern based classification in relational data mining. In: Bhowmick, S.S., Küng, J., Wagner, R. (eds.) DEXA 2008. LNCS, vol. 5181, pp. 283–296. Springer, Heidelberg (2008)
22. Boullé, M.: Optimum simultaneous discretization with data grid models in supervised classification A Bayesian model selection approach. Advances in Data Analysis and Classification 3(1), 39–61 (2009)
23. Gay, D., Boullé, M.: A bayesian approach for classification rule mining in quantitative databases. In: Flach, P.A., De Bie, T., Cristianini, N. (eds.) ECML PKDD 2012, Part II. LNCS, vol. 7524, pp. 243–259. Springer, Heidelberg (2012)
24. Lahbib, D., Boullé, M., Laurent, D.: An evaluation criterion for itemset based variable construction in multi-relational supervised learning. In: Riguzzi, F., Železný, F. (eds.) The 22nd International Conference on Inductive Logic Programming (ILP 2012), Dubrovnik, Croatia (2012)
25. Cover, T.M., Thomas, J.A.: Elements of Information Theory. Wiley, New York (1991)
26. Rissanen, J.: A universal prior for integers and estimation by minimum description length. Annals of Statistics 11(2), 416–431 (1983)
27. Shannon, C.: A mathematical theory of communication. Technical report. Bell Systems Technical Journal (1948)
28. Boullé, M.: Compression-based averaging of selective naive Bayes classifiers. Journal of Machine Learning Research 8, 1659–1685 (2007)
29. Andrews, S., Tsochantaridis, I., Hofmann, T.: Support vector machines for multiple-instance learning. In: Advances in Neural Information Processing Systems 15, pp. 561–568. MIT Press (2003)
30. Zhou, Z.H., Zhang, M.L.: Multi-instance multi-label learning with application to scene classification. In: Advances in Neural Information Processing Systems (NIPS 2006), Number i, pp. 1609–1616. MIT Press, Cambridge (2007)
31. Džeroski, S., Schulze-Kremer, S., Heidtke, K.R., Siems, K., Wettschereck, D., Blockeel, H.: Diterpene Structure Elucidation From 13C NMR Spectra with Inductive Logic Programming. Applied Artificial Intelligence 12(5), 363–383 (1998)
32. De Raedt, L.: Attribute-Value Learning Versus Inductive Logic Programming: The Missing Links (Extended Abstract). In: Page, D. (ed.) ILP 1998. LNCS, vol. 1446, pp. 1–8. Springer, Heidelberg (1998)
33. Srinivasan, A., Muggleton, S., King, R., Sternberg, M.: Mutagenesis: ILP experiments in a non-determinate biological domain. In: Proceedings of the 4th International Workshop on ILP, pp. 217–232 (1994)
34. Tomečková, M., Rauch, J., Berka, P.: STULONG - Data from a Longitudinal Study of Atherosclerosis Risk Factors. In: ECML/PKDD 2002 Discovery Challenge Workshop Notes (2002)
35. Asuncion, A., Newman, D.: UCI machine learning repository (2007)

A Declarative Modeling Language for Concept Learning in Description Logics

Francesca Alessandra Lisi

Dipartimento di Informatica, Università degli Studi di Bari "Aldo Moro", Italy
francesca.lisi@uniba.it

Abstract. Learning in Description Logics (DLs) has been paid increasing attention over the last decade. Several and diverse approaches have been proposed which however share the common feature of extending and adapting previous work in Concept Learning to the novel representation framework of DLs. In this paper we present a declarative modeling language for Concept Learning in DLs which relies on recent results in the fields of Knowledge Representation and Machine Learning. Based on second-order DLs, it allows for modeling Concept Learning problems as constructive DL reasoning tasks where the construction of the solution to the problem may be subject to optimality criteria.

1 Motivation

Machine Learning (ML) and Data Mining (DM) have gained in popularity nowadays. However, the development of applications that incorporate ML or DM techniques still remains challenging. This is due to the fact that research in ML and DM has traditionally focussed on designing high-performance algorithms for solving particular tasks rather than on developing general principles and techniques. De Raedt *et al.* [12,13] propose to overcome these difficulties by resorting to the constraint programming methodology. More precisely, they suggest to specify ML/DM problems as Constraint Satisfaction Problems (CSPs) and Optimization Problems (OPs). Their goal is to provide the user with a means for specifying declaratively what the ML/DM problem is rather than having to outline how that solution needs to be computed. This corresponds to a *model+solver*-based approach to ML and DM, in which the user specifies the problem in a *declarative modeling language* and the system automatically transforms such models into a format that can be used by a solver to efficiently generate a solution. This should be much easier for the user than having to implement or adapt an algorithm that computes a particular solution to a specific problem. The *model+solver*-based approach to ML/DM has been already investigated in De Raedt *et al.*'s work on constraint programming for itemset mining [34,20,21].

In this paper, we are interested in investigating how to model ML problems within the Knowledge Representation (KR) framework of *Description Logics* (DLs). DLs currently play a crucial role in the definition of ontology languages [1]. Learning in DLs has been paid increasing attention over the

F. Riguzzi and F. Železný (Eds.): ILP 2012, LNAI 7842, pp. 151–165, 2013.

last decade. Early work on the application of ML to DLs essentially focused on demonstrating the PAC-learnability for various terminological languages derived from the CLASSIC DL [6,8,7,19]. In particular, Cohen and Hirsh investigate the CoreCLASSIC DL proving that it is not PAC-learnable [6] as well as demonstrating the PAC-learnability of its sub-languages, such as C-CLASSIC [8], through the bottom-up LcsLEARN algorithm. These approaches tend to cast supervised Concept Learning to a structural generalizing operator working on equivalent graph representations of the concept descriptions. It is also worth mentioning unsupervised Concept Learning methodologies for DL concept descriptions, whose prototypical example is KLUSTER [24], a polynomial-time algorithm for the induction of BACK terminologies. More recently, algorithms have been proposed that follow the *generalization as search* approach [32] by extending the methodological apparatus of ILP to DL languages [3,16,17,27,28]. Systems, such as YinYang [23] and $\mathcal{DL}$-LEARNER [29], have been implemented. Based on a set of refinement operators borrowed from YinYang and $\mathcal{DL}$-LEARNER, a new version of the FOIL algorithm, named $\mathcal{DL}$-FOIL, has been proposed [18].

In spite of the increasing interest in learning in DLs, induction is still an inference little understood within the KR community. Yet, useful ontology reasoning tasks can be based on the inductive inference. Declarative modeling would be of great help to KR practitioners in specifying ML/DM problems. In this paper, we present a declarative modeling language for Concept Learning in DLs. Based on second-order DLs, it allows to model several variants of the Concept Learning problem in a declarative way. The design of the language is inspired by recent work on those non-standard inferences in DLs which support constructive reasoning, *i.e.* reasoning aimed at constructing a concept. Starting from the assumption that inductive inference is inherently constructive, the proposed language reformulates Concept Learning problem variants in terms that allow for a construction possibly subject to some optimality criteria.

The paper is structured as follows. Section 2 is devoted to preliminaries on DLs with a particular emphasis on non-standard inferences. Section 3 defines variants of the Concept Learning problem as well as variants of the solution approach for them in the DL context. Section 4 proposes a language based on second-order DLs for modeling the ML problems defined in Section 3. Section 5 discusses related work. Section 6 summarizes the contributions of the paper and outlines directions of future work.

2 Preliminaries on Description Logics

2.1 Basics

DLs are a family of decidable First Order Logic (FOL) fragments that allow for the specification of structured knowledge in terms of classes (*concepts*), instances (*individuals*), and binary relations between instances (*roles*) [1]. Complex concepts can be defined from atomic concepts and roles by means of constructors. The syntax of some typical DL constructs is reported in Table 1. A DL knowledge base (KB) Σ consists of a so-called *terminological box* (TBox) $\mathcal{T}$ and a so-called

Table 1. Syntax and semantics of some typical DL constructs

bottom (resp. top) concept	$\bot$ (resp. $\top$)	$\emptyset$ (resp. $\Delta^{\mathcal{I}}$)
atomic concept	A	$A^{\mathcal{I}} \subseteq \Delta^{\mathcal{I}}$
role	R	$R^{\mathcal{I}} \subseteq \Delta^{\mathcal{I}} \times \Delta^{\mathcal{I}}$
individual	a	$a^{\mathcal{I}} \in \Delta^{\mathcal{I}}$
concept negation	$\neg C$	$\Delta^{\mathcal{I}} \setminus C^{\mathcal{I}}$
concept intersection	$C_1 \sqcap C_2$	$C_1^{\mathcal{I}} \cap C_2^{\mathcal{I}}$
concept union	$C_1 \sqcup C_2$	$C_1^{\mathcal{I}} \cup C_2^{\mathcal{I}}$
value restriction	$\forall R.C$	$\{x \in \Delta^{\mathcal{I}} \mid \forall y \ (x,y) \in R^{\mathcal{I}} \rightarrow y \in C^{\mathcal{I}}\}$
existential restriction	$\exists R.C$	$\{x \in \Delta^{\mathcal{I}} \mid \exists y \ (x,y) \in R^{\mathcal{I}} \wedge y \in C^{\mathcal{I}}\}$
concept subsumption axiom	$C_1 \sqsubseteq C_2$	$C_1^{\mathcal{I}} \subseteq C_2^{\mathcal{I}}$
concept equivalence axiom	$C_1 \equiv C_2$	$C_1^{\mathcal{I}} = C_2^{\mathcal{I}}$
concept assertion	$C(a)$	$a^{\mathcal{I}} \in C^{\mathcal{I}}$
role assertion	$R(a,b)$	$(a^{\mathcal{I}}, b^{\mathcal{I}}) \in R^{\mathcal{I}}$

assertional box (ABox) $\mathcal{A}$. The TBox is a finite set of *axioms* which represent either is-a relations (denoted with $\sqsubseteq$) or equivalence (denoted with $\equiv$) relations between concepts, whereas the ABox is a finite set of *assertions* (or *facts*) that represent instance-of relations between individuals (resp. couples of individuals) and concepts (resp. roles). Thus, when a DL-based ontology language is adopted, an ontology is nothing else than a TBox, and a populated ontology corresponds to a whole DL KB (*i.e.*, encompassing also an ABox).

The semantics of DLs can be defined directly with set-theoretic formalizations as shown in Table 1 or through a mapping to FOL as shown in [4]. An *interpretation* $\mathcal{I} = (\Delta^{\mathcal{I}}, \cdot^{\mathcal{I}})$ for a DL KB consists of a domain $\Delta^{\mathcal{I}}$ and a mapping function $\cdot^{\mathcal{I}}$. Under the *Unique Names Assumption* (UNA)[35], individuals are mapped to elements of $\Delta^{\mathcal{I}}$ such that $a^{\mathcal{I}} \neq b^{\mathcal{I}}$ if $a \neq b$. However UNA does not hold by default in DLs. An interpretation $\mathcal{I}$ is a *model* of a KB $\Sigma = (\mathcal{T}, \mathcal{A})$ iff it satisfies all axioms and assertions in $\mathcal{T}$ and $\mathcal{A}$. In DLs a KB represents many different interpretations, i.e. all its models. This is coherent with the *Open World Assumption* (OWA) that holds in FOL semantics. A DL KB is *satisfiable* if it has at least one model. An ABox assertion α is a *logical consequence* of a KB Σ, written $\Sigma \models \alpha$, if all models of Σ are also models of α.

The main reasoning task for a DL KB Σ is the *consistency check* which tries to prove the satisfiability of Σ. This check is performed by applying decision procedures mostly based on tableau calculus. The *subsumption check* aims at proving whether a concept is included in another one according to the subsumption relationship. Another well known reasoning service in DLs is *instance check*, i.e., the check of whether an ABox assertion is a logical consequence of a DL KB. A more sophisticated version of instance check, called *instance retrieval*, retrieves, for a DL KB Σ, all (ABox) individuals that are instances of the given (possibly complex) concept expression C, i.e., all those individuals a such that Σ entails that a is an instance of C. All these reasoning tasks support so-called *standard inferences* and can be reduced to the consistency check. Besides the standard ones, additional so-called *non-standard inferences* have been investigated in DL reasoning [25]. They are treated in more detail in Section 2.2.

2.2 Non-standard Reasoning

The motivation for studying non-standard reasoning in DLs has been to support the construction and maintenance of DL KBs [31]. Indeed, although standard inferences help structuring the KB, *e.g.*, by automatically building a concept hierarchy, they are, for example, not sufficient when it comes to (automatically) generating new concept descriptions from given ones. They also fail if the concepts are specified using different vocabularies (*i.e.* sets of concept names and role names) or if they are described on different levels of abstraction. Altogether it has turned out that non-standard inferences are required for building and maintaining large DL KBs. Among them, the first ones to be studied have been the Least Common Subsumer (LCS) of a set concepts [5] and the Most Specific Concept (MSC) of an individual [33,26,2]. The LCS of a set of concepts is the minimal concept that subsumes all of them. The minimality condition implies that there is no other concept that subsumes all the concepts in the set and is less general than (subsumed by) the LCS. The notion of LCS is closely related to that of MSC of an individual, i.e., the least concept description that the individual is an instance of, given the assertions in the KB; the minimality condition is specified as before. More generally, one can define the MSC of a set of assertions about individuals as the LCS of MSC associated with each individual. Based on the computation of the MSC of a set of assertions about individuals one can incrementally construct a DL KB.

Recently, Colucci *et al.* have proposed a unified framework for non-standard reasoning services in DLs [10]. The framework is based on the use of second-order sentences in DLs [9]. It applies to so-called *constructive inferences, i.e.* those non-standard inferences that deal with finding (or constructing) a concept. More precisely, it provides a unifying definition model for all those constructive reasoning tasks which rely on specific optimality criteria to build up the objective concept. Indeed, constructive reasoning tasks can be divided into two main categories: Tasks for which we just need to compute a concept (or a set of concepts) and those for which we need to find a concept (or a set of concepts) according to some minimality/maximality criteria. In the first case, we have a set of solutions while in the second one we also have a set of sub-optimal solutions to the main problem. For instance, the set of sub-optimal solutions in LCS is represented by the common subsumers. A reformulation of LCS as optimal solution problem can be found in [10].

The work on non-standard reasoning has been more or less explicitly related to ML. E.g., LCS and MCS have been used for the bottom-up induction of CLASSIC concept descriptions from examples [8,7]. In this paper, we start from the inherently constructive nature of the inductive inference in order to extend Colucci *et al.*'s framework to Concept Learning in DLs.

3 Learning Concepts in Description Logics

3.1 Variants of the Problem

In this section, we formally define several variants of the Concept Learning problem in a $\mathcal{DL}$ setting where $\mathcal{DL}$ is any DL. The variants share the following two features:

1. The background knowledge theory is in the form of a $\mathcal{DL}$ KB $\mathcal{K}$ composed of a TBox $\mathcal{T}$ and an ABox $\mathcal{A}$, and
2. The target theory is in the form of $\mathcal{DL}$ concept definitions, *i.e.* concept equivalence axioms having an atomic concept in the left-hand side.

but differ in the requirements that an induced concept definition must fulfill in order to be considered as a correct (or valid) solution.

For the purpose, we denote:

- $\mathsf{Ind}(\mathcal{A})$ is the set of all individuals occurring in $\mathcal{A}$
- $\mathsf{Retr}_{\mathcal{K}}(C)$ is the set of all individuals occurring in $\mathcal{A}$ that are instance of a given concept C w.r.t. $\mathcal{K}$
- $\mathsf{Ind}^+_C(\mathcal{A}) = \{a \in \mathsf{Ind}(\mathcal{A}) \mid C(a) \in \mathcal{A}\} \subseteq \mathsf{Retr}_{\mathcal{K}}(C)$
- $\mathsf{Ind}^-_C(\mathcal{A}) = \{b \in \mathsf{Ind}(\mathcal{A}) \mid \neg C(b) \in \mathcal{A}\} \subseteq \mathsf{Retr}_{\mathcal{K}}(\neg C)$

These sets can be easily computed by resorting to instance retrieval inference services usually available in DL systems.

The first variant of the Concept Learning problem we consider in this paper is the supervised one. It is the base for the other variants being introduced later and is denoted with Concept Induction.

Definition 1 (Concept Induction). *Let $\mathcal{K} = (\mathcal{T}, \mathcal{A})$ be a $\mathcal{DL}$ KB. Given:*

- *a (new) target concept name C*
- *a set of positive and negative examples $Ind^+_C(\mathcal{A}) \cup Ind^-_C(\mathcal{A}) \subseteq Ind(\mathcal{A})$ for C*
- *a concept description language $\mathcal{DL}_{\mathcal{H}}$*

the Concept Induction (CI) problem is to find a concept definition $C \equiv D$ with $D \in \mathcal{DL}_{\mathcal{H}}$ such that

Completeness $\mathcal{K} \models D(a)$ $\forall a \in Ind^+_C(\mathcal{A})$ *and*
Consistency $\mathcal{K} \models \neg D(b)$ $\forall b \in Ind^-_C(\mathcal{A})$

Note that Def. 1 provides the CSP version of the supervised Concept Learning problem. However, as already mentioned, Concept Learning can be regarded also as an OP. Algorithms such as $\mathcal{DL}$-FOIL [18] testify the existence of optimality criteria to be fulfilled in CI besides the conditions of completeness and consistency. Other algorithms supporting the CI task are YINYANG [23] and $\mathcal{DL}$-LEARNER [29].

In case a previous definition D' for C is already available in $\mathcal{K}$ which however is only partially correct (*e.g.*, it is not complete) then the Concept Learning problem can be cast as a Concept Refinement problem which would amount to searching for a solution D starting from the approximation D'.

Definition 2 (Concept Refinement). *Let* $\mathcal{K} = (\mathcal{T}, \mathcal{A})$ *be a* $\mathcal{DL}$ *KB. Given:*

- *a (new) target concept name* C
- *a set of positive and negative examples* $\mathsf{Ind}_C^+(\mathcal{A}) \cup \mathsf{Ind}_C^-(\mathcal{A}) \subseteq \mathsf{Ind}(\mathcal{A})$ *for* C
- *a concept description language* $\mathcal{DL}_{\mathcal{H}}$
- *a concept* $D' \in \mathcal{DL}_{\mathcal{H}}$ *for which* $\exists a \in \mathsf{Ind}_C^+(\mathcal{A})$ *s.t.* $\mathcal{K} \not\models D'(a)$ *or* $\exists b \in \mathsf{Ind}_C^-(\mathcal{A})$ *s.t.* $\mathcal{K} \not\models \neg D'(b)$

the Concept Refinement (CR) *problem is to find a concept definition* $C \equiv D$ *with* $D \in \mathcal{DL}_{\mathcal{H}}$ *such that*

Compatibility $\mathcal{K} \models D \sqsubseteq D'$
Completeness $\mathcal{K} \models D(a) \quad \forall a \in \mathsf{Ind}_C^+(\mathcal{A})$
Consistency $\mathcal{K} \models \neg D(b) \quad \forall b \in \mathsf{Ind}_C^-(\mathcal{A})$

Note that Def. 2 differs from Def. 1 only for the further constraint (namely, compatibility with a previous concept definition) the CR problem poses on the admissible solutions.

When Concept Learning is unsupervised, the resulting problem is called Concept Formation. Typically, this problem is decomposed in two subproblems: The former is aimed at clustering individuals in mutually disjoint concepts, whereas the latter concerns the search for a definition for each of these emerging concepts.

Definition 3 (Concept Formation). *Let* $\mathcal{K} = (\mathcal{T}, \mathcal{A})$ *be a* $\mathcal{DL}$ *KB where* $\mathcal{T}$ *does not contain definitions for all the concepts with assertions in* $\mathcal{A}$. *Given a concept description language* $\mathcal{DL}_{\mathcal{H}}$, *the* Concept Formation (CF) *problem is to find (i) mutually disjoint concepts* C_i, *and (ii) for each* C_i, *a concept definition* $C_i \equiv D_i$ *with* $D_i \in \mathcal{DL}_{\mathcal{H}}$ *such that*

Completeness $\mathcal{K} \models D_i(a) \quad \forall a \in \mathsf{Ind}_{C_i}^+(\mathcal{A})$
Consistency $\mathcal{K} \models \neg D_i(b) \quad \forall b \in \mathsf{Ind}_{C_i}^-(\mathcal{A})$

Note that achieving goal (ii) in Def. 3 involves solving as many CI problems as the number of mutually disjoint concepts found in the given KB to fulfill requirement (i). Exemplars of algorithms solving the CF problem are KLUSTER [24] and CSKA [17].

3.2 Variants of the Solution

In Section 3.1, we have considered a language of hypotheses $\mathcal{DL}_{\mathcal{H}}$ that allows for the generation of concept definitions in any $\mathcal{DL}$. These definitions can be organized according to the concept subsumption relation $\sqsubseteq$. Indeed, $\sqsubseteq$ is a reflexive and transitive binary relation, *i.e.* a quasi-order. Thus, $(\mathcal{DL}_{\mathcal{H}}, \sqsubseteq)$ is a quasi-ordered set of $\mathcal{DL}$ concept definitions which defines a search space to be traversed either top-down or bottom-up by means of suitable refinement operators according to the *generalization as search* approach in Mitchell's vision [3,16].

Definition 4 (Refinement operator in DLs). *Given a quasi-ordered search space* $(\mathcal{DL}_{\mathcal{H}}, \sqsubseteq)$

- a downward refinement operator *is a mapping* $\rho : \mathcal{DL}_{\mathcal{H}} \to 2^{\mathcal{DL}_{\mathcal{H}}}$ *such that*

$$\forall C \in \mathcal{DL}_{\mathcal{H}} \quad \rho(C) \subseteq \{D \in \mathcal{DL}_{\mathcal{H}} \mid D \sqsubseteq C\}$$

- *an* upward refinement operator *is a mapping* $\delta : \mathcal{DL}_{\mathcal{H}} \to 2^{\mathcal{DL}_{\mathcal{H}}}$ *such that*

$$\forall C \in \mathcal{DL}_{\mathcal{H}} \quad \delta(C) \subseteq \{D \in \mathcal{DL}_{\mathcal{H}} \mid C \sqsubseteq D\}$$

Definition 5 (Refinement chain in DLs). *Given a downward (resp., upward) refinement operator ρ (resp., δ) for a quasi-ordered search space $(\mathcal{DL}_{\mathcal{H}}, \sqsubseteq)$, a refinement chain from $C \in \mathcal{DL}_{\mathcal{H}}$ to $D \in \mathcal{DL}_{\mathcal{H}}$ is a sequence*

$$C = C_0, C_1, \ldots, C_n = D$$

such that $C_i \in \rho(C_{i-1})$ (resp., $C_i \in \delta(C_{i-1})$) for every $1 \leq i \leq n$.

Note that, given $(\mathcal{DL}, \sqsubseteq)$, there is an infinite number of generalizations and specializations. Usually one tries to define refinement operators that can traverse efficiently throughout the hypothesis space in pursuit of one of the correct definitions (w.r.t. the examples that have been provided).

Definition 6 (Properties of refinement operators in DLs). *A downward refinement operator ρ for a quasi-ordered search space $(\mathcal{DL}_{\mathcal{H}}, \sqsubseteq)$ is*

- *(locally) finite iff $\rho(C)$ is finite for all concepts $C \in \mathcal{DL}_{\mathcal{H}}$.*
- *redundant iff there exists a refinement chain from a concept $C \in \mathcal{DL}_{\mathcal{H}}$ to a concept $D \in \mathcal{DL}_{\mathcal{H}}$, which does not go through some concept $E \in \mathcal{DL}_{\mathcal{H}}$ and a refinement chain from C to a concept equal to D, which does go through E.*
- *proper iff for all concepts $C, D \in \mathcal{DL}_{\mathcal{H}}$, $D \in \rho(C)$ implies $C \not\equiv D$.*
- *complete iff, for all concepts $C, D \in \mathcal{DL}_{\mathcal{H}}$ with $C \sqsubseteq D$, a concept $E \in \mathcal{DL}_{\mathcal{H}}$ with $E \equiv C$ can be reached from D by ρ.*
- *weakly complete iff, for all concepts $C \in \mathcal{DL}_{\mathcal{H}}$ with $C \sqsubseteq \top$, a concept $E \in \mathcal{DL}_{\mathcal{H}}$ with $E \equiv C$ can be reached from $\top$ by ρ.*

The corresponding notions for upward refinement operators are defined dually.

To design a refinement operator one needs to make decisions on which properties are most useful in practice regarding the underlying learning algorithm. Considering the properties reported in Def. 6, it has been shown that the most feasible property combination for Concept Learning in expressive DLs such as $\mathcal{ALC}$ is $\{weakly\ complete, complete, proper\}$ [27]. Only for less expressive DLs like $\mathcal{EL}$, *ideal, i.e.* complete, proper and finite, operators exist [30].

4 A Modeling Language Based on Second-Order DLs

4.1 Second-Order Concept Expressions

We assume to start from the syntax of any Description Logic $\mathcal{DL}$ where N_c, N_r, and N_o are the alphabet of concept names, role names and individual names, respectively. In order to write second-order formulas, we introduce a set $N_x = X_0, X_1, X_2, \ldots$ of concept variables, which we can quantify over. We denote by $\mathcal{DL}_X$ the language of concept terms obtained from $\mathcal{DL}$ by adding N_x.

Definition 7 (Concept term). *A concept term in $\mathcal{DL}_X$ is a concept formed according to the specific syntax rules of $\mathcal{DL}$ augmented with the additional rule $C \longrightarrow X$ for $X \in N_x$.*

Since we are not interested in second-order DLs in themselves, we restrict our language to particular existential second-order formulas of interest to this paper, *i.e.* formulas involving concept subsumptions and concept assertions.

Definition 8 (Concept expression). *Let $a_1, \ldots, a_k \in \mathcal{DL}$ be individuals, $C_1, \ldots, C_m, D_1, \ldots, D_m \in \mathcal{DL}_X$ be concept terms containing concept variables $X_0, \ldots, X_n$. A concept expression Γ in $\mathcal{DL}_X$ is a conjunction*

$$
\begin{aligned}
(C_1 \sqsubseteq D_1) \wedge \ldots \wedge (C_l \sqsubseteq D_l) \wedge (C_{l+1} \not\sqsubseteq D_{l+1}) \wedge \ldots \wedge (C_m \not\sqsubseteq D_m) \wedge \\
(a_1 : D_1) \wedge \ldots \wedge (a_j : D_l) \wedge (a_{j+1} : \neg D_{l+1}) \wedge \ldots \wedge (a_k : \neg D_m)
\end{aligned}
\tag{1}
$$

of (negated or not) concept subsumptions and concept assertions with $1 \leq l \leq m$ and $1 \leq j \leq k$.

Definition 9 (Formula). *A formula Φ in $\mathcal{DL}_X$ has the form*

$$
\exists X_0 \cdots \exists X_n.\Gamma
\tag{2}
$$

where Γ is a concept expression of the form (1) and $X_0, \ldots, X_n$ are concept variables.

We use *General Semantics*, also called Henkin semantics, for interpreting concept variables [22]. In such a semantics, variables denoting unary predicates can be interpreted only by *some subsets* among all the ones in the powerset of the domain $2^{\Delta^\mathcal{I}}$ - instead, in Standard Semantics a concept variable could be interpreted as any subset of $\Delta^\mathcal{I}$. Adapting General Semantics to our problem, the structure we consider is exactly the sets interpreting concepts in $\mathcal{DL}$. That is, the interpretation $X^\mathcal{I}$ of a concept variable $X \in \mathcal{DL}_X$ must coincide with the interpretation $E^\mathcal{I}$ of some concept $E \in \mathcal{DL}$. The interpretations we refer to in the following definition are of this kind.

Definition 10 (Satisfiability of concept expressions and formulas). *A concept expression Γ of the form (1) is satisfiable in $\mathcal{DL}$ iff there exist $n + 1$ concepts $E_0, \ldots, E_n \in \mathcal{DL}$ such that, extending the semantics of $\mathcal{DL}$ for each interpretation $\mathcal{I}$, with: $(X_i)^\mathcal{I} = (E_i)^\mathcal{I}$ for $i = 0, \ldots, n$, it holds that*

1. *for each $j = 1, \ldots, l$, and every interpretation $\mathcal{I}$, $(C_j)^\mathcal{I} \subseteq (D_j)^\mathcal{I}$ and $(a_j)^\mathcal{I} \in (D_j)^\mathcal{I}$, and*
2. *for each $j = l+1, \ldots, m$, there exists an interpretation $\mathcal{I}$ s.t. $(C_j)^\mathcal{I} \not\subseteq (D_j)^\mathcal{I}$ and $(a_j)^\mathcal{I} \notin (D_j)^\mathcal{I}$*

Otherwise, Γ is said to be unsatisfiable *in $\mathcal{DL}$. If Γ is satisfiable in $\mathcal{DL}$, then $\langle E_0, \ldots, E_n \rangle$ is a solution for Γ. A formula Φ of the form (2) is true in $\mathcal{DL}$ if there exist at least a solution for Γ, otherwise it is false.*

4.2   Modeling with Second-Order Concept Expressions

The second-order logic fragment introduced in Section 4.1 can be used as a declarative modeling language for Concept Learning problems in DLs. Hereafter, first, we show how to model MSC. This step is to be considered as functional to the modeling of the variants of Concept Learning considered in Section 3.1.

Most Specific Concept. Intuitively, the MSC of individuals described in an ABox is a concept description that represents all the properties of the individuals including the concept assertions they occur in and their relationship to other individuals. The MSC is uniquely determined up to equivalence. More precisely, the set of most specific concepts of individuals $a_1, \ldots, a_k \in \mathcal{DL}$ forms an equivalence class, and if S is defined to be the set of all concept descriptions that have $a_1, \ldots, a_k$ as their instance, then this class is the least element in $[S]$ w.r.t. a partial ordering $\preceq$ on equivalence classes induced by $\sqsubseteq$. We refer to one of its representatives by $\mathsf{MSC}(a_1, \ldots, a_k)$. The MSC need not exist. Three different phenomena may cause the non existence of a least element in $[S]$, and thus, a MSC $[S]$ might be empty, or contain different minimal elements, or contain an infinite decreasing chain $[D_1] \succ [D_2] \cdots$.

A concept E is not the MSC of $a_1, \ldots, a_k$ iff the following formula Φ_{MSC} is true in $\mathcal{DL}$:

$$\exists X.(a_1 : X) \wedge \ldots \wedge (a_k : X) \wedge (X \sqsubseteq E) \wedge (E \not\sqsubseteq X) \tag{3}$$

that is, E is not the MSC if there exists a concept X which is a most specific concept, and is strictly more specific than E.

Concept Induction. Following Def. 1, we assume that $\mathsf{Ind}_C^+(\mathcal{A}) = \{a_1, \ldots, a_m\}$ and $\mathsf{Ind}_C^-(\mathcal{A}) = \{b_1, \ldots, b_n\}$. A concept $D \in \mathcal{DL}_{\mathcal{H}}$ is a correct concept definition for the target concept name C w.r.t. $\mathsf{Ind}_C^+(\mathcal{A})$ and $\mathsf{Ind}_C^-(\mathcal{A})$ iff it is a solution for the following second-order concept expression:

$$(C \sqsubseteq X) \wedge (X \sqsubseteq C) \wedge (a_1 : X) \wedge \ldots \wedge (a_m : X) \wedge (b_1 : \neg X) \wedge \ldots \wedge (b_n : \neg X) \tag{4}$$

that is, iff D can be an assignment for the concept variable X. The CI problem can be therefore modeled with the following formula Φ_{CI}:

$$\exists X.(C \sqsubseteq X) \wedge (X \sqsubseteq C) \wedge (a_1 : X) \wedge \ldots \wedge (a_m : X) \wedge (b_1 : \neg X) \wedge \ldots \wedge (b_n : \neg X) \tag{5}$$

which covers only the CSP version of the problem. A simple OP version of CI could be modeled by asking for solutions that are compliant with a minimality criterion involving concept subsumption checks as already done for MSC. More precisely, a concept $E \in \mathcal{DL}_{\mathcal{H}}$ is not a correct concept definition for C w.r.t. $\mathsf{Ind}_C^+(\mathcal{A})$ and $\mathsf{Ind}_C^-(\mathcal{A})$ iff the following formula Φ'_{CI} is true in $\mathcal{DL}_{\mathcal{H}}$:

$$\begin{aligned} \exists X.(C \sqsubseteq X) \wedge (X \sqsubseteq C) \wedge (X \sqsubseteq E) \wedge (E \not\sqsubseteq X) \wedge \\ (a_1 : X) \wedge \ldots \wedge (a_m : X) \wedge (b_1 : \neg X) \wedge \ldots \wedge (b_n : \neg X) \end{aligned} \tag{6}$$

that is, iff there exists a concept X which is a most specific concept, and is strictly more specific than E.

Concept Refinement. Concerning Def. 2, we assume that $D' \in \mathcal{DL}_\mathcal{H}$ is a partially correct definition for the target concept name C w.r.t. $\mathsf{Ind}_C^+(\mathcal{A}) = \{a_1, \ldots, a_m\}$ and $\mathsf{Ind}_C^-(\mathcal{A}) = \{b_1, \ldots, b_n\}$. A concept $D \in \mathcal{DL}_\mathcal{H}$ is a correct concept definition for C w.r.t. $\mathsf{Ind}_C^+(\mathcal{A})$ and $\mathsf{Ind}_C^-(\mathcal{A})$ iff it makes the following formula Φ_{CR}:

$$\exists X.(C \sqsubseteq X) \wedge (X \sqsubseteq C) \wedge (X \sqsubseteq D') \wedge \\ (a_1 : X) \wedge \ldots \wedge (a_m : X) \wedge (b_1 : \neg X) \wedge \ldots \wedge (b_n : \neg X) \tag{7}$$

true in $\mathcal{DL}_\mathcal{H}$. Note that Φ_{CR} is similar in the spirit to formula (6).

Concept Formation. As regards Def. 3, the problem decomposition of the CF problem allows us to model the first subproblem with Φ_{MSC} and the second subproblem with a formula which substantially relies on Φ_{CI} as many times as the number of the mutually disjoint concepts found as a result of the solution to the first subproblem.

4.3   An Illustrative Example

For illustrative purposes, we choose the New Testament Names (shortly, NTN) ontology developed within the Semantic Bible project[1] because it is medium-sized (51 concepts, 29 roles, 723 individuals), contains rich background knowledge (52 concept subsumption axioms) and is still manageable by a DL reasoner as a whole. The structure of the NTN ontology is illustrated in Figure 1 and is adequately represented with the very expressive $\mathcal{SHOIN}(D)$ DL.

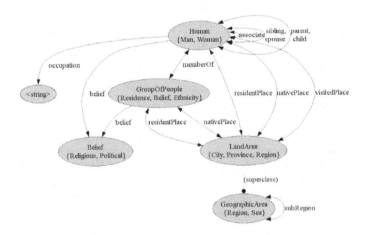

Fig. 1. Structure of the New Testament Names ontology

With reference to NTN (which therefore will play the role of $\mathcal{K}$ as in Def. 1), we might want to induce a concept definition for the target concept name $C = $ Brother from the following positive and negative examples:

- $\mathsf{Ind}^+_{\texttt{Brother}}(\mathcal{A}) = \{\texttt{Archelaus}, \texttt{HerodAntipas}\}$
- $\mathsf{Ind}^-_{\texttt{Brother}}(\mathcal{A}) = \{\texttt{Almighty}, \texttt{Gabriel}, \texttt{Michael}, \texttt{Satan}, \texttt{Jesus}\}$

We remind the reader that the examples are chosen from the sets $\mathsf{Retr}_\mathcal{K}(\texttt{Brother})$ and $\mathsf{Retr}_\mathcal{K}(\neg\texttt{Brother})$, respectively.

According to (5), the intended CI problem can be modeled as follows:

$$\begin{aligned}
&\exists X.(\texttt{Brother} \sqsubseteq X) \wedge (X \sqsubseteq \texttt{Brother}) \wedge \\
&(\texttt{Archelaus} : X) \wedge (\texttt{HerodAntipas} : X) \wedge \\
&(\texttt{Almighty} : \neg X) \wedge (\texttt{Gabriel} : \neg X) \wedge (\texttt{Michael} : \neg X) \wedge \\
&(\texttt{Satan} : \neg X) \wedge (\texttt{Jesus} : \neg X)
\end{aligned} \tag{8}$$

Also the solutions for one such problem can be simply the concepts that can be generated from a $\mathcal{SHOIN}(\mathrm{D})$ language built upon the atomic concept and role names occurring in NTN (except, of course, for the target concept name).

Among the concepts (instantiations of X) satisfying the formula (8), there is

$$\exists \texttt{siblingOf}.(\exists \texttt{enemyOf}.\texttt{Human}) \tag{9}$$

which describes the set of individuals who are sibling of some enemy of some human being. It provides a correct concept definition for Brother w.r.t. the given examples, i.e. the following concept equivalence axiom

$$\texttt{Brother} \equiv \exists \texttt{siblingOf}.(\exists \texttt{enemyOf}.\texttt{Human}) \tag{10}$$

is a solution for the CI problem in hand.

5   Related Work

The study reported in this paper opens an interesting direction of research at the intersection of ML/DM and KR. For this research we have taken inspiration from recent results in both areas, namely De Raedt et al.'s work on declarative modeling for ML/DM [12] and Colucci et al.'s work on non-standard reasoning in DLs [10]. Interestingly, both works pursue a unified view on the inferential problems of interest to the respective fields of research. This match of research efforts in the two fields has motivated the work presented in this paper which, therefore, moves a step towards bridging the gap between KR and ML/DM in areas such as the maintenance of KBs where the two fields have already produced promising results though mostly independently from each other. A further source of inspiration has been the general framework for data mining proposed by Dzeroski [14]. He discusses the requirements that such a framework should fulfill: It should elegantly handle different types of data, of data mining tasks, and of patterns/models. He also discusses the design and implementation of data mining algorithms, as well as their composition into nontrivial multi-step knowledge discovery scenarios relevant for practical application. He then proceeds by laying out some basic concepts, starting with (structured) data and generalizations (*e.g.*, patterns and models) and continuing with data mining tasks and basic components of data mining algorithms (*i.e.*, refinement operators, distances, features and kernels). He next discusses how to use these concepts to formulate constraint-based data mining tasks and design generic data mining algorithms. He finally discusses how these components would fit in the overall framework and in particular into a declarative modeling language for data mining and knowledge discovery which incorporates some features from both functional programming and logic programming. However, the language is only sketched.

New questions and challenges are raised by the cross-fertilization of these results. Notably, the choice of the solver is a critical aspect in any *model+solver*-based approach. De Raedt *et al.*'s work shows that off-the-shelf constraint programming techniques can be applied to various ML/DM problems, once reformulated as CSPs and OPs [34,20,21]. On the KR side, the issue is still open. In [10], Colucci *et al.*'s have provided a sound and complete procedure for solving constructive reasoning problems in DLs, which combines a tableaux calculus for DLs with rules for the substitution of concept variables in second-order concept expressions. However, the procedure does not terminate in all cases, since some of the above problems are known to be undecidable. More recently, Colucci and Donini [11] have presented a modular Prolog prototype system aimed at proving the feasibility of their unified framework for non-standard reasoning in DLs. The prototype supports only some constructive reasoning problems, *e.g.* LCS, and two simple DLs ($\mathcal{EL}$ and $\mathcal{ALN}$). The authors plan to extend the approach to different DLs, and to implement, for each investigated DL, further non-standard reasoning services in the prototype. Also, they intend to improve the system efficiency. However, their approach has some theoretical limitations. In particular, it relies on so-called structural subsumption for checking concept inclusion axioms. This feature restricts the applicability of the approach to DLs for which structural subsumption is complete. The authors claim that, for more expressive DLs, the fixpoint mechanism could still be exploited, but using some higher-order tableaux methods that are still to be defined and whose correctness and termination should be proved.

6   Summary and Directions of Future Work

In this paper we have provided a formal characterization of Concept Learning in DLs according to a declarative modeling language which abstracts from the specific algorithms used to solve the problem. To this purpose, we have defined a fragment of second-order logic under the general semantics which extends the one considered in Colucci *et al.*'s framework [10] with (negated or not) concept assertions. One such fragment allows to model the MSC problem, which was left out as future work in [10], and different variants of Concept Learning (Concept Induction, Concept Refinement and Concept Formation). Furthermore, as a minor contribution, we have suggested that Mitchell's *generalization as search* approach to Concept Learning is just that unifying framework necessary for accompanying the declarative modeling language proposed in this paper with a way of computing solutions to the problems declaratively modeled with this language. More precisely, the computational method we refer to is based on the iterative application of suitable refinement operators. Since many refinement operators for DLs are already available in the literature, the method can be designed such that it can be instantiated with a refinement operator specifically defined for the DL in hand.

In the future, we intend to investigate how to express optimality criteria such as the information gain function within the second-order concept expressions and how to integrate the *generalization as search* approach with second-order calculus. Concerning the integration issue, a simple way of addressing the issue would be to develop a software module able to transform the ML problem statements expressed as second-order concept expressions into a format acceptable by existing DL learning systems. However, in order to improve the efficiency of Concept Learning algorithms for the DL case, a more sophisticated way would be to rely on an appropriate solver. Rather than logic programming and constraint programming, answer set programming extended with higher-order reasoning and external evaluations [15] seems to be more promising in our case due to the notorious links between DLs and disjunctive Datalog.

Acknowledgements. The author would like to thank Luc De Raedt for the fruitful discussion on the open issues of declarative modeling for ML/DM, not only in general but also in the specific case of interest to this paper, and Simona Colucci for her precious advice on the use of second-order DLs for modeling constructive reasoning problems.

References

1. Baader, F., Calvanese, D., McGuinness, D., Nardi, D., Patel-Schneider, P. (eds.): The Description Logic Handbook: Theory, Implementation and Applications, 2nd edn. Cambridge University Press (2007)
2. Baader, F.: Least common subsumers and most specific concepts in a description logic with existential restrictions and terminological cycles. In: Gottlob, G., Walsh, T. (eds.) IJCAI 2003: Proceedings of the 18th International Joint Conference on Artificial Intelligence, pp. 319–324. Morgan Kaufmann Publishers Inc., San Francisco (2003)
3. Badea, L., Nienhuys-Cheng, S.-H.: A refinement operator for description logics. In: Cussens, J., Frisch, A. (eds.) ILP 2000. LNCS (LNAI), vol. 1866, pp. 40–59. Springer, Heidelberg (2000)
4. Borgida, A.: On the relative expressiveness of description logics and predicate logics. Artificial Intelligence 82(1-2), 353–367 (1996)
5. Cohen, W.W., Borgida, A., Hirsh, H.: Computing least common subsumers in description logics. In: Proc. of the 10th National Conf. on Artificial Intelligence, pp. 754–760. The AAAI Press / The MIT Press (1992)
6. Cohen, W.W., Hirsh, H.: Learnability of description logics. In: Haussler, D. (ed.) Proceedings of the Fifth Annual ACM Conference on Computational Learning Theory, COLT 1992, Pittsburgh, PA, USA, July 27-29. ACM (1992)

7. Cohen, W.W., Hirsh, H.: The learnability of description logics with equality constraints. Machine Learning 17(2-3), 169–199 (1994)
8. Cohen, W.W., Hirsh, H.: Learning the CLASSIC description logic: Thoretical and experimental results. In: Proc. of the 4th Int. Conf. on Principles of Knowledge Representation and Reasoning (KR 1994), pp. 121–133. Morgan Kaufmann (1994)
9. Colucci, S., Di Noia, T., Di Sciascio, E., Donini, F.M., Ragone, A.: Second-order description logics: Semantics, motivation, and a calculus. In: Haarslev, V., Toman, D., Weddell, G.E. (eds.) Proceedings of the 23rd International Workshop on Description Logics (DL 2010), Waterloo, Ontario, Canada, May 4-7. CEUR Workshop Proceedings, vol. 573. CEUR-WS.org (2010)
10. Colucci, S., Di Noia, T., Di Sciascio, E., Donini, F.M., Ragone, A.: A unified framework for non-standard reasoning services in description logics. In: Coelho, H., Studer, R., Wooldridge, M. (eds.) Proceedings of the 19th European Conference on Artificial Intelligence, ECAI 2010, Lisbon, Portugal, August 16-20. Frontiers in Artificial Intelligence and Applications, vol. 215, pp. 479–484. IOS Press (2010)
11. Colucci, S., Donini, F.M.: Inverting subsumption for constructive reasoning. In: Kazakov, Y., Lembo, D., Wolter, F. (eds.) Proceedings of the 2012 International Workshop on Description Logics, DL 2012, Rome, Italy, June 7-10. CEUR Workshop Proceedings, vol. 846. CEUR-WS.org (2012)
12. De Raedt, L., Guns, T., Nijssen, S.: Constraint programming for data mining and machine learning. In: Fox, M., Poole, D. (eds.) Proceedings of the Twenty-Fourth AAAI Conference on Artificial Intelligence, AAAI 2010, Atlanta, Georgia, USA, July 11-15. AAAI Press (2010)
13. De Raedt, L., Nijssen, S., O'Sullivan, B., Van Hentenryck, P.: Constraint programming meets machine learning and data mining (dagstuhl seminar 11201). Dagstuhl Reports 1(5), 61–83 (2011)
14. Džeroski, S.: Towards a general framework for data mining. In: Džeroski, S., Struyf, J. (eds.) KDID 2006. LNCS, vol. 4747, pp. 259–300. Springer, Heidelberg (2007)
15. Eiter, T., Ianni, G., Schindlauer, R., Tompits, H.: A uniform integration of higher-order reasoning and external evaluations in answer-set programming. In: IJCAI, pp. 90–96 (2005)
16. Esposito, F., Fanizzi, N., Iannone, L., Palmisano, I., Semeraro, G.: Knowledge-intensive induction of terminologies from metadata. In: McIlraith, S.A., Plexousakis, D., van Harmelen, F. (eds.) ISWC 2004. LNCS, vol. 3298, pp. 441–455. Springer, Heidelberg (2004)
17. Fanizzi, N., Iannone, L., Palmisano, I., Semeraro, G.: Concept formation in expressive description logics. In: Boulicaut, J.-F., Esposito, F., Giannotti, F., Pedreschi, D. (eds.) ECML 2004. LNCS (LNAI), vol. 3201, pp. 99–110. Springer, Heidelberg (2004)
18. Fanizzi, N., d'Amato, C., Esposito, F.: DL-FOIL concept learning in description logics. In: Železný, F., Lavrač, N. (eds.) ILP 2008. LNCS (LNAI), vol. 5194, pp. 107–121. Springer, Heidelberg (2008)
19. Frazier, M., Pitt, L.: CLASSIC learning. Machine Learning 25(2-3), 151–193 (1996)
20. Guns, T., Nijssen, S., De Raedt, L.: Evaluating pattern set mining strategies in a constraint programming framework. In: Huang, J.Z., Cao, L., Srivastava, J. (eds.) PAKDD 2011, Part II. LNCS, vol. 6635, pp. 382–394. Springer, Heidelberg (2011)
21. Guns, T., Nijssen, S., De Raedt, L.: Itemset mining: A constraint programming perspective. Artificial Intelligence 175(12-13), 1951–1983 (2011)
22. Henkin, L.: Completeness in the theory of types. Journal of Symbolic Logic 15(2), 81–91 (1950)

23. Iannone, L., Palmisano, I., Fanizzi, N.: An algorithm based on counterfactuals for concept learning in the semantic web. Applied Intelligence 26(2), 139–159 (2007)
24. Kietz, J.U., Morik, K.: A polynomial approach to the constructive induction of structural knowledge. Machine Learning 14(1), 193–217 (1994)
25. Küsters, R. (ed.): Non-Standard Inferences in Description Logics. LNCS (LNAI), vol. 2100. Springer, Heidelberg (2001)
26. Küsters, R., Molitor, R.: Approximating most specific concepts in description logics with existential restrictions. AI Communications 15(1), 47–59 (2002)
27. Lehmann, J., Hitzler, P.: Foundations of Refinement Operators for Description Logics. In: Blockeel, H., Ramon, J., Shavlik, J., Tadepalli, P. (eds.) ILP 2007. LNCS (LNAI), vol. 4894, pp. 161–174. Springer, Heidelberg (2008)
28. Lehmann, J., Hitzler, P.: Concept learning in description logics using refinement operators. Machine Learning 78(1-2), 203–250 (2010)
29. Lehmann, J.: DL-Learner: Learning Concepts in Description Logics. Journal of Machine Learning Research 10, 2639–2642 (2009)
30. Lehmann, J., Haase, C.: Ideal Downward Refinement in the $\mathcal{EL}$ Description Logic. In: De Raedt, L. (ed.) ILP 2009. LNCS, vol. 5989, pp. 73–87. Springer, Heidelberg (2010)
31. McGuinness, D.L., Patel-Schneider, P.F.: Usability issues in knowledge representation systems. In: Mostow, J., Rich, C. (eds.) Proceedings of the Fifteenth National Conference on Artificial Intelligence and Tenth Innovative Applications of Artificial Intelligence Conference, AAAI 1998, IAAI 1998, Madison, Wisconsin, USA, July 26-30, pp. 608–614. AAAI Press/The MIT Press (1998)
32. Mitchell, T.: Generalization as search. Artificial Intelligence 18, 203–226 (1982)
33. Nebel, B. (ed.): Reasoning and Revision in Hybrid Representation Systems. LNCS, vol. 422. Springer, Heidelberg (1990)
34. Nijssen, S., Guns, T., De Raedt, L.: Correlated itemset mining in ROC space: a constraint programming approach. In: Elder IV, J.F., Fogelman-Soulié, F., Flach, P.A., Zaki, M.J. (eds.) Proceedings of the 15th ACM SIGKDD International Conference on Knowledge Discovery and Data Mining, Paris, France, June 28-July 1, pp. 647–656. ACM (2009)
35. Reiter, R.: Equality and domain closure in first order databases. Journal of ACM 27, 235–249 (1980)

Identifying Driver's Cognitive Load Using Inductive Logic Programming

Fumio Mizoguchi[1,2], Hayato Ohwada[1], Hiroyuki Nishiyama[1],
and Hirotoshi Iwasaki[*]

[1] Faculty of Sci. and Tech. Tokyo University of Science,
2641 Yamazaki, Noda-shi,
CHIBA, 278-8510, Japan
[2] WisdomTex Co. Ltd.,
1-17-3 Meguro-ku Meguro, Tokyo 153-0063, Japan
Denso IT Laboratory
mizo@wisdomtex.com, ohwada@rs.tus.ac.jp, hiroyuki@rs.noda.tus.ac.jp,
hiwasaki@d-itlab.co.jp

Abstract. This paper uses inductive logic programming (ILP) to identify a driver's cognitive state in real driving situations to determine whether a driver will be ready to select a suitable operation and recommended service in the next generation car navigation systems. We measure the driver's eye movement and collect various data such as braking, acceleration and steering angles that are qualitatively interpreted and represented as background knowledge. A set of data about the driver's degree of tension or relaxation regarded as a training set is obtained from the driver's mental load based on resource-limited cognitive process analysis. Given such information, our ILP system has successfully produced logic rules that are qualitatively understandable for rule verification and are actively employed for user-oriented interface design. Realistic experiments were conducted to demonstrate the learning performance of this approach. Reasonable accuracy was achieved for an appropriate service providing safe driving.

1 Introduction

The next-generation driving support system requires both highly sophisticated functions and user-friendly interfaces (telephone or e-mail communication, audio-visual service, etc.) according to the driver's situation. This support system must therefore recognize the driver's cognitive state to determine whether he or she will be ready to select a suitable operation and recommended service. Obviously, it is better to provide information when the user is relaxed rather than tense.

The next-generation services should judge the driver's mental load by considering both the driver's individual characteristics and operations, and the driver mental load considered that driver is unconscious. The mental load has been

---

[*] Corresponding author.

F. Riguzzi and F. Železný (Eds.): ILP 2012, LNAI 7842, pp. 166–177, 2013.
© Springer-Verlag Berlin Heidelberg 2013

analyzed by mathematical equations, stochastic models, and engineering control models in the fields of bionics and brain science[6,4,7]. Although those were based on the biological understanding and clarification of the living body itself, the important point here is how the human cognitive state can be specified in terms of the states of living body. However, such specification may be hard to investigate because of the ambiguity induced by individual differences or unconscious reactions.

In this paper, we focus on how a user's cognitive load can be determined, analyzed, and used from vital reactions (eye movement) and user operations (driving a car). To do this, we measure the driver's eye movement and gather driving data such as accelerator use, braking, and steering. Eye movement is used in the field of physiological psychology for clarifying control[3] and is directly related to perception that can be considered an indication of cognitive load. Driving a car requires cognition and prediction of the surrounding environment and is influenced by situations. Furthermore, it includes reflex action interruptions and resource competition for performing two or more operations simultaneously[12].

This paper takes an ILP approach to the above cognitive state identification problem in a realistic car-driving task. We set up binary cognitive loads (relaxed or tense) that are analyzed by interviewing the driver to determine his cognitive state with regard to a driving video review and use the obtained data as a set of training examples. For background knowledge, we identify an object the driver watches and collect driving data obtained from an in-vehicle LAN; those data are then processed as qualitative data. Given such information, our ILP system has successfully produced logic rules that are qualitatively understandable for rule verification and refinement. We conducted realistic experiments to demonstrate the learning performance of this approach, and obtained reasonable accuracy for designing really useful user-interfaces.

This paper is organized as follows. Section 2 presents raw data obtained from an in-vehicle LAN. Section 3 arranges these data that are transformed into qualitative descriptions in Section 4. Section 5 describes background knowledge, and Section 6 provides a way to set up training data. Section 7 introduces learned rules, and Section 8 includes performance evaluation. The final section provides conclusions.

2   Raw Data

To collect data about the driver's eye movement, the system EMR-8 is used [1]. The device measures horizontal and vertical viewing angles in degrees. For the purpose of this study, we consider 60 data points per second.

The Controller Area Network [2] (CAN) is an in-vehicle LAN used to gather driving data. We can obtain the accelerator depression rate (0% to 100%),

---

[1] EMR-8 is produced by NAC Image Teck., Inc. (www.eyemark.jp)

[2] This is a standard used for the data transfer between in-vehicle equipment. The International Organization for Standardization is standardizing it as ISO 11898 and ISO 11519.

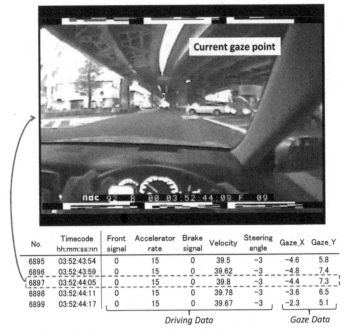

No.	Timecode hh:mm:ss:nn	Front signal	Accelerator rate	Brake signal	Velocity	Steering angle	Gaze_X	Gaze_Y
6895	03:52:43:54	0	15	0	39.5	−3	−4.6	5.8
6896	03:52:43:59	0	15	0	39.62	−3	−4.8	7.4
6897	03:52:44:05	0	15	0	39.8	−3	−4.4	7.3
6898	03:52:44:11	0	15	0	39.78	−3	−3.6	6.5
6899	03:52:44:17	0	15	0	39.67	−3	−2.3	5.1

Driving Data   *Gaze Data*

Fig. 1. Eye movement and driving data

braking signal (0 or 1), steering signal (-450 to 450 degrees), a signal representing the gear (0 to 4), the front separation (in meters), and so on. To measure these, we modified a Toyota Crown and obtained 10 data points per second.

The eye-movement data and the car-driving data must be synchronized as indicated in Fig. 1. The third row data (No. 6897) states that the car's velocity is 39.8km/h with an accelerator depression rate of 15% and is running almost straight (steering angle -3); the driver is gazing at the center (the coordinate is (-4.4,7.3)). All of the data can be used to produce background knowledge as demonstrated in the next section.

3   Data Arrangement

Eye movement indicates where a person is looking. It also expresses caution and concern, and is used for evaluating software usability [11].

When a person looks at something every day, the eyes repeat cycles of rapid movement and stationary periods. The rapid movements are called "saccade," and the stationary periods are called "fixation" [8]. In saccade, the eyes rotate to position the target at the central fovea of the retina. Saccade is defined as the movement that rotates the eye at speeds of 100 degrees/second or more. Movement that rotates the eyes at 100 degrees/second or less is called "pursuit" because the eye pursues a smoothly moving target. Because saccade is related with capturing a target, it is closely related to the observer's motivation and cognitive process.

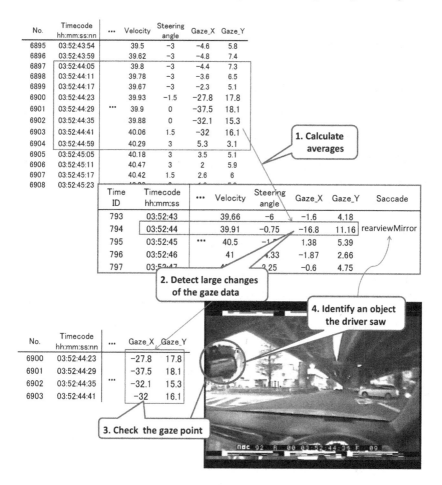

Fig. 2. Data arrangement process for each second. Driving data is averaged, and saccade events are detected.

We identify a driver's cognitive load induced by saccade in one second, and average raw data for each second. By tracking saccade events, we detect targets such as the rearview mirror, car navigation device, road signs, and pedestrians. This raw data arrangement consists of the following steps:

Step 1. Collect a set of raw data measured in a second, and average each attribute value of driving data (Fig. 2, Step 1).

Step 2. Detect a large change of eye movement that reveals a saccade event, and obtain a set of time codes (Fig. 2, Step 2).

Step 3. Calculate the gaze point area corresponding to saccade from raw data (Fig. 2, Step 3).

Step 4. Identify an object (*e.g.* rearview mirror) within the gaze point through video analysis (Fig. 2, Step 4). If the object is identified, it is used as the value of "saccade" attribute.

Step 4 is performed manually (not automatically) at the current stage. In Fig. 2, rearview mirror is identified as the saccade target.

4   Data Transformation

The data-transformation phase gathers time-series saccade events and generates qualitative expressions of driving data. This is inspired by the research of qualitative reasoning that focuses on the change of variables such as an increase or decrease and the relation between variables such as proportionality and differentiation [9]. Although qualitative reasoning has also been employed in ILP literature [1,2,14], we use qualitative reasoning to deal with driver's individual characteristics, unconscious reactions, and human situated cognition [5].

A driver's cognitive state can be qualitatively characterized by a resource-limited model [12]. In driving a car, a resource-required action corresponds to an accelerator operation making the vehicle speed up, and a resource-free action, to an operation to slow the vehicle. Although this model is simple and abstract, qualitatively describing driving data may help to capture the driver's cognitive state [13].

Some averaged real data are mapped to a small number of categories. In contrast to braking signals and front signals (indicating whether a car exists in front) that are binary attributes, accelerator rate and steering angle are handled as four qualitative values (zero, low, middle, high) based on the quantiles on the standard normal distribution. A value for velocity is divided to the ordered set (zero, <10, <20, <30, <40, <50, >=50) according to normal town-street driving. Moreover, parameter changes in qualitative reasoning are extended to the four categories [3] (low, middle, high, veryHigh) with two directions of change (up, down) [4]. This means that a qualitative state difference between the current state and the next state takes one of the eight forms.

Although a driver's cognitive state can be characterized by a sequence of saccades, we employ only adjacent saccades in a short time period (5 to 10 seconds) to predict the cognitive state for recognition and decision-making. This includes driving data after five seconds because some operations are intended by the driver and are considered as predictable changes for the driver. Thus, driving data for a small number of saccade events are considered for background knowledge.

Figure 3 illustrates the process of data transformation consisting of the following steps:

Step 1. Collect an ordered set of saccade event data in which "saccadeID" is inserted for each data.

Step 2. Add new attributes indicating the differences in the short time period (five seconds before and after) where each difference is represented by Δ-second.

---

[3] They are calculated by using the quantiles on the standard normal distribution.

[4] noChange is used to indicate that there is no change.

Saccade ID	Time ID	...	Velocity	Steering angle	Gaze_X	Gaze_Y	Saccade
121	794		39.91	−0.75	−16.8	11.16	rearviewMirror
122	799	...	38.89	−2.06	−10.5	7.88	rearviewMirror
123	801		33.22	2	−26.5	−9.23	carNavigation

Add difference data

Saccade ID	Time ID	...	Velocity	Velocity △−5s	Velocity △−1s	Velocity △+5s	Steering angle	Steering △−5s	Steering △−1s	Steering △+5s	...	Saccade
121	794		39.91	−1.7	0.3	−1	−0.75	28.1	5.3	−1.3		rearviewMirror
122	799	...	38.89	−1	−1.2	−13.5	−2.06	−1.3	−5.1	0.7	...	rearviewMirror
123	801		33.22	−7.8	−3.3	−13.7	2	6.3	7.3	−3.8		carNavigation

Translating into qualitative information

Saccade ID	Time ID	...	Velocity	Velocity △−5s	Velocity △−1s	Velocity △+5s	...	Saccade
121	794		under40	downLow	upLow	downLow		rearviewMirror
122	799	...	under40	downLow	downLow	downMiddle	...	rearviewMirror
123	801		under40	downMiddle	downLow	downMiddle		carNavigation

Fig. 3. Data transformation

Step 3. Translate the data obtained at Step 2 into the corresponding qualitative data using the categories (upLow, upMiddle, upHigh, downLow, downMiddle and downHigh).

5   Background Knowledge

Table 1 presents a set of predicate types and their mode declarations given to the background knowledge. The first type corresponds to qualitative values for each eye movement and driving data. This is described by the saccade ID and a parameter value. The second type is a qualitative state difference in a short time period and is described as *parameter_diff*(ID,Time,Val) where Time indicates the time difference in seconds. The third one (before_event and after_event) is used to get information of adjacent saccades. Note that multiple adjacent saccades exist within five seconds in some cases.

Table 1. Predicates and their mode declarations in background knowledge. Mode + indicates input variable, - output variable, and # constant.

Types	Predicates
qualitative value	accele(+ID, #Val), brake(+ID, #Val), velocity(+ID, #Val), steering(+ID, #Val), front(+ID, #Val), gazeX(+ID, #Val), gazeY(+ID, #Val), lookAt(+ID, #Target)
qualitative state difference	accele_diff(+ID,#Time,#Val), brake_diff(+ID,#Time,#Val), velocity_diff(+ID,#Time,#Val), front_diff(+ID,#Time,#Val), steering_diff(+ID,#Time,#Val), gazeX_diff(+ID,#Time,#Val), gazeY_diff(+ID,#Time,#Val)
adjacent saccades	before_event(+ID, -ID), after_event(+ID, -ID)

6   Training Examples

We set up binary cognitive loads (relaxed or tense) that are analyzed by interviewing the driver to determine his cognitive state with regard to a driving video review. A set of data about the driver's degree of tension or relaxation is regarded as a training set and is obtained by also considering the driver's mental load based on the resource-limited cognitive process model [12].

The resource-limited model indicates the mechanism used to coordinate the resource consumption through the repeated execution of driver operations and eye movements. For example, the accelerator increases speed, and the used resource will increase. If too much of the resource is used, a resource-free action like braking is invoked. A decision about the cognitive load is based mainly on an interview with the driver, but the resource-limited model is used to check the decision under the following criteria:

tense: If saccade occurs and time resource-free actions are executed, the used resource is full. In this case, the driver's cognitive load is heavy.

relaxed: If saccade occurs and the resource is stable or resource-required actions are executed, the resource is available for use. This case indicates that the driver is relaxed.

A cognitive state is represented as `class(+ID,#Class)` where `Class` is either `tense` or `relaxed`. These two classes are mutually exclusive; a positive example `class(1,tense)` is regarded as the negative example of `class(1,relaxed)`.

7   Learned Rules

We measured driving and eye-movement data of a skilled driver for about 20 minutes driving on two types of road, an urban road with much traffic (Shibuya area) and a road with moderate traffic (Noda area). Table 2 presents the statistics of raw data, arranged data, and saccade-event data.

Table 2. Used data for the experiments

Area	Measured time (second)	Raw data	Arranged data for each second	Saccade event data
Shibuya	1345	80780	9415	220
Noda	1310	72429	9170	240

Saccade event data is classified as either **tense** or **relaxed**, and Table 3 indicates its class distribution.

Our ILP system (GKS[10]) is employed to learn rules for each driving area. The machine we used is Windows 7 OS with two 2.40GHz Intel(R) Core(TM) i5 CPUs with 4GB memory. The system generated 33 rules (17 for tense and 16 for relaxed) for the Shibuya area and 39 rules (14 for tense and 25 for relaxed)

Table 3. Class distribution of the saccade event data

Area	tense	relax	Total
Shibuya	105	115	220
Noda	117	123	240

for the Noda area. Learning times were 533 seconds (Shibuya) and 2400 seconds (Noda).

We present typical rules below. "{T,F}" denotes the number of positive examples (T) and the number of negative examples (F) the rule covers.

```
{21,0} class(A, tense) :-
       before_event(A, B), lookAt(B, corner), accele(B, zero).
```

The first rule *naturally* describes our understanding in which a driver becomes tense when he looks at the corner of the front window without pressing the accelerator. In this case, he is making a turn.

The following rule indicates another tense state:

```
{20,0} class(A, tense) :-
@@@@ before_event(A, B), before_event(B, C), brake(B, on),
@@@@ brake_diff(C, -1, noChange), gazeY(C, front).
```

This rule indicates that multiple saccades occur in a short time period (denoted by the variables A and B) with continued braking regardless of watching the front direction. This is not a trivial rule represented in terms of *non-determinate* predicates (before_event), indicating the advantage of an ILP-based learner.

The above two rules indicate that the driver's cognitive load is heavy, and thus additional service should not be provided. In contrast, the following rules indicate when service can be provided:

```
{35,0} class(A, relax) :-
       brake(A, off), brake_diff(A, +5, noChange),
       steering(A, straight).
```

This rule *naturally* describes a normal driving situation in which a driver becomes relaxed when going straight without continued braking. The following is another rule indicating a relaxed driver:

```
{15,0} class(A, relax) :-
       after_event(A, B), brake(B, off),
       front_diff(B, -5, noChange), steering_diff(A, +5, rightLow),
       steering_diff(B, -5, rightLow).
```

This rule is more complicated because it predicts the driver's action. The predicate after_event is used for a predictable state, representing that the driver will release the brake and go straight while driving smoothly. The rule cannot be applied in our current car navigation system, but it may be useful for next-generation systems.

We obtained unexpected rule:

```
{10,0} class(A, tense) :-
@@@@ velocity(A, under10), velocity_diff(A, -1, upLow),
@@@@ velocity_diff(A, +5, upLow).
```

This rule says that the driver is tense even when the driving speed is low and slightly accelerating. Although the driver seems to have available resources, he carefully looked in front of the car. Our rule is based on saccade events, but in this case there is only one saccade, so missing some critical intentional action executions.

8 Performance Evaluation

Experiments were conducted to achieve the following.

Aim1 seeks to describe the difference of accuracy among three types of background knowledge. Background knowledge B1 includes only qualitative parameter values of driving data. Background knowledge B2 adds qualitative differences of parameter changes to B2. Background knowledge B3 contains all information including the information about adjacent saccade events.

Aim2 attempts to derive the learning curve based on the assumption of real-time learning. We divide training examples into ten subsets in time sequence. In this setting, training examples progressively increase in time order.

Aim3 is designed to describe the accuracy of rules in applying another new road (driver is the same person, using the same vehicle).

In Aim1, we conducted a 10-fold cross validation assessment for each item of background knowledge. Performance measures are *accuracy*, *recall* and *precision* defined in the appendix. The result is listed in Table 4 in which B3 is the most effective and B2 is the second-most effective. Case B2 indicates that information about qualitative state differences is very important for cognitive state classification due to the effectiveness of the qualitative reasoning approach. Case B3 indicates that information within a short time period is useful for discriminating important factors of cognitive states. As indicated in the previous section, such

Table 4. Performance measures of the 10-fold cross-validation for B1, B2 and B3

Area	Data set	Accuracy	Precision	Recall
	B1	75.5 ± 1.3%	73.8%	75.2%
Shibuya	B2	84.5 ± 1.3%	85.1%	81.9%
	B3	85.9 ± 0.6%	84.9%	85.7%
	B1	72.5 ± 1.0%	71.1%	73.5%
Noda	B2	81.3 ± 0.6%	81.6%	79.5%
	B3	84.2 ± 0.8%	83.8%	83.8%

information is used to produce a first-order version of rules. This means that the driver's cognitive load depends on a time-series of action executions rather than a single action execution.

We then experimented with Aim2 using progressively incremental training data. In contrast to a 10-fold cross validation, we assign the saccade events in the set to 10 equal partitions in time order. In the first state, the first partition is a set of training examples, and the remaining partitions construct a set of test examples. The second stage takes the first two partitions as the training data set and the remaining eight partitions as the test data set. This process repeats nine times.

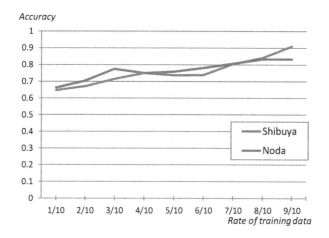

Fig. 4. Accuracies of incremental training data

Figure 4 illustrates how accuracy changes with a series of incremental training data. At every stage, predictive accuracy exceeds the corresponding default accuracy. It exceeds 70% in the third stage and 80% in the eighth stage. This result is quite reasonable in accordance with the cross validation result.

The last aim is achieved by applying learned rules to another driving situation. Figure 5 presents the test results in which all driving data in another situation construct a set of test data. The test data set is equally divided into 10 partitions to capture the change of accuracy.

The accuracy of Shibuya's rule set is 78.3%, and that of Noda's rule set is 80.9% (average). These coincide with the accuracies for the third to the sixth stages in Fig. 4. However, variances among different partitions cannot be ignored due to the driving load. More detailed analysis on load information is needed.

9   Conclusions

This paper applied ILP to identify a driver's cognitive state using real driving data. We described how to acquire raw data such as eye-movement and driving

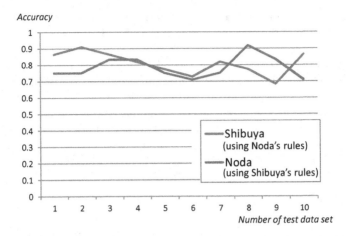

Fig. 5. Accuracies using a set of test data constructed from another driving situation

data, arrange the raw data, and transform the data into qualitative data. We then constructed background knowledge and training examples, and produced rules classifying the driver's cognitive state. We also conducted realistic experiments to demonstrate the learning performance of this approach and obtained reasonable accuracy. The results indicate that ILP could capture the cognitive mental load based on measurable data, and we hope that ILP will be useful for designing user-interfaces for next-generation car navigation systems.

Future work includes investigating the applicability of learned rules to other drivers. We selected a skilled driver, but we have to consider driving data for inexperienced drivers as well. Furthermore, we should demonstrate how to clarify individual differences among drivers compared with other approaches of brain science.

References

1. Bain, M., Sammut, C.: A framework for behavioural cloning. Machine Intelligence 15, 103–129 (1998)
2. Bratko, I., Muggleton, S., Varsek, A.: Learning Qualitative Models of Dynamic Systems. In: Inductive Logic Programming, pp. 437–452 (1992)
3. Carpenter, R.H.S.: Movements of the Eyes. Pion Ltd. (1988)
4. Chapin, J.K., Moxon, K.A., Markowitz, R.S., Nicolelis, M.A.L.: Real-time control of a robot arm using simultaneously recorded neurons in the motor cortex. Nature Neuroscience 2(7), 664–670 (1999)
5. Clancey, W.J.: Situated Cognition: On Human Knowledge and Computer Representations. Cambridge University Press (1997)
6. Enderle, J., Blanchard, S.M., Bronzino, J.: Introduction to Biomedical Engineering. Academic Press (2005)
7. Findlay, J.M., Walker, R.: A model of saccade generation based on parallel processing and competitive inhibition. Behavioral and Brain Science 22, 661–721 (1999)

8. Findlay, J.M., Gilchrist, I.D.: Active Vision. Oxford University Press (2003)
9. Kuipers, B.: Qualitative Reasoning: Modeling and Simulation with Incomplete Knowledge. The MIT Press (1994)
10. Mizoguchi, F., Ohwada, H.: Constrained Relative Least General Generalization for Inducing Constraint Logic Programs. New Generation Computing 13, 335–368 (1995)
11. Nielsen, J., Pernice, K.: Eyetracking Web Usability. New Riders Press (2009)
12. Norman, D.A., Bobrow, D.G.: On data-limited and resource-limited processes. Cognitive Psychology 7(1), 44–64 (1975)
13. Sega, S., Iwasaki, H., Hiraishi, H., Mizoguchi, F.: Qualitative Reasoning Approach to a Driver's Cognitive Mental Load. International Journal of Software Science and Computational Intelligence 3(4), 18–32 (2011)
14. Suc, D., Vladusic, D., Bratko, I.: Qualitatively faithful quantitative prediction. Artificial Intelligence 158(2), 189–214 (2004)

Appendix A: Used Formula

Formula for accuracy, precision and recall is as follows (TP: true positives, FP: false positives, FN: false negatives, TN: true negative):

$$Accuracy = \frac{TP + TN}{TP + FP + FN + TN}$$

$$Precision = \frac{TP}{TP + FP} \qquad Recall = \frac{TP}{TP + FN}$$

Opening Doors: An Initial SRL Approach

Bogdan Moldovan[1], Laura Antanas[1], and McElory Hoffmann[1,2]

[1] Department of Computer Science, Katholieke Universiteit Leuven, Belgium
[2] Department of Mathematical Sciences, Stellenbosch University, South Africa

Abstract. Opening doors is an essential task that a robot should perform. In this paper, we propose a logical approach to predict the action of opening doors, together with the action point where the action should be performed. The input of our system is a pair of bounding boxes of the door and door handle, together with background knowledge in the form of logical rules. Learning and inference are performed with the probabilistic programming language ProbLog. We evaluate our approach on a doors dataset and we obtain encouraging results. Additionally, a comparison to a propositional decision tree shows the benefits of using a probabilistic programming language such as ProbLog.

Keywords: robotics, mobile manipulation, door opening and grasping, logical rules, ProbLog.

1 Introduction

In the context of the EU-project on *Flexible Skill Acquisition and Intuitive Robot Tasking for Mobile Manipulation in the Real World*[1] (FIRST-MM), one of the goals of autonomous robots is to perform mobile manipulation tasks in indoor environments. In this setting, an essential condition is that the robot can operate doors during navigation. A complete solution to this general problem requires a system that can solve several tasks: detecting and localising the door and its handle, recognising the grasping points, finding the right actionable point and action and finally, performing the action on the handle. In this work we focus on two of these tasks: detecting the actionable point and the action movement itself. We assume that the door and handle are detectable by the robot. This is an object detection problem that has been addressed previously in the literature using several approaches, using either 2D [1] or 3D information [3,4,19].

Detecting the action and action points is a challenging manipulation task. It depends on the sensorimotor control of the robot, the gripper, the type of handle and the properties of the door. To be opened, each door requires different actions depending on the side of the door that the robot is approaching. The action also depends on the type of the handle, its relative position to the door and sometimes even the objects around the door. Usually if hinges are detected on the side of the door and the light switch on the opposite side next to the handle, the door needs to be pulled. Similarly, while the shape of the handle can be quite a good

---

[1] More information available at: http://www.first-mm.eu

F. Riguzzi and F. Železný (Eds.): ILP 2012, LNAI 7842, pp. 178–192, 2013.
© Springer-Verlag Berlin Heidelberg 2013

indicator of a suitable action point (i.e., knob), sometimes it cannot be detected reliably, for example when an L-shaped handle is confused with a horizontal one. This may directly influence valid graspable points, given the robot hand type, and limit the actionable handle points. In this case, the relative position of the contact points with the door, the relative positions of the candidate action point to the door sides or other points in the point cloud of the handle may play a key role for grasping and performing the action [18,11].

In the cases mentioned above, generalisations over opening directions, point positions in the point cloud and types of handles are needed. Additionally, it is highly likely that a robot will encounter a door or handle that it has not seen before. Hence it is not possible to enumerate all possible doors and handles. Moreover, the objects in the surrounding environment, such as the hinges, play a role in predicting the correct action and action point for opening the door. Following this idea, Rosman and Ramamoorthy [16] introduced approaches to learn the symbolic spatial relationships between objects which allow performing various robotic tasks. Thus, they take the relations in the environment into account. Similarly, Bereson et al. [2] takes contextual information into account when planning grasps.

In this work we propose the use of statistical relational learning (SRL) [5,14] to generalise over doors and handles in order to predict door opening actions and handle actionable points. SRL combines logical representations, probabilistic reasoning mechanisms and machine learning and thus is suitable for our task. Although several existing SRL approaches could be used to solve our problem [10,15], in this work we consider *probabilistic logic programming languages* (PPLs). Our choice is mainly motivated by the fact that PPLs are specially designed to describe and infer probabilistic relational models or, more generally, logic theories. In particular, we use Problog [6], a PPL, that allows us to specify our problem declaratively, using symbolic rules, while being able to deal with noisy data due to its probabilistic language. Other SRL frameworks could be used as well, however ProbLog is very convenient for encoding world knowledge due to: i) its declarative nature; ii) the fact that weights can be directly viewed as probabilities; iii) its capability to handle numerical values, which is something to consider when dealing with point clouds (for future developments of this work).

Thus, the main contribution of this paper is the use of a ProbLog theory to predict the actions to open doors and well as the action points. We use as input solely extracted properties of door images. In our logical representation of the domain, every visual scene is mapped to a logical interpretation. However, ProbLog could also model the noisy nature of the detection aspects and the uncertainty of the environment where the robot operates. As a key element, we encode background knowledge about the domain, in a natural way, using a set of logical rules which reduce the number of parameters of the theory. Finally, we use a learning from interpretations setting [7] to learn the parameters of our ProbLog theory. The approach is general enough to be able to deal with point clouds as well as 2D visual images.

We evaluated our approach on a dataset containing 60 images of doors. The results are promising and motivate us to continue this work with a real robot scenario. Additionally we compare against a random classifier, a majority classifier and a propositional decision tree. We report superior results using our relational approach, which shows the role of background knowledge when solving the opening door problem. Some work on predicting how to best open a door by a robot setting exists [12]. However, it does not make use of logical representations as we do.

The outline of this paper is as follows: in Section 2 we introduce the problem and the approach used to solve it, and in Section 3 we present our learning and inference setting. We show experimental results in Section 4, after which we conclude and mention future directions in Section 5.

2   Problem Modeling

2.1   Problem Description

We first introduce an initial setting for a high-level logical and relational reasoning system that can be used by a robot for opening doors. We assume the robot is able to detect doors and door handles, so we assume to have access to bounding boxes in the image for both the door frame and the door handle. Figure 1 presents two such examples of detected door frame (in red) and handle (in blue) with their bounding boxes. Later, we can add prior probabilities on the positions of the frame and handle for a more realistic scenario which involves object detection uncertainty. The setting can be expanded to include other detected objects in the environment to help us identify the action needed to open the door by providing additional relational contextual cues.

Fig. 1. Annotated doors: **(a)** push down action, **(b)** push in

In this setting, we are interested in predicting the high-level (initial) action the robot needs to perform in order to open the door, and where this action should be applied (action point). Once these are determined, the robot can grasp the handle and execute the action. In a more advanced setting it can be imagined that we can generalise over possible grasping points depending on the specific robot hand. Here, we are just interested in predicting the action/action point pairs.

We assume that the robot can open a door by pushing it in any of the six 3D directions, labelled as *in, out, left, right, up, down*, and turning the handle in two directions: *clockwise* and *counterclockwise*. In total, there are eight possible high-level actions. At a high-level, we think of the action point in terms of which end of the handle needs to be acted upon, so we will discretise this into 5 different values: *up, down, left, right* and *centre*. For future work, we plan to upgrade our model with continuous actionable points in the handle point cloud.

Although currently the proposed model does not include explicit relations between objects or object points in the scene, it can be easily extended to incorporate such information. For example, if the object is represented by a point cloud, knowing that two points are very close spatially allows us to exploit symetries and redundancies in the domain. By using such a relation we can generalize over points in the cloud, and thus produce a more compact model.

2.2   Approach

From the bounding boxes for the door frame and handle we obtain a set of positions in the $x - y$ plane for both the door and handle: $(x_{min}, y_{min}, x_{max}, y_{max})$. Based on these we define five features $(F_1, ..., F_5)$, namely: the handle aspect ratio, the handle width relative to the door width (or handle relative width), the handle height relative to the door height (or handle relative height), the position of the centre of the handle relative to the door frame in the x-axis and in the y-axis (or handle relative width/height positions).

We assume that these features are independent and additionally, we discretize them. Handle aspect ratio can take the values {*big-width, small-width, square, small-height, big-height*}. Handle relative width and height can take the values {*small, medium, large*}. Finally, the position of the handle relative to the door can take the values {*center-pos, up-pos, down-pos, left-pos, right-pos*} on the x-axis or y-axis.

Action Prediction. An initial intuition is that we can use a Naive Bayes classifier [17] in order to predict the action based on these features. The Bayesian Network for our Naive Bayes model is illustrated in Figure 2. Given our computed features $F_1, ..., F_5$ from the observed x and y positions of the bounding boxes of the door frame and handle and using our independence assumption, we can compute the conditional probability [17] of an action A as:

$$P(A|F_1, ..., F_5) = \frac{P(A) * P(F_1, ..., F_5|A)}{P(F_1, ..., F_5)} = \frac{P(A) * P(F_1|A) * ... * P(F_5|A)}{P(F_1) * ... * P(F_5)}.$$

Then, in order to predict A, we compute the maximum a posteriori (MAP) probability estimate as: $\arg\max_A P(A) * \prod_{i=1}^{5} P(F_i|A)$. However, in a fully propositional setting this requires the learning of many parameters, even in such a small domain with five features, taking values from a small discretised set. We propose to go towards a relational setting, where background knowledge can be used as a set of logical rules to reduce the number of parameters that need to be learnt, and thus the number of learning examples that need to be used, and at a later stage to generalise over our setting.

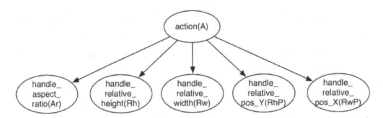

Fig. 2. The Bayesian Network used for the Naive Bayes classifier for action prediction

Action Point Prediction. The action point can be determined both by the type of the already predicted action and the features $F_1, ..., F_5$. Since the action and these features are not independent, we cannot use our previous approach that we used for predicting the action. For the purpose of predicting the action point we define a Bayesian Network and learn its parameters. The Bayesian Network is illustrated in Figure 3. In order to predict the position, we need to compute: $\arg\max_{Pos} P(Pos|A, F_1, ..., F_5)$.

Similarly to the action prediction task, we augment our model with background knowledge in the form of logical rules which constrain the action point based on the action and related features. For example, a *push in* or *out* on a handle with a big width or height (or big aspect ratio) should be done at the centre of the handle. This also helps us reduce the number of parameters that we need to learn. In practice, our experiments show that the action point prediction is influenced only by the action, relative position of the centre of the handle on the x-axis and handle relative height.

For both tasks we can use ProbLog to compute the probabilities. In the next section we describe our learning and inference setting with ProbLog.

3   Learning and Inference with ProbLog

ProbLog is a probabilistic extension of the Prolog programming language. We first review the main principles and concepts underlying ProbLog. Afterwards, we explain how we perform learning and inference with our model within the ProbLog language.

In ProbLog – as in Prolog – an atom is an expression of the form $a(t_1, ..., t_k)$ where a is a predicate of arity k with $t_1, ..., t_k$ terms. A term can be a variable, a constant, or a functor applied to other terms. An expression is called ground if it

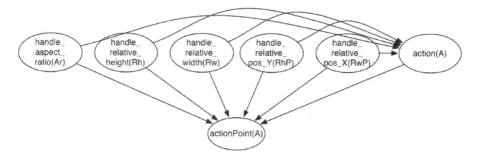

Fig. 3. The Bayesian network for action point prediction

does not contain variables. Definite clauses are universally quantified expressions of the form $h : -b_1, \ldots, b_n$, where h and b_i are atoms. A fact is a clause without the body. Additionally, in ProbLog a clause c_i may be labeled with a probability p_i in the form $p_i :: h : -b_1, \ldots, b_n$. Similarly, a ProbLog labeled fact signifies that the fact is true with probability p_i.

A ProbLog theory T consists of a set of facts F and a set of definite clauses BK which express the background knowledge. The semantics of a set of definite clauses is given by its least Herbrand model, the set of all ground facts entailed by the theory. If $T = \{p_1 :: c_1, \ldots, p_n :: c_n\}$, ProbLog defines a probability distribution over logic theories $L \subseteq L_T = \{c_1, \ldots, c_n\}$ in the following way: $P(L|T) = \prod_{c_i \in L} p_i \prod_{c_i \in L_T \setminus L} (1 - p_i)$. Additionally, in ProbLog we are interested in computing the probability that a query q succeeds in the theory T. The success probability $P(q|T)$ corresponds to the probability that the query q has a proof, given the distribution over logic programs.

For our prediction tasks, we first build a ProbLog theory of the Bayesian Networks, which we augment with logical rules reflecting our background knowledge. The theory then supports inference for predicting both tasks by answering probabilistic queries. In our case the end goal query is the conditional probability of a particular action (or action point) given the set of observations made about the world, that is about the handle in relation with the door. The logical rules are (general) background knowledge and they are able generalize over different properties of the handle relative to the door, actions or action points. In this way, we generalise by reducing the number of parameters of a fully ground model. Because our goal is to solve both action prediction and action point prediction, our ProbLog theory will contain two models, one for each task.

We next present an excerpt from our ProbLog theory used to answer probabilistic queries.

```
%theory for action prediction

%parameters for the Naive Bayes model
0.0:: action(push_up).
0.43:: action(push_in).
```

```
0.05:: action(push_out).
...
0.74:: har(small_height,left).
0.11:: har(big_height,left).
...
0.31:: hrw(small,in).
0.13:: hrw(medium,in).
0.56:: hrw(large,in).
...
0.998:: hrh(small,turnc).
0.997:: hrh(small,turncc).
0.001:: hrh(large,turnc).
...
0.60:: hrwp(left_point,down).
0.004:: hrwp(center_pos,down).
0.24:: hrwp(center_pos,right).
...
0.017:: hrhp(down_pos,up).
0.34:: hrhp(center_pos,up).
0.001:: hrhp(up_pos,out).
...
%definite clauses for the Naive Bayes (action prediction)
handleAspectRatio(square):- action(turnc), har(square,turnc).
handleAspectRatio(square):- action(turncc), har(square,turncc).
handleAspectRatio(Ar):- action(Ac), har(Ar,Ac), Ar\=square.
handleAspectRatio(Ar):- action(Ac), har(Ar,Ac),
                        Ac\=turnc, Ac\=turncc.

handleRelativeWidth(large):- action(in), hrw(large,in).
handleRelativeWidth(large):- action(out), hrw(large,out).
handleRelativeWidth(Rw):- action(Ac), hrw(Rw,Ac), Rw\=large.
handleRelativeWidth(Rw):- action(Ac), hrw(Rw,Ac), Ac\=in, Ac\=out.

%similar relations for relative height
...
handleRelativePosition\_X(RwP):- action(Ac), hrhp(HrWp,Ac).
handleRelativePosition\_Y(RhP):- action(Ac), hrhp(HrHp,Ac).

%extended theory with action point prediction

%parameters for the Bayesian Network model
0.09:: ap(center,up,right_pos).
0.06:: ap(center,left,right_pos).
0.17:: ap(center,up,left_pos).
...
```

```
0.09:: ap(down,left,left_pos).
```

```
%definite clauses for the Bayesian Network
%(action point prediction)
actionPoint(center):- handleRelativeWidth(large).
actionPoint(center):- handleRelativeHeight(large).
actionPoint(center):- handleRelativeWidthPosition_X(center_pos).
actionPoint(center):- action(turnc).
actionPoint(center):- action(turncc).
actionPoint(center):- action(in).
actionPoint(center):- action(out).
actionPoint(center):- action(Ac), ap(center,Ac,HrWp),
                handleRelativeWidthPosition_X(RwP), Ac\=turnc,
                Ac\=turncc, Ac\=in, Ac\=out, RwP\=center_pos.
actionPoint(left):- action(Ac), ap(left,Ac,HrWp),
                handleRelativeWidthPosition_X(RwP), Ac\=turnc,
                Ac\=turncc, Ac\=in, Ac\=out, RwP\=center_pos.
...
%similarly for the rest of action points
```

The Naive Bayes model for action prediction can be directly encoded in ProbLog using rules such as:

$\text{handleAspectRatio}(Ar) \leftarrow \text{action}(Ac), \text{har}(Ar, Ac).$

$\text{handleRelativeWidth}(Rw) \leftarrow \text{action}(Ac), \text{hrw}(Rw, Ac).$

where $\text{har}(Ar, Ac)$ is a probabilistic fact representing the conditional probability of the handle aspect ratio given the action and $\text{hrw}(Rw, A)$ is a probabilistic fact representing the conditional probability of the handle relative width given the action. The probabilities of its groundings (e.g., $0.74 :: \text{har}(\text{small\_height}, \text{left})$, $0.11 :: \text{har}(\text{big\_height}, \text{left}), \dots$) are parameters of the model. However, to reduce them, we augment the rules with additional background knowledge. For example, the rule encoding the handle relative width feature can be extended to encode that if the handle relative width or height is large, the action that needs to be performed is either a push *in* or a push *out* (as illustrated in Figure 1(b)). We encode this in ProbLog using two rules the following way:

$\text{handleRelativeWidth}(\text{large}) \leftarrow \text{action}(Ac), Ac = \text{in}, \text{hrw}(1, Ac).$

$\text{handleRelativeWidth}(\text{large}) \leftarrow \text{action}(Ac), Ac = \text{out}, \text{hrw}(1, Ac).$

Similarly, if the handle is a knob, it needs to be turned in one of the two directions *clockwise* or *counterclockwise*. In this case the handle is characterized by a handle aspect ratio close to 1 (i.e. a bounding box close to a *square*) and we encode this case in ProbLog using two other rules:

$\text{handleAspectRatio}(\text{square}) \leftarrow \text{action}(Ac), Ac = \text{turn\_clock}, \text{har}(\text{square}, Ac).$

$\text{handleAspectRatio}(\text{square}) \leftarrow \text{action}(Ac), Ac = \text{turn\_clockwise},$
$$\text{har}(\text{square}, Ac).$$

Similar rules are defined for the action prediction task for:

`handleRelativeHeight(Rh)`

`handleRelativePosition_Y(RhP)`

`handleRelativePosition_X(RwP).`

The model can also be extended later with background contextual knowledge gathered from the environment, like the presence of other objects near the door which could give an indication about how to open it. Furthermore, ProbLog allows us to add priors (e.g., Gaussian) on the x and y axis positions of the detected door frame and handle to model uncertainty in object detection.

The theory defined for the action prediction task can be extended with extra rules to encode the ProbLog model of the BN associated with the action point task. Similarly, it will contain the respective background knowledge in the form of logical rules, which enables us to generalise.

The BN model for action point prediction can be encoded in ProbLog using a rule such as:

`actionPoint(center) : −action(Ac), ap(center, Ac, RwP),`
                    `handleRelativeWidthPosition_X(RwP).`

where `ap(Point, Ac, RwP)` is a probabilistic predicate representing the conditional probability of the action point given the action and the relative position of the centre of the handle on the x-axis. These probabilities are the model parameters. Similarly, to reduce their number, we augment the rules with additional background knowledge. From experience we know that any turn action requires the robot to perform a caging grasp of the knob, which means grabbing the handle at the centre. This can be encoded as definite rules as follows:

`actionpoint(centre) ← action(Ac), A = turn_counterclock.`

`actionpoint(centre) ← action(Ac), A = turn_clockwise.`

Once these background rules are added, we would not need the model parameters `ap(center, turn_counterclock, RwP)` and `ap(center, turn_clockwise, RwP)` anymore. Similar rules one can define for the left, right, up and down action points. A clear advantage of such models is in the context of predicting exact points in a point cloud. Then, such rules will generalize over similarly spatially displayed points in the cloud.

Now our model is encoded via probabilistic facts and logical clauses. ProbLog can be used both to learn the parameters or answer probabilistic queries [9].

3.1 Learning

In order to perform either task, additionally to the theory we need examples, which in this case are facts characterizing the world, and thus each example defines an interpretation of the image. Interpretations are used for parameter learning in the training phase and as evidence to query the theory in the inference phase. An interpretation and example for learning the parameters of the door action prediction model is illustrated in Example 1.

Example 1.

`example(1).`

known(1, handleAspectRatio(small_height), true).
known(1, handleRelativeHeight(long), true).
known(1, handleRelativeWidth(medium), true).
known(1, handleRelativePosition_Y(up_point), true).
known(1, handleRelativePosition_X(left_point), true).
known(1, action(in), true).

We are in a learning from interpretation setting, thus we learn the parameters of our models using ProbLog LFI (or the learning from partial interpretations setting within ProbLog) [7]. Given a ProbLog program $T(p)$ where the parameters $p = \langle p_1, ..., p_n \rangle$ of the probabilistic labeled facts $p_i :: c_i$ in the program are unknown, and a set of M (possibly partial) interpretations $D = I_1, ..., I_M$, known as the training examples, ProbLog LFI estimates using a Soft-EM algorithm the maximum likelihood probabilities $\hat{p} = \langle \hat{p_1}, ..., \hat{p_n} \rangle$ such that $\hat{p} = \arg\max_P P(D|T(p)) = \arg\max_P \prod_{m=1}^{M} P_w(I_m|T(p))$, where $P_w(I)$ is the probability of a partial interpretation $I = (I^+, I^-)$ with the set of all true atoms I^+ and the set of all false atoms I^-. ProbLog LFI is also able to learn parameters in the case of partial observations, which is useful to generalise over the cases when the door or handle is not fully observed.

In essence, the algorithm constructs a propositional logic formula for each interpretation that is used for training and uses a Soft-EM to estimate the marginals of the probabilistic parameters. For more details please see [7][2].

3.2 Inference

After the learning phase setting, we performed inference in order to do either action recognition or action point prediction. It is assumed that the robot can observe the properties of the world (door and handle in our case) and it needs to infer which action needs to be performed to open the door and which point to action. This resumes to querying the ProbLog theory for the conditional probabilities $P(A|F_1, \ldots, F_5)$ for each of the 8 possible actions and $P(Pos|A, F_1, \ldots, F_5)$ for each of the 5 possible discretized action points. We then compute the MAP probability estimate of the action, and afterwards of the action point.

4   Experiments and Results

For the purpose of our initial experimental setup, we collected a set of 60 door images. Most of these were taken from the Caltech and Pasadena Entrances 2000 dataset[3] and from the CCNY Door Detection Dataset for Context-based Indoor Object Recognition [20]. To increase variation in the different types of doors and handles we also added a few images from a Google image search. The images were manually annotated with bounding boxes for the door and handles, as well

---

[2] In practice we have used the implementation available at:
http://dtai.cs.kuleuven.be/problog/tutorial-learning-lfi.html.

[3] Available at: http://www.vision.caltech.edu/html-files/archive.html

as the action needed to open the door and the action point. We randomly split this dataset into two sets of 30 images, one to be used for training the ProbLog model by running ProbLog LFI to learn parameters, and one for testing by running inference to make first predictions about the action and then about the action point for the predicted action. Each experiment was run five times with different train and test sets and the results averaged.

In Table 1 we can see some examples of doors and the predictions produced by our logical approach. The correct predictions are shown in green, while the incorrect ones in red, with the ground truth value shown between brackets.

Table 1. Example doors and predictions: correct predictions are in green, incorrect ones in red. The ground truth value is between brackets

Action: turncc (out)
Position: center (down)

Action: in (right)
Position: center (up)

Action: in (in)
Position: center (center)

Action: out (out)
Position: center (right)

Action: in (in)
Position: center (center)

Action: in (in)
Position: center (center)

Action: in (in)
Position: center (center)

Action: down (down)
Position: left (left)

Action: turncc (turncc)
Position: center (center)

On this dataset, we compared our approach against three other baseline approaches. We compared against the following approaches on our collected door dataset: a random classifier, a majority classifier, and a propositional learner setting, for which we used a decision tree. These approaches are intended as a baseline to justify the benefit of using our logical approach. We will show that we obtain better prediction rates for the two prediction tasks of action prediction and action point prediction on the dataset.

We first set up a random classifier. We obtained the prior probabilities for the action and action point for each of the five training sets of 30 images, and then use a random classifier for the two prediction tasks. The results are summarized in Table 2 for the action prediction task, and Table 3 for the action point prediction task.

Similarly, we then set up a majority classifier. We determine from the training sets which are the action and action point majority classes and use this information in a majority classifier on the test sets. The majority class for the action is *out* in three of the training data sets, and *in* in the other two. The majority class for the action point is *center* in all the training datasets. The results of the two prediction tasks are also summarized in Table 2 and Table 3.

For the propositional learner setting baseline, we modelled a decision tree classifier in Weka [8]. We used the J48 decision tree algorithm in Weka, a version of the C4.5 [13] decision tree algorithm. We trained the decision tree on the training sets data. An example of a learned decision tree from a training set for predicting the action needed to open the door can be seen in Figure 4. An example of a learned decision tree from a training set for predicting the action position needed to open the door can be seen in Figure 5.

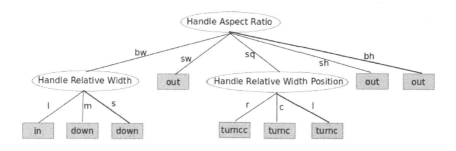

Fig. 4. Example obtained decision tree for predicting the action

We used the learned decision trees on the tests sets. The results of the two prediction tasks are also summarized in Table 2 and Table 3.

The results of our preliminary experiments against the three mentioned baselines are promising. The results show that our relational approach outperforms the three baseline approaches and so that even our initial prior rules add a benefit and thus justify the benefit of using a relational approach. We plan to extend

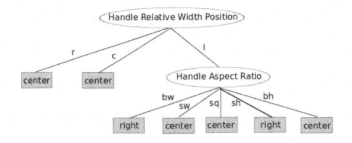

Fig. 5. Example obtained decision tree for predicting the action position

Table 2. Action prediction

Method	Total experiments	Avg. Success	Percentage
Random Classifier	30	8.6	28.67%
Majority Classifier	30	9.2	30.67%
Decision Tree	30	21.4	71.33%
Relational Approach	30	23.6	78.67%

Table 3. Action point prediction

Method	Total experiments	Avg. Success	Percentage
Random Classifier	30	16	53.33%
Majority Classifier	30	21.2	70.67%
Decision Tree	30	22.6	75.33%
Relational Approach	30	23	76.67%

our model with the ideas suggested in this paper and perform more extensive experiments that would extend the relational domain of our initial setting.

5   Conclusion and Future Work

We described an initial approach that uses SRL, and in particular ProbLog, to predict the action a robot needs to perform in order to open doors and the point it needs to action. The experiments showed that our relational approach produced better results than a random classifier, a majority classifier, and a J48 decision tree.

Although the initial model is limited, there are future ideas on adding more context by considering other objects in the environment and extending the action point prediction to consider multiple interest points on the handle which are relationally related. These can be obtained by automatically detecting different grasping points for the handle [11]. Additionally, we plan to include probabilistic priors on the doorframe and handle positions to model real-world detection uncertainty. Furthermore, our model can be extended with a temporal relational aspect to generalise over opening doors that need a sequence of actions (e.g., first push the handle down, then pull the door).

Our final goal is to test the algorithms with a realistic simulator – a common evaluation setup in the robotic community, as well as with real robots. However, this is a rather challenging task, given that even obtaining a dataset for training is difficult and error-prone. However, some solutions exist. One of them is introduced by Saxena et al. [18], where they employ an automated technique to generate a realistic, synthetic dataset. In particular, the use 3D-models of objects together with computer graphics algorithms to construct 2D-images with known ground truth seems feasible and works well in practice. Therefore, as feature work we plan to extend our work with synthetic generated datasets to perform experiments with real-world data. Finally, we will use a realistic simulator to extend our work to multiple robotic hands opening different doors.

Acknowledgements. Bogdan Moldovan is supported by the IWT (agentschap voor Innovatie door Wetenschap en Technologie). Laura Antanas is supported by the European Community's 7th Framework Programme, grant agreement First-MM-248258.

References

1. Antanas, L., van Otterlo, M., Oramas, J.M., Tuytelaars, T., De Raedt, L.: A relational distance-based framework for hierarchical image understanding. In: ICPRAM (2), pp. 206–218 (2012)
2. Berenson, D., Diankov, R., Nishiwaki, K., Kagami, S., Kuffner, J.: Grasp planning in complex scenes. In: 7th IEEE-RAS International Conference on Humanoid Robots, pp. 42–48. IEEE (2007)
3. Blum, M., Springenberg, J.T., Wulfing, J., Riedmiller, M.: A learned feature descriptor for object recognition in rgb-d data. In: 2012 IEEE International Conference on Robotics and Automation (ICRA), pp. 1298 –1303 (May 2012)
4. Bo, L., Lai, K., Ren, X., Fox, D.: Object recognition with hierarchical kernel descriptors. In: CVPR, pp. 1729–1736 (2011)
5. De Raedt, L., Kersting, K.: Probabilistic inductive logic programming. In: Prob. Ind. Log. Progr., pp. 1–27 (2008)
6. De Raedt, L., Kimmig, A., Toivonen, H.: ProbLog: A probabilistic Prolog and its application in link discovery. In: IJCAI, pp. 2462–2467 (2007)
7. Gutmann, B., Thon, I., De Raedt, L.: Learning the parameters of probabilistic logic programs from interpretations. In: Gunopulos, D., Hofmann, T., Malerba, D., Vazirgiannis, M. (eds.) ECML PKDD 2011, Part I. LNCS, vol. 6911, pp. 581–596. Springer, Heidelberg (2011)
8. Hall, M., Frank, E., Holmes, G., Pfahringer, B., Reutemann, P., Witten, I.H.: The WEKA data mining software: an update. SIGKDD Explorations 11(1) (2009)
9. Kimmig, A., Demoen, B., De Raedt, L., Costa, V.S., Rocha, R.: On the implementation of the probabilistic logic programming language ProbLog (2011)
10. Landwehr, N., Kersting, K., De Raedt, L.: nFOIL: Integrating Naïve Bayes and FOIL (2005)

11. Moreno, P., Hörnstein, J., Santos-Victor, J.: Learning to grasp from point clouds. VisLab Technical Report (2011)
12. Quigley, M., Batra, S., Gould, S., Klingbeil, E., Le, Q.V., Wellman, A., Ng, A.Y.: High-accuracy 3D sensing for mobile manipulation: Improving object detection and door opening. In: ICRA, pp. 2816–2822. IEEE (2009)
13. Quinlan, R.: C4.5: Programs for Machine Learning. Morgan Kaufmann, San Mateo (1993)
14. De Raedt, L.: Logical and Relational Learning. Springer (2008)
15. De Raedt, L., Thon, I.: Probabilistic rule learning. In: Frasconi, P., Lisi, F.A. (eds.) ILP 2010. LNCS, vol. 6489, pp. 47–58. Springer, Heidelberg (2011)
16. Rosman, B., Ramamoorthy, S.: Learning spatial relationships between objects. The International Journal of Robotics Research 30(11), 1328–1342 (2011)
17. Russel, S., Norvig, P.: Artificial Intelligence: A Modern Approach. Prentice-Hall (1995)
18. Saxena, A., Driemeyer, J., Kearns, J., Ng, A.Y.: Robotic grasping of novel objects. In: Neural Information Processing Systems, pp. 1209–1216 (2006)
19. Steder, B., Rusu, R.B., Konolige, K., Burgard, W.: Point feature extraction on 3D range scans taking into account object boundaries. In: ICRA, pp. 2601–2608 (2011)
20. Yang, X., Tian, Y., Yi, C., Arditi, A.: Context-based indoor object detection as an aid to blind persons accessing unfamiliar environments. In: ACM Multimedia. ACM (2010)

Probing the Space of Optimal Markov Logic Networks for Sequence Labeling

Naveen Nair[1,2,3], Ajay Nagesh[1,2,3], and Ganesh Ramakrishnan[2,1]

[1] IITB-Monash Research Academy, Old CSE Building, IIT Bombay
[2] Department of Computer Science and Engineering, IIT Bombay
[3] Faculty of Information Technology, Monash University
{naveennair,ajaynagesh,ganesh}@cse.iitb.ac.in

Abstract. Discovering relational structure between input features in sequence labeling models has shown to improve their accuracies in several problem settings. The problem of learning relational structure for sequence labeling can be posed as learning Markov Logic Networks (MLN) for sequence labeling, which we abbreviate as Markov Logic Chains (MLC). This objective in propositional space can be solved efficiently and optimally by a Hierarchical Kernels based approach, referred to as StructRELHKL, which we recently proposed. However, the applicability of StructRELHKL in complex first order settings is non-trivial and challenging. We present the challenges and possibilities for optimally and simultaneously learning the structure as well as parameters of MLCs (as against learning them separately and/or greedily). Here, we look into leveraging the StructRELHKL approach for optimizing the MLC learning steps to the extent possible. To this end, we categorize first order MLC features based on their complexity and show that complex features can be constructed from simpler ones. We define a self-contained class of features called absolute features ($\mathcal{AF}$), which can be conjoined to yield complex MLC features. Our approach first generates a set of relevant $\mathcal{AF}$s and then makes use of the algorithm for StructRELHKL to learn their optimal conjunctions. We demonstrate the efficiency of our approach by evaluating on a publicly available activity recognition dataset.

Keywords: Feature Induction, Hierarchical Kernel Learning, Markov Logic Networks, Sequence labeling.

1 Introduction

Learning and prediction problems in real world have to deal with complex relationships among entities combined with uncertainties in these relationships. These complex relationships are quite often represented compactly in the form of first order logical statements. The uncertainties are typically captured in the form of probabilities or probabilistic weights. The systems capable of handling such complex logical relationships and their uncertainties are generally classified as Statistical Relational Learning (SRL) systems [1,2]. One of the most popular

F. Riguzzi and F. Železný (Eds.): ILP 2012, LNAI 7842, pp. 193–208, 2013.

SRL frameworks is the Markov Logic Network (MLN) [3,4,5,6], which combines the expressive representation formalism of first order logic and the ability of probabilistic graphical models to handle uncertainty [3,4]. Our focus, in this paper, is learning MLNs for the specific problem of sequence labeling. We first briefly introduce MLNs in the following paragraph and then proceed to define our sequence labeling problem.

Markov Logic Networks: MLNs [3] extend first order logical systems by incorporating probabilistic information to the clauses/rules. They are typically represented as a collection of first order clauses with real valued weights attached to each clause. Each vertex in the graphical representation is a first order predicate. The edges between these predicates represent the logical connectives in a first order formula. Therefore, a clique in the graph represents a first order clause. A grounded[1] MLN is a Markov Network in the propositional space.

MLNs define a probability distribution over a possible world I as,

$$P(I|H,B) = \frac{1}{Z} \prod_{C \in H \cup B} (\phi_C)^{n_C(I)} \tag{1}$$

where H is the hypothesis, B is the background knowledge, $\phi_C = e^{f_C}$, f_C is the weight attributed to the clause C, $n_C(I)$ is the number of true groundings of C in I and Z is the normalization constant. Therefore, an ideal MLN should have hypothesis,

$$H^* = \arg\max_H \prod_{I \in \hat{I}} P(I|H,B) \tag{2}$$

where $\hat{I}$ is the set of true interpretations.

Conventional MLN systems tend to learn the structure and parameters separately. There have been a few approaches recently to learn the structure and parameters simultaneously [7,8,9]. However since the feature space is exponential, all the MLN structure learning approaches are greedy and thus cannot guarantee optimal models. Therefore, learning optimal MLNs is a hard task. In this paper, we propose an approach that optimizes a substantial part of MLN structure learning, wherein we learn the final set of features and their parameters simultaneously from simpler features by leveraging an optimal feature learning algorithm.

We now briefly discuss the sequence labeling problem. The contributions of this paper can be extended to other acyclic structured output classification settings. Since we derive our approach for structured output spaces, which is more general, the derivations trivially cover simpler MLN settings.

Sequence Labeling: Sequence labeling is the task of assigning a class label to each instance in a sequence of observations. Typical sequence labeling algorithms learn probabilistic information about the neighboring states along with

---

[1] Grounding is replacing variables with constants/objects.

the probabilistic information about the observations. Hidden Markov Models (HMM) [10] and Conditional Random Fields (CRF) [11] are two traditional systems used popularly for sequence labeling problems. A HMM makes use of the assumption that, in a sequence, the label y_t at sequence step t is independent of all previous labels given y_{t-1} at time $t-1$ and the observation $\mathbf{x}_t$ at time t is independent of all other variables given y_t, and factorize the joint probability distribution in the form of the transition (label-label) distribution and the emission (label-observation) distribution. During the training phase, parameters that maximize the joint probability of the input and output sequences in the training data are learned. Whereas, a CRF maximizes the conditional probability of the output sequence given the input sequence to learn parameters. These parameters are later used to identify the (hidden) label sequence that best explains a given sequence of observations. Inference is typically performed efficiently by a dynamic programming algorithm called the Viterbi algorithm [12]. Recently, Tsochantaridis *et al.* proposed a maximum margin framework for structured output spaces such as sequence labeling, which is referred to as StructSVM [13]. It generalizes the standard Support Vector Machines (SVM) with the margin defined as the difference in the score of the original output sequence with any other possible output sequence.

Recent works, including ours, have looked into the problem of learning better sequence labeling models by discovering the structure in the input space in the form of conjunctions [14,15]. However since these approaches employ a greedy search to discover useful conjunctions, an optimal model is not guaranteed. In [16], we proposed a Hierarchical Kernels based learning approach for Structured Output Spaces (StructRELHKL) for learning optimal conjunctive (propositional) features for sequence labeling. In this work, we look into leveraging the StructRELHKL framework to learn first order features. Before going into the details of our approach, we now give a brief introduction to one of the application areas of sequence labeling called the activity recognition domain, which is our motivating problem.

Activity Recognition: Activity recognition systems are ubiquitous in the modern era of smart systems. One specific example is the use of activity recognition systems and approaches to monitor the activities of users in domicile environments; for instance, to monitor the activities of daily living of elderly people living alone, for estimating their health condition [17,18,19]. Such non-intrusive settings typically have on/off sensors installed at various locations in a home. Binary sensor values are recorded at regular time intervals. The joint state of these sensor values at a particular time forms our observation. The user activity at a particular time forms the hidden state or label. The history of sensor readings and (manually) annotated activities can be used to train prediction models such as the Hidden Markov Model (HMM) [10], the Conditional Random Field (CRF) [11] or StructSVM [13], which could be later used to predict activities based on sensor observations [18,15].

Activity recognition datasets tend to be sparse; that is, one could expect very few sensors to be on at any given time instance. Moreover, in a setting such as activity recognition, one can expect certain combinations of (sensor) readings to be directly indicative of certain activities. While HMMs, CRFs and StructSVM attempt to capture these relations indirectly, Nair *et al.* illustrate that discovering activity specific conjunctions of sensor readings (as emission features in HMM) can improve the accuracy of prediction [15]. McCallum [20] follows a similar approach for inducing features for a CRF model to solve Natural Language Processing tasks. Both these approaches, since they employ a greedy search for discovering features, have the limitation that an optimal model is not guaranteed. An exhaustive search for optimal features is not feasible in real world settings. We, in [16], proposed a hierarchical kernels based learning approach (StructRELHKL) for learning optimal feature conjunctions for sequence labeling, which we build on for learning first order relational features in this work. We now present the sequence labeling problem in a Markov Logic framework.

Sequence Labeling as Markov Logic Network: The training objective in sequence labeling can be posed as learning features that make the score,

$$F : \mathcal{X} \times \mathcal{Y} \to \mathbb{R} \tag{3}$$

of the original output sequence Y greater than any other possible output sequence, given an input sequence X. The score is defined as,

$$F(X, Y; \mathbf{f}) = \langle \mathbf{f}, \psi(X, Y) \rangle \tag{4}$$

where ψ is the feature vector (features describing observation structure and transitions), and $\mathbf{f}$ is the weight vector. Inference is performed by the decision function $\mathcal{F} : \mathcal{X} \to \mathcal{Y}$ defined by

$$\mathcal{F}(X; \mathbf{f}) = \underset{Y \in \mathcal{Y}}{\arg\max} \ F(X, Y; \mathbf{f}) \tag{5}$$

In this work, we intend to learn first order features for the emission relationships and their weights, while learning the weights for all the transition relationships. Therefore, our ψ will include label specific emission features in first order along with the transition features. For example, rules of the form *activity(T, prepare-BreakFast) ← microwave(T), previousItemMoved(T, X), breakFastItem(X)* can be an emission rule/feature. Whereas transition rules can include rules such as *activity(T, eatBreakFast) ← activity(T-1, prepareBreakFast)*.

This can be viewed as a Markov Logic Network. We refer to this MLN for sequence labeling as a Markov Logic Chain (MLC). In this context, HMM, CRF, StructSVM, *etc.* can be viewed as special instances of MLC. Learning first order MLC (or MLN) is challenging due to the huge feature space. Conventional MLN systems learn the structure using greedy search algorithms and thus could lead to local optimum models.

We have recently proposed a hierarchical kernels based approach for structured output classification to learn MLC features in propositional space, which is a sub-instance of general MLCs. The Hierarchical kernel learning approach, originally proposed by Bach [21] for binary classification problems, learns an optimal set of sparse and simple features that has an ordering defined. We have extended the HKL approach in [16] to structured output spaces, which we call Rule Ensemble Learning using Hierarchical Kernels in Structured Output Spaces (StructRELHKL). StructRELHKL, in sequence labeling, exploits the hierarchical structure in the (exponential) space of emission features and efficiently learns a sparse set of simple propositional features and their weights. One of the fundamental requirements of these approaches is the summability of kernels over descendants in polynomial time. Since our focus here is to learn features in first order settings, that has inherent complexities such as huge number of groundings, variable sharing, background knowledge, refinement operators using unifications and anti-unifications, subsume equivalence *etc.*, it is hard to sum the descendant kernels in polynomial time. Therefore, the first order extension of StructRELHKL is non-trivial and is a challenging problem. In this paper, we investigate the possibility of leveraging StructRELHKL in discovering first order features. We briefly introduce the intuitions for our approach next.

Leveraging StructRELHKL for Learning First Order MLCs: We intend to learn first order features for sequence labeling. An example of first order feature is a feature that capture input relationships across sequence steps. Gutmann and Kersting recently proposed a first order extension of CRF (TildeCRF) that captures input relationships across sequence steps [22]. However, this approach also employs a greedy search to discover useful features. To the best of our knowledge, no previous work has explored the possibility of incorporating optimal feature learning in any step of MLC structure learning.

As discussed in the previous paragraphs, it is not feasible to apply StructRELHKL directly to learn first order MLCs. Although, grounding the first order predicates with all possible constants (according to the language bias) and leveraging StructRELHKL to learn optimal conjunctions of them seems to solve the problem, it is not feasible in large settings due to the huge search spaces. We therefore propose to employ StructRELHKL for learning optimal conjunctions of a powerful subclass of features, in the process of learning MLC features. To this end, we categorize first order MLN features based on their complexity and identify the class of features that can be efficiently constructed from simpler ones by StructRELHKL[2]. We identify a self-contained class of features called the absolute features ($\mathcal{AF}$) as the building blocks, whose unary/multiple conjunctions result in the final model. Our approach first generates $\mathcal{AF}$s that cover a threshold number of examples (weak relevance) and employs the StructRELHKL algorithm to simultaneously learn optimal conjunctions of $\mathcal{AF}$s and their weights

---

[2] In this work, we restrict our discussion to function-free first order definite features. A class specific feature can be constructed by conjoining the body literals of a definite clause whose head depicts the class label.

(as against learning them separately and/or greedily). We learn optimal MLCs with respect to weakly relevant $\mathcal{AF}$s. To summarize, we present the challenges and possibilities for optimally and simultaneously learning the structure as well as parameters of MLCs. We evaluate the efficiency of our approach on a publicly available activity recognition dataset and compare our results against Tilde-CRF [22], the state-of-the-art relational sequence tagging tool. A brief summary of some of the related approaches is given in the next paragraph.

Related Work: Huynh and Mooney in [23] propose an online weight learning algorithm for MLNs using a max-margin framework. The approach is for learning only the weights, whereas, our focus is on learning the structure as well as parameters for sequence labeling problems. In [24], Huynh and Mooney discusses their online algorithm for learning the structure and parameters of MLNs. In each iteration of their approach, the current model is used to predict the output, and if the prediction is wrong, the incorrect prediction is treated as a negative example, new clauses are learned that differentiate the true and false examples. The new clauses are learned by searching for atoms that are in the true example and not in the false one. This is performed by a relational path finding algorithm. Weight learning is then performed by an L1 regularized formulation, which also nullifies many non relevant clauses in the current structure. In contrast, our approach employs a batch learning algorithm. It learns a large number of candidate clauses called absolute clauses, which are then conjoined optimally to learn the final model. Our approach optimizes a convex formulation to learn structure and parameters with respect to the absolute clauses (features). A Logical Hidden Markov Model is discussed in [25], which deals with sequences over logical atoms. A model selection approach for Logical Hidden Markov Model is proposed in [26], which is based on Expectation Maximization algorithm and Inductive Logic Programming principles. Our approach differs from their approach in the sense that our objective is to explore the relationships among multiple observations at a sequence step to improve efficiency of sequence labeling. Thon *et al.* in [27] and [28] elaborate on relational markov processes which are concerned with efficient parameter learning and inference. They assume that a structure has been provided upfront. Similarly, a relational bayesian network learning is discussed in [29] with the goal of learning the parameters given the structure of the bayes-net.

TildeCRF [22] has an objective similar to our approach, where the relational structure and parameters of a CRF for sequence labeling are learned. TildeCRF uses relational regression trees and gradient tree boosting for learning the structure and parameters. The main difference of our approach with TildeCRF is that in our approach, unlike in TildeCRF, we derive convex formulations for a significant portion of learning steps.

Paper Organization: In section 2, the complexity based categorization of features is discussed. We discuss our approach in section 3. Experimental setup and results are discussed in section 4 and we conclude the paper in section 5.

2   First Order Definite Features for Markov Logic Chains

Most of the Inductive Logic Programming (ILP) systems and their statistical extensions (SRL systems) learn clauses by searching a space (often a lattice) of clauses in the domain. The search space is typically controlled by language restrictions, which define the type of clauses to be learned. One common way to solve a classification problem is to learn definite clauses (clauses having one head predicate conditioned on the values of zero or more body predicates). Since we are interested in such a setup, we confine our discussion to the space of definite clauses. We use the terms first order definite clause and first order feature interchangeably, as one can be derived from the other. We start by defining categories of predicates and then discuss the complexity based classification of features.

Similar to *structural* and *property* predicates in 1BC clauses [30], we define two types of predicates, *viz. (inter) relational* and *quality* predicates. A *relational* predicate is a binary predicate that represents the relationship between *types* or between a *type* and its parts, where a *type* is an entity described by its attributes. A *quality* predicate is a predicate that reveals a quality/property of a part (or subset) of a *type*. From the example clauses given below, microwave(_,_), and before(_,_) are *relational* predicates and all other predicates are *quality* predicates.

1. prepareDinner(T) :- microwave(T,X1), soak(X1)
2. prepareDinner(T) :- microwave(T,X1)
3. prepareLunch(T) :- microwave(T,X1), powdered(X1), cereal(X1)
4. prepareLunch(T) :- microwave(T,X1), dry(X1), microwave(T,X2),
                                     before(X1,X2), wet(X2), cereal(X2)
5. prepareDinner(T) :- microwave(T,X1), soak(X1), microwave(T,X2),
                                     nonSoak(X2)

We now categorize first order definite features based on complexity into Absolute Features ($\mathcal{AF}$), Primary Features ($\mathcal{PF}$), Composite Features ($\mathcal{CF}$) and Definite Features ($\mathcal{DF}$). The definitions of $\mathcal{AF}$ and $\mathcal{CF}$ are used in this paper, while the other categories are presented for supporting the definitions for these.

Absolute Features ($\mathcal{AF}$):
In absolute features (clauses), new local variables can only be introduced in a *relational* predicate, where a local variable is a variable not present in the head predicate. Unlike in 1BC clauses, any number of new local variables can be introduced in a *relational* predicate. Any number of *relational* and *quality* predicates can be conjoined to form a $\mathcal{AF}$ such that the resultant $\mathcal{AF}$ is minimal and the local variables introduced in *relational* predicates are consumed by some other *relational* or *quality* predicates. Here a minimal clause is one which cannot be constructed from smaller clauses that share no common variables other than that in the head. So clauses 1, 3 and 4 above are $\mathcal{AF}$s whereas clauses 5 (since it is not minimal) and 2 (since variable X1 is not consumed) are not.

Primary Features ($\mathcal{PF}$):
Primary features (clauses) are absolute clauses (features) that have at-most one *quality* predicate for every new local variable introduced. This is similar to elementary features in [30] except that elementary features allow only one new local variable in a *structural* predicate. Clause 1 is a $\mathcal{PF}$ whereas the other clauses do not conform to the restrictions imposed.

Composite Features ($\mathcal{CF}$):
Composite Features (clauses) are definite clauses that are formed by the conjunction of one or more $\mathcal{AF}$s without unification of body literals. Only the head predicates are unified. As in $\mathcal{AF}$s, every local variable introduced in a relational predicate should be consumed by other relational or quality predicates. Clauses 1, 3, 4 and 5 are $\mathcal{CF}$s where as 2 is not.

First Order Definite Features ($\mathcal{DF}$):
First order definite features (clauses) are features with none of the above restrictions. Therefore, all the given examples are $\mathcal{DF}$s.

We now state some of the relationships between these categories of features.

Claim 1. The set of primary features is a proper subset of the set of Absolute Features. That is, $\mathcal{PF} \subset \mathcal{AF}$.

Proof. From definition, $\mathcal{PF}$s are $\mathcal{AF}$ with the restriction that a new local variable introduced should be transitively consumed by a single *quality* predicate. Hence $\mathcal{PF} \subseteq \mathcal{AF}$. Now, consider the clause 3 above, which is an $\mathcal{AF}$ but not a $\mathcal{PF}$. Hence, $\mathcal{PF} \neq \mathcal{AF}$.

Claim 2. The set of absolute features is a proper subset of the set of composite features. That is, $\mathcal{AF} \subset \mathcal{CF}$.

Proof. From definition, $\mathcal{CF}$s are conjunctions of one or more $\mathcal{AF}$s. Therefore, all $\mathcal{AF}$s are $\mathcal{CF}$s (unary conjunctions). Now, consider the clause 5 above, which is a $\mathcal{CF}$ but not a $\mathcal{AF}$. Hence, $\mathcal{AF} \neq \mathcal{CF}$.

Claim 3. The set of composite features is a proper subset of the set of full first order definite features. $\mathcal{CF} \subset \mathcal{DF}$.

Proof. From definition, $\mathcal{DF}$s are first order definite clauses without any restrictions imposed for $\mathcal{CF}$s. Therefore, $\mathcal{CF} \subseteq \mathcal{DF}$. Now consider the clause 2 above, which is a $\mathcal{DF}$ but not a $\mathcal{CF}$. Therefore, $\mathcal{CF} \neq \mathcal{DF}$.

Claim 4. Every $\mathcal{AF}$ can be constructed from $\mathcal{PF}$s using unifications.

Proof. The difference an $\mathcal{AF}$ has with $\mathcal{PF}$ is that it can have more than one *quality* predicates for each local variable introduced. Let l_p be a *relational* literal in the body of a $\mathcal{AF}$ clause which introduces only one local variable. Let $l_1, l_2, \ldots, l_{p-1}$ be the set of *relational* literals in the body, which l_p depends on. Let there be $n \geq 0$ number of dependency chains starting from l_p to some *quality* predicates, each of which is represented as $l^i_{p+1}, \ldots, l^i_k$. We define l_p as a pivot literal if $n > 1$. For simplicity, we assume there is only one pivot in a clause. Now, we can construct n $\mathcal{PF}$ clauses from this with the body of the ith clause as $l_1, l_2, \ldots, l_{p-1}, l_p, l^i_{p+1}, \ldots, l^i_k$, where l^i_k is a *quality* predicate. It is trivial to see that these n clauses can be unified to construct the original $\mathcal{AF}$. For multiple pivot literal clauses, the above method can be applied recursively until $\mathcal{PF}$ clauses are generated. The proof can be extended to pivot literals with multiple new local variables by using a dependency tree structure in place of chain.

Claim 5. Every $\mathcal{CF}$ can be constructed from $\mathcal{AF}$s by conjunctions.

Proof. By definition.

We briefly discuss below some of the existing complexity based categorization of first order definite features in the following paragraph.

1BC clauses and elementary clauses introduced in [30] are similar to $\mathcal{AF}$s and $\mathcal{PF}$s respectively with the restriction that a structural predicate can have only one new local variable. Simple clauses are defined in [31] as the clause with at-most one sink literal, where a sink literal is one which has no other literal dependent on it. Simple clauses need not have a sink for a local variable and thus differs from $\mathcal{PF}$s. We present the proposed approach for learning first order features for sequence labeling in the next section.

3   Towards Optimal Feature Learning for Markov Logic Chains

We now formalize our intutions explained in the introduction section for leveraging StructRELHKL to learn first order MLC features.

As explained in the introduction section, we focus on learning first order features for sequence labeling. An example class of first order features in sequence labeling domain is the features that capture input relationships across sequence steps. TildeCRF [22] is an existing approach that explores such feature space. However the approach pursued is greedy. Although the Hierarchical Kernels based feature learning approach for structured output spaces (StructRELHKL) [16], that we proposed recently, learns optimal conjunctive features in propositional settings, it has limitations in exploring first order space. The main limitation is that, due to the complex refinement operators using unification and anti-unification, any ordering of the first order features restricts the requirement of StructRELHKL to sum the descendant kernels in polynomial time. Alternatively, grounding the first order predicates with all possible constants and then leveraging StructRELHKL to learn optimal features is not feasible due to the huge feature space. Moreover, such settings could lead to less effective models, due to the redundant information present in the models learned. Here, we explore the space of first order features, wherein, our optimal conjunctive feature learning algorithm is leveraged to the extent possible in the learning step.

MLC Features: Our objective is to learn label specific relational features as observation features for sequence labeling. These types of features can be represented in the form of definite clauses. As discussed in section 2, Composite Features ($\mathcal{CF}$) are first order definite features with the restriction that local variables introduced in relational predicates have to be used by other relational or quality predicates. This category of features is particularly interesting in our case, because they are powerful to represent a large part of definite clauses. $\mathcal{CF}$s are not defined to include clauses such as clause 2 in section 2. In predicting smaller granular activities such as *prepareBreakFast, prepareLunch, etc.*, what is being cooked in the microwave oven becomes interesting. Absence of such information doesn't give meaning to the rule. On the other hand, if one wants to predict larger granular activities such as whether the current activity is cooking or not, then what is being cooked inside microwave is less important. To this effect, a new predicate *microwaveOn(T)* can be defined and we can construct $\mathcal{CF}$s from this. We have also proved that composite features ($\mathcal{CF}$) can be constructed from absolute features ($\mathcal{AF}$) with unary/multiple conjunctions without unifications and $\mathcal{AF}$s can be constructed from primary features ($\mathcal{PF}$) with unifications. Therefore, the space of $\mathcal{CF}$s can be defined as a partial order over $\mathcal{PF}$s with unifications and conjunctions. As discussed before, the optimal feature learning approach (StructRELHKL) is not applicable in a partial order with unifications. However, since the space of $\mathcal{CF}$s is a partial order over conjunctions of $\mathcal{AF}$s, it is possible to leverage the StructRELHKL framework to learn $\mathcal{CF}$s from $\mathcal{AF}$s. With proper language restrictions, $\mathcal{AF}$s can be generated by ILP methods. Our approach is to generate $\mathcal{AF}$s using ILP techniques and employ StructRELHKL to find optimal conjunctions of $\mathcal{AF}$s. The next paragraph gives some insight into the optimality of the resultant MLC.

We define an $\mathcal{AF}$ as strongly relevant if it is constructed from optimal unifications of $\mathcal{PF}$s, which is a hard task. On the other hand, we consider a feature to be weakly relevant if it covers at-least a threshold percentage of examples. We are interested in $\mathcal{AF}$s that are at-least weakly relevant and our approach learns optimal MLCs with respect to the weakly relevant $\mathcal{AF}$s. We now derive our approach in the following paragraph.

Optimizing MLC Feature Learning Steps: The objective is to first learn self contained weakly relevant $\mathcal{AF}$s by employing ILP methods and then learn optimal conjunctions of $\mathcal{AF}$s and their weights simultaneously by leveraging the StructRELHKL approach. Since there is a plethora of literature available on ILP approaches for searching clauses with language restrictions [2,1], we skip the discussion on generating $\mathcal{AF}$s and move on to give an overview of StructRELHKL [16] framework to construct $\mathcal{CF}$s.

An MLC should have features defining transition relationships between the labels as well as the observation relationships. Let ψ_T represent the feature vector corresponding to all transition features and ψ_{CF} represent the feature vector corresponding to all observation features (space of all possible conjunctions of $\mathcal{AF}$s). For the sake of visualization, lets assume there is a partial order of $\mathcal{CF}$s for each label in a multi-class setup. As defined in the introduction section, ψ represents a combination of ψ_T and ψ_{CF}. We assume that both ψ_{CF} and ψ_T are of dimension equal to the dimension of ψ with zero values for all elements not in their context. In similar spirit, the feature weight vector $\mathbf{f}$ can be constructed from $\mathbf{f_{CF}}$ and $\mathbf{f_T}$. Similarly, $\mathcal{V}$, the indices of the elements of ψ, can be constructed from $\mathcal{V_{CF}}$ and $\mathcal{V_T}$. Our objective is to simultaneously select a sparse set of CFs and their weights along with all the transition feature weights. To achieve this we build on the StructSVM framework [13] and employ a sparsity inducing hierarchical regularizer [21,32] on emission features and the standard 2-norm regularizer for transition features (as sparsity is desired only in the observation space). The margin of this Support Vector Machine (SVM) setup is defined as the difference in scores (defined in (4)) of the original output sequence and any other possible output sequence. Therefore, the objective can be stated as learning a sparse and simple set of observation features and all transition features that maximize the difference in scores of original sequences with any other possible sequence in the training data. These features can later be used during inference to find the best sequence among the set of possible sequences for a given observation. The SVM objective can be stated as,

$$\min_{\mathbf{f},\boldsymbol{\xi}} \frac{1}{2}\Omega_{CF}(\mathbf{f_{CF}})^2 + \frac{1}{2}\Omega_T(\mathbf{f_T})^2 + \frac{C}{m}\sum_{i=1}^{m}\xi_i,$$

$$\forall i, \forall Y \in \mathcal{Y} \setminus Y_i : \langle \mathbf{f}, \psi_i^\delta(Y)\rangle \geq 1 - \frac{\xi_i}{\Delta(Y_i, Y)}$$

$$\forall i : \xi_i \geq 0 \tag{6}$$

where $\Omega_{CF}(\mathbf{f_{CF}})$ is the hierarchical $(1,\rho)$ norm regularizer [32] defined as, $\sum_{v \in \mathcal{V_{CF}}} d_v \| \mathbf{f_{CF}}_{D(v)} \|_\rho$, $\rho \in (1,2]$, $d_v \geq 0$ is a prior parameter showing usefulness of the feature conjunctions, $D(v)$ represents the set of descendants (including itself) of node v in the partial order (similarly $A(v)$ represents the set of ancestors of node v), $\mathbf{f_{CF}}_{D(v)}$ is the vector with elements as $\| f_{CFw} \|_2$, $\forall w \in D(v)$, and $\| . \|_\rho$ represents the ρ-norm , $\Omega_T(\mathbf{f_T})$ is the 2-norm regularizer $(\sum_i f_{Ti}^2)^{\frac{1}{2}}$, m is the number of examples, C is the regularization parameter, ξ's are the slack variables introduced to allow errors in the training set in a soft margin SVM formulation, $X_i \in \mathcal{X}$ and $Y_i \in \mathcal{Y}$ represent the ith input and output sequences respectively, $\langle \mathbf{f}, \psi_i^\delta(Y)\rangle$ represents the value $\langle \mathbf{f}, \psi(X_i, Y_i)\rangle - \langle \mathbf{f}, \psi(X_i, Y)\rangle$, and $\Delta(Y, \hat{Y})$ represents the loss when the true output is Y and the prediction is $\hat{Y}$.

The 1-norm in $\Omega_{CF}(\mathbf{f_{CF}})$ forces many of the $\| \mathbf{f_{CF}}_{D(v)} \|_\rho$ to be zero. Even in cases where $\| \mathbf{f_{CF}}_{D(v)} \|_\rho$ is not forced to zero, the ρ-norm forces many of node v's descendants to zero. This ensures a sparse and simple set of features.

The above SVM setup has to deal with two exponential spaces. The first is that of the exponential space of features and the second problem is the exponential number of constraints for the objective. The next paragraph outlines solution to this.

By solving (6), most of the emission feature weights are expected to be zero. As illustrated in [32,16], the solution to the problem when solved with the original set of features is the same but requires less computation when solved only with features having non zero weights at optimality. Therefore, an active set algorithm can be employed to incrementally find the optimal set of non zero weights. In each iteration of the active set algorithm, since the constraint set in (6) is exponential, a cutting plane algorithm has to be used to find a subset of constraints of polynomial size so that the corresponding solution satisfies all the constraints with an error not more than ϵ. The MLC objective is solved by deriving a dual for (6) with the feature set reduced to active features and a sufficient condition to check for optimality. The active set algorithm starts with the top nodes in the partial order as active set and at each iteration, solves the dual of (6), checks a sufficient condition for optimality, adds those nodes in the sources (subset of a set of nodes in a partial order that have no parent in the set) of the active set's complement that violate the sufficient condition to the active set. The process continues until there are no nodes violating the sufficient condition. The dual of (6) is derived as,

$$\min_{\eta \in \Delta_{|\mathcal{V}|,1}} g(\eta) \tag{7}$$

where $g(\eta)$ is defined as,

$$\max_{\alpha \in S(\mathcal{Y},C)} \sum_{i,Y \neq Y_i} \alpha_{iY} - \frac{1}{2}\boldsymbol{\alpha}^\top \boldsymbol{\kappa_T}\boldsymbol{\alpha} - \frac{1}{2}\Big(\sum_{w \in \mathcal{V}} \zeta_w(\eta)(\boldsymbol{\alpha}^\top \boldsymbol{\kappa_{FCw}}\boldsymbol{\alpha})^{\bar{\rho}} \Big)^{\frac{1}{\bar{\rho}}}, \tag{8}$$

$$\zeta_w(\eta) = \Big(\sum_{v \in A(w)} d_v^\rho \eta_v^{1-\rho} \Big)^{\frac{1}{1-\rho}}, \qquad \bar{\rho} = \frac{\rho}{2(\rho-1)}, \text{ and}$$

$$S(\mathcal{Y},C) = \{\boldsymbol{\alpha} \in \mathbb{R}^{m(n^l-1)} \mid \alpha_{i,Y} \geq 0, \, m\sum_{Y \neq Y_i} \frac{\alpha_{iY}}{\Delta(Y,Y_i)} \leq C, \, \forall i, Y\}$$

A sufficient condition for optimality, with the current active set $\mathcal{W}$, with a duality gap less than ϵ is derived as

$$\max_{u \in sources(\mathcal{W}^c)} \sum_{i,Y \neq Y_i} \sum_{j,Y' \neq Y_j} \boldsymbol{\alpha}_{WiY}^\top \sum_{p=1}^{l_i} \sum_{q=1}^{l_j} 2\Big(\prod_{k \in u} \frac{\psi_{FCk}(\mathbf{x}_i^p)\psi_{FCk}(\mathbf{x}_j^q)}{b^2} \Big)$$

$$\Big(\prod_{k \notin u} \Big(1 + \frac{\psi_{FCk}(\mathbf{x}_i^p)\psi_{FCk}(\mathbf{x}_j^q)}{(1+b)^2}\Big) \Big) \boldsymbol{\alpha}_{WjY'}$$

$$\leq \Omega_{FC}(\mathbf{f_{FCW}})^2 + \Omega_T(\mathbf{f_{TW}})^2 + 2(\epsilon - e_\mathcal{W}) \tag{9}$$

Inference is performed by applying dynamic programming methods [12]. We now discuss the experimental setup in the following section.

4   Experiments

Our approach first learns a weakly relevant set of absolute features and then learns their optimal conjunctions using StructRELHKL. We use Warmr [33,34], an ILP data mining algorithm that learns frequent patterns reflecting one to many and many to many relationships, to learn absolute features. Warmr uses an efficient level wise search through the pattern space. With proper language bias, absolute features can be generated by Warmr. The absolute features learned by Warmr are then input to StructRELHKL code to learn the structure and parameters of the final Markov Logic Chain. Our StructRELHKL is a java program, which implements the active set algorithm. In each iteration of the active set algorithm, the dual objective is solved using the current model, a sufficiency condition for optimality is checked on the nodes in the complement of active set that has parents only in the active set, and the violating nodes are added to the active set. The sub-problem of solving the dual is performed by a cutting plane algorithm, which starts with an empty set of constraints. In each iteration, the algorithm solves the objective with the current set of constraints and adds new constraints that violate the margin (defined by the current model) at-least by a threshold value.

Our experiments are carried out on a publicly available activity recognition dataset. This data, provided by Kasteren *et al.* [18], has been extracted from a household fitted with 14 binary sensors at *bedroomDoor*, *bathroomDoor*, *microwave*, *fridge*, *cupboards*, etc.. Eight activities have been annotated for 4 weeks by a subject. Activities are daily house hold activities like *sleeping*, *usingToilet*, *preparingDinner*, *preparingBreakfast*, *leavingOut*, etc. There are 40006 data instances. The dataset is skewed. For instance, the activity *leavingOut* occurs about 56.4% of time, while the activity *preparingBreakfast* occurs only 0.3% of time. Since the authors of the dataset are from the University of Amsterdam, we will refer to the dataset as the UA data.

The data is split into different sequences and each sequence is treated as an example. We perform our experiments in a four fold cross-validation setup. On each fold, we train our model on 25% of data and test on the remaining 75%[3]. We report performance in terms of micro-average and macro-average labeling accuracies. The micro-average accuracy, referred to as time-slice accuracy in [18], is the weighted average of per-class accuracies, weighted by the number of instances of the class. Macro-average accuracy, referred to as class accuracy in [18], is the average of the per-class accuracies.

We compare our approach with the TildeCRF [22]. The TildeCRF is the state-of-the-art ILP approach for learning relational features for sequence labeling, and works in the same feature space that we are interested in. The comparison is outlined in Table 1.

It can be observed that our results are competitive to the state-of-the-art sequence labeling approach, TildeCRF. Our approach returned better micro-average accuracy than TildeCRF, while reporting lesser macro-average accuracy. Micro-averaged accuracy is typically used as the performance evaluation measure. However in data that is biased towards some classes, macro-average

---

[3] Since in real world problems such as activity recognition, a trained model has to be used for a period much longer than the period training data is collected, here we considered training on a small part of the data and testing on the rest.

Table 1. Micro average accuracy (%), and macro average accuracy (%) using TildeCRF and the proposed MLC approach on UA dataset

	Micro avg.(%)	Macro avg.(%)
TildeCRF	56.22 ($\pm$12.08)	35.36 ($\pm$6.55)
MLC	60.36 ($\pm$6.99)	30.39 ($\pm$4.31)

accuracy being too low is considered to be a poor performance. Our standard deviation values are also less compared to the competitor algorithm. Our approach on average took about 25 hours for training while TildeCRF took 2.5 hours.

5   Conclusion

Recent works have shown the importance of learning the input relational structure (features) in sequence labeling problems [14,15]. Most of the existing feature learning approaches employ greedy search techniques to discover relational features. We have recently proposed a hierarchical kernels based approach for structured output classification (StructRELHKL) that is capable of learning optimum feature conjunctions to build propositional sequence labeling models [16]. However, StructRELHKL works in propositional domain and has limitations in exploring first order space. In this paper, we presented the challenges and possibilities of leveraging StructRELHKL to optimize first order feature learning steps. To this end, we categorized first order features based on complexity and identified the class of features that can be constructed using StructRELHKL from simpler ones. We therefore learn a simple and powerful sub-class of first order features called the absolute features using Inductive Logic Programming techniques and learn their optimal conjunctions using StructRELHKL to build the final model. Our experiments show competitive performance compared to the state-of-the-art relational sequence labeling tool, the TildeCRF.

References

1. Getoor, L., Taskar, B.: Statistical relational learning. MIT Press (2006)
2. Nienhuys-Cheng, S.H., de Wolf, R.: Foundations of Inductive Logic Programming. Springer-Verlag New York, Inc., Secaucus (1997)
3. Richardson, M., Domingos, P.: Markov logic networks. Mach. Learn. 62(1-2), 107–136 (2006)
4. Domingos, P., Kok, S., Poon, H., Richardson, M., Singla, P.: Unifying logical and statistical AI. In: Proceedings of the 21st National Conference on Artificial Intelligence, AAAI 2006, vol. 1, pp. 2–7. AAAI Press (2006)
5. Muggleton, S., De Raedt, L., Poole, D., Bratko, I., Flach, P., Inoue, K., Srinivasan, A.: ILP turns 20. Mach. Learn. 86(1), 3–23 (2012)
6. Zhuo, H.H., Yang, Q., Hu, D.H., Li, L.: Learning complex action models with quantifiers and logical implications. Artif. Intell. 174(18), 1540–1569 (2010)

7. Kok, S., Domingos, P.: Learning the structure of markov logic networks. In: Proceedings of the 22nd International Conference on Machine Learning, ICML 2005, pp. 441–448. ACM, New York (2005)
8. Biba, M., Ferilli, S., Esposito, F.: Structure learning of markov logic networks through iterated local search. In: Proceedings of the 2008 Conference on ECAI 2008: 18th European Conference on Artificial Intelligence, pp. 361–365. IOS Press, Amsterdam (2008)
9. Khot, T., Natarajan, S., Kersting, K., Shavlik, J.: Learning markov logic networks via functional gradient boosting. In: Proceedings of the 2011 IEEE 11th International Conference on Data Mining, ICDM 2011, pp. 320–329. IEEE Computer Society, Washington, DC (2011)
10. Rabiner, L.R.: Readings in speech recognition, pp. 267–296. Morgan Kaufmann Publishers Inc., San Francisco (1990)
11. Lafferty, J.D., McCallum, A., Pereira, F.C.N.: Conditional random fields: Probabilistic models for segmenting and labeling sequence data. In: Proceedings of the Eighteenth International Conference on Machine Learning, ICML 2001, pp. 282–289. Morgan Kaufmann Publishers Inc., San Francisco (2001)
12. Forney, G.J.: The viterbi algorithm. Proceedings of IEEE 61(3), 268–278 (1973)
13. Tsochantaridis, I., Hofmann, T., Joachims, T., Altun, Y.: Support vector machine learning for interdependent and structured output spaces. In: Proceedings of the Twenty-First International Conference on Machine Learning, ICML 2004, pp. 104–111. ACM, New York (2004)
14. McCallum, A., Li, W.: Early results for named entity recognition with conditional random fields, feature induction and web-enhanced lexicons. In: Proceedings of the Seventh Conference on Natural Language Learning at HLT-NAACL 2003, CONLL 2003, vol. 4, pp. 188–191. Association for Computational Linguistics, Stroudsburg (2003)
15. Nair, N., Ramakrishnan, G., Krishnaswamy, S.: Enhancing activity recognition in smart homes using feature induction. In: Cuzzocrea, A., Dayal, U. (eds.) DaWaK 2011. LNCS, vol. 6862, pp. 406–418. Springer, Heidelberg (2011)
16. Nair, N., Saha, A., Ramakrishnan, G., Krishnaswamy, S.: Rule ensemble learning using hierarchical kernels in structured output spaces. In: Twenty-Sixth AAAI Conference on Artificial Intelligence (2012)
17. Wilson, D.H.: Assistive intelligent environments for automatic health monitoring. PhD Thesis, Carnegie Mellon University (2005)
18. van Kasteren, T., Noulas, A., Englebienne, G., Kröse, B.: Accurate activity recognition in a home setting. In: Proceedings of the 10th International Conference on Ubiquitous Computing, UbiComp 2008, pp. 1–9. ACM, New York (2008)
19. Gibson, C., van Kasteren, T., Krose, B.: Monitoring homes with wireless sensor networks. In: Proceedings of the International Med-e-Tel Conference (2008)
20. McCallum, A.K.: Efficiently inducing features of conditional random fields. In: Proceedings of the Nineteenth Conference Annual Conference on Uncertainty in Artificial Intelligence (2003)
21. Bach, F.: High-dimensional non-linear variable selection through hierarchical kernel learning. Technical report, INRIA, France (2009)
22. Gutmann, B., Kersting, K.: TildeCRF: Conditional random fields for logical sequences. In: Fürnkranz, J., Scheffer, T., Spiliopoulou, M. (eds.) ECML 2006. LNCS (LNAI), vol. 4212, pp. 174–185. Springer, Heidelberg (2006)
23. Huynh, T.N., Mooney, R.J.: Online max-margin weight learning with markov logic networks. In: Proceedings of the AAAI 2010 Workshop on Statistical Relational AI (Star-AI 2010), Atlanta, GA, pp. 32–37 (July 2010)

24. Huynh, T.N., Mooney, R.J.: Online structure learning for markov logic networks. In: Gunopulos, D., Hofmann, T., Malerba, D., Vazirgiannis, M. (eds.) ECML PKDD 2011, Part II. LNCS, vol. 6912, pp. 81–96. Springer, Heidelberg (2011)

25. Kersting, K., De Raedt, L., Raiko, T.: Logical hidden markov models. Journal of Artificial Intelligence Research 25 (2006)

26. Kersting, K.: Say em for selecting probabilistic models for logical sequences. In: Proceedings of the Twenty First Conference on Uncertainty in Artificial Intelligence, pp. 300–307. Morgan Kaufmann (2005)

27. Thon, I.: Don't fear optimality: Sampling for probabilistic-logic sequence models. In: De Raedt, L. (ed.) ILP 2009. LNCS, vol. 5989, pp. 226–233. Springer, Heidelberg (2010)

28. Thon, I., Landwehr, N., De Raedt, L.: Stochastic relational processes: Efficient inference and applications. Mach. Learn. 82(2), 239–272 (2011)

29. Schulte, O., Khosravi, H., Kirkpatrick, A., Man, T., Gao, T., Zhu, Y.: Modelling relational statistics with bayes nets. In: Proceedings of 22nd International Conference on Inductive Logic Programming (ILP 2012). Springer (2012)

30. Flach, P., Lachiche, N.: 1BC: A first-order bayesian classifier. In: Džeroski, S., Flach, P. (eds.) ILP 1999. LNCS (LNAI), vol. 1634, pp. 92–103. Springer, Heidelberg (1999)

31. McCreath, E., Sharma, A.: LIME: A system for learning relations. In: Richter, M.M., Smith, C.H., Wiehagen, R., Zeugmann, T. (eds.) ALT 1998. LNCS (LNAI), vol. 1501, pp. 336–374. Springer, Heidelberg (1998)

32. Jawanpuria, P., Nath, J.S., Ramakrishnan, G.: Efficient rule ensemble learning using hierarchical kernels. In: Getoor, L., Scheffer, T. (eds.) ICML, pp. 161–168. Omnipress (2011)

33. Dehaspe, L., Toivonen, H.: Discovery of frequent datalog patterns. Data Min. Knowl. Discov. 3(1), 7–36 (1999)

34. Dehaspe, L., Toironen, H.: Relational data mining, pp. 189–208. Springer-Verlag New York, Inc., New York (2000)

What Kinds of Relational Features Are Useful for Statistical Learning?

Amrita Saha[1], Ashwin Srinivasan[2], and Ganesh Ramakrishnan[1]

[1] Department of Computer Science and Engineering,
Indian Institute of Technology,
Bombay, India
[2] Department of Computer Science, Indraprastha Institute of Technology,
New Delhi, India

Abstract. A workmanlike, but nevertheless very effective combination of statistical and relational learning uses a statistical learner to construct models with features identified (quite often, separately) by a relational learner. This form of model-building has a long history in Inductive Logic Programming (ILP), with roots in the early 1990s with the LINUS system. Additional work has also been done in the field under the categories of propositionalisation and relational subgroup discovery, where a distinction has been made between *elementary* and *non-elementary* features, and statistical models have been constructed using one or the other kind of feature. More recently, constructing relational features has become an essential step in many model-building programs in the emerging area of Statistical Relational Learning (SRL). To date, not much work—theoretical or empirical—has been done on what kinds of relational features are sufficient to build good statistical models. On the face of it, the features that are needed are those that capture diverse and complex relational structure. This suggests that the feature-constructor should examine as rich a space as possible, in terms of relational descriptions. One example is the space of all possible features in first-order logic, given constraints of the problem being addressed. Practically, it may be intractable for a relational learner to search such a space effectively for features that may be potentially useful for a statistical learner. Additionally, the statistical learner may also be able to capture some kinds of complex structure by combining simpler features together. Based on these observations, we investigate empirically whether it is acceptable for a relational learner to examine a more restricted space of features than that actually necessary for the full statistical model. Specifically, we consider five sets of features, partially ordered by the subset relation, bounded on top by the set F_d, the set of features corresponding to definite clauses subject to domain-specific restrictions; and bounded at the bottom by F_e, the set of "elementary" features with substantial additional constraints. Our results suggest that: (a) For relational datasets used in the ILP literature, features from F_d may not be required; and (b) Models obtained with a standard statistical learner with features from subsets of features are comparable to the best obtained to date.

F. Riguzzi and F. Železný (Eds.): ILP 2012, LNAI 7842, pp. 209–224, 2013.

1   Introduction

The emerging area of statistical relational learning (SRL) is characterised by
a number of distinct strands of research. Especially prominent is research concerned with the construction of models from data that use representations based
on either first-order logic programs augmented with probabilities or probabilistic graphical models. The concerns here are the usual ones to do with expressive power, estimation, and inference: what kinds of probabilistic models can
be constructed with one representation or the other; how can we estimate the
structure and parameters in the model; how do we answer queries exactly, given
data that is observed, and perhaps missing; and so on. The combination of relational learning with statistical modeling, however, has a longer history within
Inductive Logic Programming (ILP), with origins at least as early as 1990, with
the LINUS system [14]. Since then, there are have been regular reports in the
literature on the use of ILP systems, as a tool for constructing relational features
for use in statistical modeling [22].

An argument can be made that construction of relational features must necessarily require some form of first-order learning, of which ILP is an instance
(for example, see [13]). Arguments in-principle aside, the literature also suggests
that augmenting any existing features with ILP-constructed relational ones can
substantially improve the predictive power of a statistical model [24,4,22]. There
are thus good practical reasons to persist with this variant of statistical and
logical learning. On the other hand there has been some work done on comparing the different kinds of propositionalisation techniques used to transform the
search space from the space of first-order hypothesis to the space of propositional features which can be handled by more scalable propositional/statistical
learners. [12] claims that of the two main kinds of propositionalization methods,
namely logic oriented and database-oriented, both have their specific advantages.
While logic-oriented methods can handle complex relational structures in the
form of background knowledge and provide more expressive relational models,
database-oriented models are much more scalable. According to their empirical findings, a combination of features from these two groups are necessary for
learning good models.

Even within this well-trodden corner of statistical relational learning (no more,
perhaps, than a "poor man's SRL"), there are some issues that remain unad-
dressed. To date, not much work—theoretical or empirical—has been done on
what kinds of relational features are sufficient to build good statistical models.
On the face of it, the features that are needed are those that capture diverse and
complex relational structure. This suggests that the feature-constructor should
examine as rich a space as possible, in terms of relational descriptions. One example is the space of all possible features in first-order logic, given constraints of
the problem being addressed. Practically, it may be intractable for a relational
learner to search such a space effectively for features that may be potentially
useful for a statistical learner. Additionally, the statistical learner may also be
able to capture some kinds of complex structure by combining simpler features
together. For example, a statistical learner like a support vector machine or

logistic regression that may also be able to approximate the effect of a conjunction of these features using their weighted sum. Reports in the ILP literature suggest at least 5 kinds of relational feature classes: (1) F_d: (the set of) features from definite clauses with no restrictions other than those of the domain [22,24]; (2) F_i: features from "independent" clauses that place restrictions on the sharing of existential variables [2]; (3) F_r: features denoting a class of relational subgroup that place additional restrictions on the use of existential variables in independent clauses [15]; (4) F_s: features from "simple" clauses in the sense described in [17]; and (5) F_e: features developed from the class of "elementary" clauses described in [16]. In this paper, we show certain subset relationships hold between these sets. These are shown diagrammatically in Fig. 1. When exploring whether smaller feature-spaces are adequate, we use these relationships to investigate empirically whether exploring larger sets of features adds any significant predictivity to a statistical learner. Several Statistical Relational Learning approaches in the past have focused on learning from specific classes of features and construct more complicated ones if necessary by boosting. In [10,19] it has been empirically shown in various learning settings, that boosting of weak features performs well. The same technique of boosting applied on different settings, by using differnt variants of the loss functions have been discussed repeatedly in SRL literature, for example, in [5,8,9,21]. There has also been attempt at posing the relational feature construction as a problem of combining macro-operators by statistical learner and delegating search to them [1]. In this macro-operator paradigm, it has been shown that by following a particular propositionalization technique, the trade-off between feature construction cost and learning cost can be better handled and infact the propositionalized dataset becomes PAC-learnable.

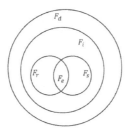

Fig. 1. Relationships between the sets of features considered in this paper

The rest of the paper is organised as follows. In Section 2 we describe the mapping between clauses and relational features. We also describe in greater detail the five feature classes F_d–F_e. In Section 3 we derive the relationships shown in Fig. 1 and enumerate some consequences that follow directly from them. Section 4 describes an empirical investigation, using standard statistical learners of the smallest size feature class that appears to be useful for constructing predictive models for several ILP benchmark datasets. Section 5 concludes the paper.

2   Feature Classes

In this paper, we will take first-order relational features simply to be first-order clauses that either maximize some objective either collectively or individually (effectively, a restatement of the distinction in [7] between strong and weakly relevant features, in optimisation terms). Specifically, we will assume that is a one-to-one correspondence between first-order relational features and definite clauses (alphabetic variants of a clause are treated as being the same clause). Formally, we adopt the same notation as in [22] for a function that maps a first-order definite clause to a feature. The set of examples provided to an ILP system can be defined as a binary relation which is a subset of the Cartesian product $\mathcal{X} \times \mathcal{Y}$ where $\mathcal{X}$ denotes the set of individuals (*i.e.* structured objects consisting of first-order predicates) and $\mathcal{Y}$ denotes the finite set of classes. The definite clauses obtained as an output of the ILP system can be represented in the form $h_j(x, c) : Class(x, c) \leftarrow Cp_j(x)$ where $Cp_j : \mathcal{X} \to \{0, 1\}$ is a nonempty conjunction of predicates on the variable $x \in \mathcal{X}$ and c is the class variable specified in the head predicate. For convenience, we will say $Head(h_j(x, c)) = Class(x, c)$ and $Body(h_j(x, c)) = Cp_j(x)$. Given a clause $h_j(x, c)$, a first-order feature can be defined as $f_j(x) = 1$ iff $Body(h_j(x, c)) = 1$ and 0 otherwise. Constraints on $Body(h_j(x, c))$ allow us to define several kinds of features. Following [17] (with a small difference) we distinguish between "source" predicates and "sink" predicates in the language of clauses allowed in the domain. The former are those that contain at least one output argument and any number of input arguments (in the sense used by the mode declarations in [18]) and the latter are those that contain no output arguments. In a clause, thus all new existentially quantified variables should be introduced in the source literals and not in the sink literals. [1] Additionally, following [2], for a clause $h_j(x, c)$, the independent components in $Body(h_j(x, c))$ are partitions of the literals in $Body(h_j(x, c))$ into sets, such that each partition consists of a "connected" set of literals (once again, using input and output variables in the sense of [18]), and that literals across partitions only share the head variable x. We are then able to define the following kinds of relational feature classes:

The class F_d: This consists of first-order features obtained from definite clauses $h_j(x, c)$ in the functional manner described above. That is, no constraints are placed on $Body(h_j(x, c))$. Features used in [24], for example, belong to this class.

The class F_i: This is a restricted version of the class F_d, consisting of features from clauses $h_j(x, c)$ such that $Body(h_j(x, c))$ consists of exactly 1 independent component.

The class F_r: This is a restricted version of the class F_i, consisting of features obtained from clauses $h_j(x, c)$ such that $Body(h_j(x, c))$ consists of exactly 1 independent component, and with an additional constraint that all new existential variables introduced by a source literal appear in source or sink literals in $Body(h_j(x, c))$. The features in [15] are from this class, and the first-order

---

[1] "Structural" predicates used in [16] are thus binary source literals that introduce a single new variable, and "property" predicates are sink literals.

features described in [16] are a special case of features in F_r (the special case arising from restrictions on sources and sinks to structural and property predicates as described earlier).

The class F_s: This consists of features obtained from clauses $h_j(x, c)$ such that h_j is a simple clause in the sense defined by [17]. That is, $Body(h_j(x, c))$ contains exactly 1 sink literal.

The class F_e: This is a restricted version of the class F_r, consisting of features obtained from clauses $h_j(x, c)$ such that $Body(h_j(x, c))$ contains exactly 1 sink literal. "Elementary" features described in [16] are a special case of features in F_e.

3   Relationships between Feature Classes

Some of the relationships between feature classes are evident from the definitions. That is: $F_r \subseteq F_i \subseteq F_d$. The following additional statements hold

- $F_s \subseteq F_i$
- $F_s \not\subseteq F_r$ and $F_r \not\subseteq F_s$
- $F_e = F_r \cap F_s$

Thus, the feature classes exhibit the following hierarchical structure:
$$F_e = (F_r \cap F_s) \subseteq F_r \subseteq F_i \subseteq F_d$$
$$F_e = (F_r \cap F_s) \subseteq F_s \subseteq F_i \subseteq F_d$$

Given these subset relationships between feature classes, it is also of some importance to consider whether there exists a way of reconstructing, using logical operations, every feature in a superset class by combining features from a subset (or smaller) class (clearly, the reverse is always possible: a feature in a subset class can always be constructed from a feature in the superset class). The interest here is of course that if such logical relationships hold, then features in the larger class may be approximated by statistical learners using weighted combinations of features from the smaller class. The following logical relationships hold between the feature classes:

- Every feature in F_i can be constructed from features in F_s
- Every feature in F_d can be constructed from features in F_i

Thus, every feature in F_d can be constructed from F_s. In addition:

- Not every feature in F_i can be constructed from features in F_r
- Not every feature in F_r can be constructed from features in F_e
- Not every feature in F_s can be constructed from features in F_e

The proofs of the relations have been elaborated in Appendix A. We now evaluate empirically the utility of features from the different classes, proceeding from the smallest (F_e) to the largest (F_d).

4   Empirical Evaluation

4.1   Aims

Our aim is to obtain empirical evidence of the smallest feature class that is found to yield good statistical models. Specifically, we mean that we wish to find using a set of well-studied benchmark data sets and popular statistical learners, whether there is a feature class such that adding features from larger classes yields no significant increases in predictive accuracy.

4.2   Materials

Domains. We use biochemical datasets that have been used in a range of papers in the ILP literature. These are tabulated below. These datasets have been used widely: see, for example

Dataset	No. of instances for positive class	No. of instances for negative class
Alz (Amine) [25]	343	344
Alz (Acetyl) [25]	443	443
Alz (Memory) [25]	321	321
Alz (Toxic) [25]	443	443
Carcin [25]	182	155
DSSTox [25]	220	356
Mut(188) [25]	125	63

Fig. 2. Datasets used in the paper

4.3   Algorithms and Machines

The statistical learners used in the paper are these:

- A support vector machine(SVM). Specifically, we examine SVMs with both the L1-norm, also known in literature as L1-SVM (the LibLINEAR implementation) and the L2-norm, known as L2-SVM,(the LibSVM implementation) on the weight vector used as the regularizer. The L1-norm is known to induce sparsity on the feature space, forcing a form of feature selection which is important when the number of features is large.
- Logistic regression. Specifically, we consider a standard version of this technique and a faster and more efficient variant called SMLR (Sparse Multinomial Logistic Regression) [11]
- Rule ensemble learning using maximum likelihood estimation (MLRules) [3]. This employs a greedy approach of maximising likelihood to construct an ensemble of rules from the features. Like logistic regression, the weights of the rules are derived from the conditional probability distribution learnt. So it can be thought of as a generalization of Logistic Regression, where the space explored to construct rule ensembles can be considered as an approximation of conjunctions of rules/features.

All statistical learners here, compute the decision function as a weighted linear combination of the features (which can be thought of as an approximation to a logical conjunction). Features in each feature class are constructed using the ILP system Aleph [23]. It is possible to enforce the constraints associated with each feature class in a straightforward manner as part of background knowledge provided to this system.

All experiments were conducted on a machine equipped with a 8-core Intel i7 2.67GHz processors and 8 gigabytes of random access memory.

4.4   Method

The methodology adopted for providing the statistical learners features from feature classes F_e to F_d is as follows:

1. Select a total ordering on the classes $F_e \prec \cdots \prec F_d$ that is consistent with the partial ordering imposed by subset relationships between these sets.
2. For each dataset and each statistical learner:
   (a) For sets S_j from smallest (F_e) to the largest (F_d), in the total ordering $\prec$:
   (b) Let $F_0 = \emptyset$
       i. Repeat R times:
           A. Obtain a set of features F from S_j
           B. $F_j = F_{j-1} \cup F$
           C. Obtain an estimate of the predictive accuracy A_j of the statistical learner with features F_j
       ii. Obtain the mean predictive accuracy $\overline{A_j}$ across the R repeats
3. Determine the smallest set S_k after which there are no significant changes in mean predictive accuracy $\overline{A_k}$

The following details are relevant:

1. There are only 2 total orderings possible, given the subset relationships between the feature classes: $F_e \prec_1 F_r \prec_1 F_s \prec_1 F_i \prec_1 F_d$ and $F_e \prec_2 F_s \prec_2 F_r \prec_2 F_i \prec_2 F_d$. According to our incremental algorithm, here F_s in the 5th column, F_s in $\prec_1$ and F_r in $\prec_2$ actually refer to sets of features from $F_r \cup F_s$, because of their order of appearance. The results we report here are with $\prec = \prec_1$. Our conclusions do not change with either $\prec_1$ or $\prec_2$.
2. All predictive accuracies are estimated using 10-fold cross-validation.
3. The results are averaged over $R = 4$ repetitions. Larger values of R would result in smaller standard errors of the mean estimate.
4. Statistical learners have parameters that require optimisation. It has been shown elsewhere [25] that using default values of parameters can result in suboptimal models, which can clearly confound the conclusions that can be drawn here. For each set of training data in a cross-validation run, we set aside some small part of the training data as a "validation" set, and use this to tune the parameters of the statistical learner. The "best" parameter value that results is then used to construct a model on the training data; and then evaluated on

the test dataset of that cross-validation run. This does not necessarily yield the best model possible, but accuracies are usually higher than those obtained with the default settings for the parameters.

5. The feature-construction procedure employed by the ILP engine requires an upper-bound k on the number of features produced for each class label in the example data. For experiments reported here, $k = 500$. There is a randomised element to the feature-construction procedure in Aleph, which can been seen as selecting (non-uniformly) upto k features, from the set of all possible features allowed. The selection of features is controlled by parameters that correspond to precision and recall in the data mining literature. For experiments here, these values have been deliberately left low (and hence, easy to obtain for the ILP engine). The intent is that the statistical learner should be able to combine these to obtain higher values.

6. We have elected to assess the utility of features from each feature class by examining the improvements in mean accuracy by augmenting features already present from feature classes earlier in the ordering $\prec$. The alternative of simply comparing accuracies with a new set of features drawn from the from each feature class X in the ordering would have confounded matters, since this set could contain features from a class $Y \subseteq X$. It would not then be known whether the increases in accuracy, if any, are due to features from a class X, or those from class Y.

7. It is known that for large R, mean accuracies are distributed normally. For small R, it is known that the sampling distribution of the mean is a t-distribution with $R - 1$ degrees of freedom. This is used in statistical comparisons when needed (as will be seen, it is often evident whether differences are significant, and no undue statistical testing is needed).

4.5   Results and Discussion

The comparative performance of the models is shown in Fig. 3. The principal result from this tabulation is this: broadly, there is little value in including features from F_d (the largest feature class considered here). Examining the table in greater detail, we are able to obtain the number of "wins" for each feature class (that is, the number of times the highest predictive accuracy results from using features in that class). A cross-tabulation of this against the learners is shown in Fig. 4:

The data in Figs. 3 and 4 suggest that it is sufficient to consider features from F_i (that is, features from clauses containing one independent component). Now, while it is possible to reconstruct every feature in F_d exactly as a simple logical conjunction, of some features in F_i, it is also possible to reconstruct several (but not all) features in F_d by simple logical conjunction of select features in F_r, F_s and F_e. This appears to be exploited by all of the statistical learners here, since it is by no means necessary for any of them that features from F_d are consistently required to produce the best results. This is reinforced further, if rather than considering outright wins, we consider a model as being good enough if there no (statistically) significant difference to the best model. Then, the performance tallies can be summarized as in Fig. 5.

Data	Learner	Using Features From				
		F_e	F_r	F_s	F_i	F_d
Alz(Amine)	L2-SVM	66.14±0.84	65.71±0.69	67.23±0.86	**81.62±0.92**	81.45±0.61
	L1-SVM	65.50±0.15	65.84±0.14	64.91±1.65	78.34±4.18	**80.31±1.57**
	LR	65.42±0.0	65.42±0.0	64.38±0.0	79.32±3.93	**81.28±0.31**
	SMLR	65.42±0.0	66.05±0.22	67.12±0.38	81.20±2.56	**81.63±2.77**
	MLRules	64.39±0.0	64.39±0.0	68.67±0.0	**82.32±1.18**	82.22±0.35
Alz(Acetyl)	L2-SVM	69.42±0.33	68.52±0.70	68.59±2.11	**74.40±0.80**	72.97±0.47
	L1-SVM	69.85±0.05	68.55±0.0	66.20±0.0	**74.06±0.91**	72.78±1.44
	LR	69.82±0.0	69.82±0.0	71.93±0.0	73.87±0.44	**74.16±0.24**
	SMLR	70.36±0.11	70.32±0.14	72.17±0.12	**73.34±1.38**	73.20±0.65
	MLRules	69.59±0.0	69.59±0.0	70.95±0.0	71.67±0.47	**72.44±0.52**
Alz(Memory)	L2-SVM	59.80±0.22	61.09±0.45	63.46±1.07	**68.39±1.15**	68.07±1.75
	L1-SVM	56.99±1.29	62.89±1.07	64.33±0.25	**71.29±1.04**	67.96±0.41
	LR	59.07±0.08	59.07±0.08	65.87±0.0	**70.31±1.54**	69.30±0.55
	SMLR	59.69±0.0	58.44±1.32	66.34±0.73	70.10±1.26	**71.83±1.67**
	MLRules	58.83±0.19	58.83±0.19	64.76±0.08	**70.13±1.14**	67.91±0.64
Alz(Toxic)	L2-SVM	70.72±1.0	71.97±1.99	77.17±0.82	**83.01±0.73**	82.04±2.40
	L1-SVM	70.06±0.0	71.73±0.0	80.94±3.15	82.59±0.70	**82.83±2.21**
	LR	72.84±0.0	72.09±1.51	77.91±0.0	81.65±0.79	**83.32±0.87**
	SMLR	74.50±2.35	74.07±0.79	76.76±0.16	82.59±0.39	**84.21±0.69**
	MLRules	73.18±0.0	73.18±0.0	77.44±0.0	**84.50±0.44**	84.19±0.56
Carcin	L2-SVM	60.59±1.58	**62.15±1.75**	59.49±1.75	62.03±1.28	59.85±1.67
	L1-SVM	**61.21±0.0**	59.71±0.0	58.66±3.10	60.26±0.71	59.21±1.29
	LR	**59.00±0.0**	57.71±0.16	54.01±2.79	51.80±1.37	52.52±3.49
	SMLR	**60.88±0.0**	60.17±0.38	59.51±2.91	58.80±3.42	57.94±2.84
	MLRules	**59.39±0.25**	59.20±0.15	58.83±1.96	57.94±2.31	56.64±0.94
DSSTox	L2-SVM	69.36±1.15	69.86±1.19	72.21±1.22	**73.12±0.94**	70.90±3.15
	L1-SVM	69.25±0.25	69.00±0.25	70.77±1.60	**71.91±1.77**	69.21±1.34
	LR	70.49±0.0	70.49±0.0	**72.18±1.71**	71.01±0.87	70.32±0.47
	SMLR	72.24±0.0	**72.29±0.09**	70.71±2.62	70.95±2.94	71.02±1.38
	MLRules	71.55±0.0	71.55±0.0	**72.85±1.43**	72.73±1.52	72.16±1.09
Mut(188)	L2-SVM	–	65.80±1.25	85.84±1.36	**86.49±0.28**	84.63±1.58
	L1-SVM	–	63.70±0.30	84.92±0.65	85.27±0.31	**85.52±0.48**
	LR	–	67.12±0.0	74.85±1.74	77.48±3.38	**77.61±2.88**
	SMLR	–	73.44±0.0	**88.06±1.57**	85.83±1.81	84.10±2.02
	MLRules	–	71.86±0.0	85.04±0.76	**86.50±1.40**	85.70±1.55

Fig. 3. Mean predictive accuracies of statistical models including features from F_e to F_d. "–" indicates no features in this class were possible given the domain constraints. Here, refers to the F_s in the incremental order $\prec_1$, hence it actually denotes the set of features from $F_r \cup F_s$. Same notation has been followed in Fig. 4. and Fig. 5.

Feature Class	Number of Wins					Total Wins
	L1-SVM	L2-SVM	LR	SMLR	MLRules	
F_e	1	0	1	1	1	4/35
F_r	0	1	1	1	0	3/35
F_s	0	0	1	1	1	3/35
F_i	3	6	1	1	4	15/35
F_d	3	0	4	3	1	11/35

Fig. 4. Number of outright wins for a feature class. This is number of occasions out of the total number of possible occasions (*i.e.* 35) on which a statistical learners achieves the highest mean predictive accuracy using features from that class.

Feature Class	Number of Good Enough Models					Total No. of Good Models
	L1-SVM	L2-SVM	LR	SMLR	MLRules	
F_e	2	1	2	2	2	9/35
F_r	2	2	1	4	4	13/35
F_s	3	2	1	4	4	14/35
F_i	7	7	6	7	7	34/35
F_d	4	5	6	7	7	25/35

Fig. 5. Number of good enough models (out of all possible models *i.e.* 35), using a feature class. A model is taken to be good enough if its predictive accuracy is not statistically different to the model with the highest predictive accuracy.

Data	Statistical Model	ILP Model With Parameter Selection & Optimization
Alz (Amine)	82.32±1.18	80.20
Alz (Acetyl)	74.16±0.24	77.40
Alz (Memory)	71.83±1.67	67.40
Alz (Toxic)	84.50±0.44	87.20
Carcin	62.15±1.75	59.10
DSSTox	73.12±0.94	73.10
Mut(188)	88.06±1.57	88.30

Fig. 6. Comparison of mean predictive accuracies of statistical models against the ILP models constructed with parameter selection and optimisation (see [25])

These tabulations do suggest that of the classes $F_e \dots F_d$, the class F_i may be the most useful. But how good are the models obtained with F_i, when compared against the ILP models reported in the literature? Fig. 6 shows a comparison of the best statistical models against the predictive accuracies of the ILP models obtained after parameter selection and optimisation [25].

Finally, although not relevant to the aims of the experiment here, we note some exceptional behaviour on the "Carcin" dataset, in which there is a fairly consistent trend of decreasing predictive accuracies as we progress from F_e to F_d. An examination of re-substitution (training-set) accuracies for this data shows the opposite trend, suggesting that these data may be especially prone to overfitting.

We also experimented with $\prec = \prec_2$ and found that the results were quite similar with the ones with $\prec = \prec_1$ tabulated here. In interest of space, those results have been omitted.

The methodology of experimentation discussed above, constructs a total ordering of the feature classes by cumulatively augmenting the features already obtained from the feature classes earlier in the ordering $\prec$. But this process of incrementally adding features can increase the size of the hypothesis space. Increasing the size of the hypothesis space naturally increases the possibility that models perform better simply due to chance effects. We investigate for this by sampling a fixed number of features from each feature- class only, without cumulatively augmenting the already present features generated by its subsumed feature classes. The second set of experiments based on this setting has been elaborated in Table 7 and the comparative study of the performance of the different feature

Data	Learner	Using Features From					
		F_e	F_r	F_s	$F_{r\cup s}$	F_i	F_d
Alz(Amine)	L2-SVM	65.27	66.15	79.54	78.29	**79.56**	77.82
	L1-SVM	63.95	68.19	76.91	**78.12**	77.38	74.61
	LR	65.56	66.45	79.98	**80.91**	79.13	79.47
	SMLR	67.12	67.75	79.10	78.76	**79.33**	79.17
	MLRules	65.42	66.0	80.73	80.62	**80.74**	78.74
Alz(Acetyl)	L2-SVM	66.83	67.02	69.37	67.55	**73.53**	72.54
	L1-SVM	65.13	65.28	69.22	67.55	**73.53**	72.54
	LR	66.75	67.17	67.55	68.01	**73.53**	72.54
	SMLR	66.68	68.14	68.09	**71.01**	69.88	70.2
	MLRules	66.30	67.56	69.96	**72.22**	69.60	70.35
Alz(Memory)	L2-SVM	59.8	61.21	64.94	63.38	67.46	**69.3**
	L1-SVM	59.8	61.21	64.94	63.38	67.46	**69.3**
	LR	59.8	61.21	64.94	63.38	67.46	**69.3**
	SMLR	60.88	59.17	67.73	**71.01**	68.04	67.17
	MLRules	60.411	59.78	69.91	**72.56**	68.52	66.63
Alz(Toxic)	L2-SVM	70.47	74.86	77.08	76.52	**82.06**	81.59
	L1-SVM	75.67	75.55	80.28	79.4	**84.68**	81.40
	LR	70.47	74.86	77.08	76.52	**82.06**	81.59
	SMLR	71.27	58.85	78.94	**82.78**	79.03	81.43
	MLRules	70.33	55.90	80.33	**82.22**	79.79	81.43
Carcin	L2-SVM	58.81	**63.87**	59.29	58.41	63.23	57.58
	L1-SVM	58.81	**63.87**	59.29	58.41	63.23	57.58
	LR	58.81	**63.87**	59.29	58.41	63.23	57.58
	SMLR	**61.16**	60.50	59.07	55.43	59.37	59.32
	MLRules	**60.52**	55.26	56.64	54.23	57.29	55.67
DSSTox	L2-SVM	70.32	69.69	71.25	71.25	**73.65**	66.87
	L1-SVM	70.32	69.69	71.25	71.25	**73.65**	66.87
	LR	70.32	69.69	71.25	71.25	**73.65**	66.87
	SMLR	71.77	**73.67**	72.01	71.06	69.63	71.48
	MLRules	66.98	69.10	71.84	71.11	**72.48**	72.47
Mut(188)	L2-SVM	–	–	85.14	85.14	**86.66**	82.45
	L1-SVM	–	–	85.14	85.14	**86.77**	82.45
	LR	–	–	85.14	85.14	**86.66**	82.45
	SMLR	–	–	**73.43**	**73.43**	64.95	72.72
	MLRules	–	–	74.43	74.43	65.47	**76.93**

Fig. 7. Mean predictive accuracies of statistical models including features from F_e to F_d. "–" indicates no features in this class were possible given the domain constraints. Here, refers to the $F_{r\cup s}$ actually denotes the set of features from $F_r \cup F_s$. Same notation has been followed in Fig. 4. and Fig. 5.

Feature Class	Total Number of Outright Wins	Total Number of Good Models
F_e	2/35	3/35
F_r	4/35	3/35
F_s	1/35	9/35
$F_{r\cup s}$	8/35	16/35
F_i	**16/35**	**27/35**
F_d	4/35	17/35

Fig. 8. Number of outright wins for a feature class and the Number of good enough models generated from features obtained from those classes

classes in terms of number of outright wins and number of good enough statistical models constructed, has been presented in 8.

5   Concluding Remarks

In this paper we have explored the relationship between several feature spaces that have been reported in ILP literature and examined empirically whether it is possible to construct good statistical models using features from smaller spaces. The intuition underlying this is that statistical models may often be able to approximate the effect of more elaborate features by weighted combinations of simpler features. Our results suggests that the class F_i, consisting of features constructed from clauses containing exactly one independent component seems to be particularly useful. This makes some sense: a linear combination of multiple features from F_i can approximate the reconstruction of a full first-order feature, since no variable sharing is required between such features. In fact, this leads us to hypothesize that statistical learners like [6] that perform conjunctive feature learning will not perform any better than learners using weighted combinations of features from F_i, and will incur a greater computational cost. Further, this also leads us to believe that weighted linear combinations of first order clauses in relational learning models such as MLN [20], Relational Markov Networks [26] and the boosting techniques like [10,19] could be efficiently and effectively approximated by weighted linear combinations of clauses from simpler classes such as F_i and is part of our ongoing work. On the other hand the main focus of this paper has been directed at learning a discriminative classification model whereas the general Statistical Relational Learning also addresses a generative model learning setting. It would be interesting to see how this study of learning from specific feature classes can make the generative learning more efficient.

Acknowledgements. Ashwin Srinivasan is supported by a Ramanujan Fellowship of the Government of India. He is also a Visiting Professor at the Department of Computer Science, Oxford University; and a Visiting Professorial Fellow at the School of CSE, University of NSW, Sydney.

A   Appendix A

A.1   Proofs of Results from Section 3

Relationship between F_e and F_s : $F_e \subseteq F_s$.

> *Proof.* Every feature from the class F_e also belongs to the class F_s, since it is minimal (*i.e.* cannot be decomposed into smaller features that share only global variable), and contains a single sink (*i.e.* cannot be decomposed into smaller features that share only a local variable). On the other hand, a feature from the class F_s may not belong to the class F_e, since the former may have some unused output variables, For example, in the trains problem originally proposed by Ryzhard Michalski, $eastbound(A) \leftarrow hasCar(A, B)$ is a valid clause from F_s but it cannot belong to class F_e.

Relationship between F_s and F_r : $F_s \not\subset F_r$. And the reverse is neither true, *i.e.* $F_r \not\subset F_s$.

Proof. Every feature from the class F_s may not belong to the class F_r, attributed to the same reason of local-variable reusal property that the latter must satisfy, a property that need not hold for the former. The same clause mentioned above as an example, *viz.*, $eastbound(A) \leftarrow hasCar(A, B)$, will not yield a feature from F_r (or for that matter, from F_e), though it is a valid clause from F_s. Also the reverse is not true *i.e.* every feature from F_r may not belong to F_s, since the former can have any number of property nodes *i.e.* sinks, which is not possible in the case of the latter.

Relationship between F_e, F_r and F_s : $F_e = (F_r \cap F_s)$.

Proof. By the very definition of a feature from class F_e, it must belong to class F_r and since it cannot have more than one sink, it must also be a valid feature from class F_s.

Relationship between F_r and F_i : $F_r \subseteq F_i$

Proof. Every feature from class F_r also belongs to class F_i because of the minimal property, but the reverse is not true *i.e.* not every feature from class F_i is a valid feature from class F_r again because of the variable reusal property. For example, $eastbound(A) \leftarrow hasCar(A, B)$ is a valid feature from class F_i but cannot belong to class F_r because of the unused output variable C introduced by the structural predicate, $hasCar(A, B)$. Also there may not exist any property predicate in the clause from class F_i as in this example.

Relationship between F_s and F_i : $F_s \subseteq F_i$.

Proof. By the very definitions of features from class F_s and F_i, it can be seen that every feature from class F_s is a valid feature from class F_i but the reverse is not true.

Relationship between F_i and F_d : *i.e.* $F_i \subseteq F_d$

Proof. It is obvious that every feature from class F_i is a first-order definite feature. But the reverse is not true, since there exist clauses such as the following one from the trains problem,
$eastbound(A) \leftarrow hasCar(A, B), hasCar(A, C), short(B), closed(C)$
that are not independent (since they can be decomposed further into independent components).

A.2   Reconstruction Property of Feature Classes

1. Every full first-order feature from definite clauses *i.e.* from F_d can be constructed from features from F_s.

   Proof. The one-to-one mapping from clauses to features allows any feature f_d from F_d to be inverted to a definite clause c_d. Also [17] states that every definite clause can be constructed from simple causes. So given a set of simple clauses S_s that can reconstruct the first order definite clause c_d, the mapping can be used to construct the set of features corresponding to the set S_s of simple clauses used to reconstruct c_d and hence feature f_d.

2. Every feature from F_i can be constructed from features from F_s.

   Proof. Since every feature from F_i is also a full first-order feature (owing to the subset relation) and every full first-order feature from F_d can be reconstructed from features from F_s, it follows that every feature from F_i can be reconstructed from features from F_s by performing a combination of logical conjunction and (possibly) variable unification.

3. Every feature from F_r can be constructed from features from F_s.

   Proof. Similar to the above justification of (2), since each feature from F_r is also a feature from F_d, it can be constructed from features from F_s by performing their logical conjunction and (possibly) variable unification.

4. Every feature from F_e can be constructed from features from F_s.

   Proof. This follows from the subset relation $F_e \subseteq F_s$. In general, any feature of a subclass can be constructed from its superclass. This is just stating the obvious.

5. Every full first-order feature from F_d can be constructed from features from F_i.

   Proof. This follows from (1), since every full first-order feature from F_d can be constructed from features from F_s and the latter is a subset of F_i. In this case the reconstruction is much simpler since it is equivalent to logical conjunction without any variable unification required.

6. Not every feature from F_r can be constructed from features from F_e.

   Proof. Because of the redefined structural predicates that can introduce any non-zero number of new variables, there can be features from F_r that reuse these newly introduced variables separately in different property predicates. In that case no single property predicate will be sufficient to satisfy the variable-reusal property. The clauses corresponding to these features from F_r cannot be reconstructed from any number of clauses obtained by the inverse-mapping from features from F_e. For example, the feature corresponding to the clause $eastbound(A) \leftarrow hasCarFollowsCar(A, B, C), short(B), closed(C)$ is a valid feature belonging to F_r which cannot be constructed from features obtained from F_e class alone.

7. Not every full first-order feature from F_d can be constructed from features from F_r.

   Proof. Clauses corresponding to features from F_d that do not satisfy the variable-reusal property cannot be reconstructed from any number of clauses inverse-mapped from features from F_r.

   For the same reason as above, the following two properties additionally hold.

8. Not every feature from F_i can be constructed from features from F_e.
9. Not every full first-order feature from F_d can be constructed from features from F_e.

References

1. Alphonse, É.: Macro-operators revisited in inductive logic programming. In: Camacho, R., King, R., Srinivasan, A. (eds.) ILP 2004. LNCS (LNAI), vol. 3194, pp. 8–25. Springer, Heidelberg (2004)
2. Santos Costa, V., Srinivasan, A., Camacho, R., Blockeel, H., Demoen, B., Janssens, G., Van Laer, W., Cussens, J., Frisch, A.: Query transformations for improving the efficiency of ILP systems. Journal of Machine Learning Research 4, 491 (2002)
3. Dembczynski, K., Kotlowski, W., Slowinski, R.: Maximum likelihood rule ensembles. In: ICML, pp. 224–231 (2008)
4. Gottlob, G., Leone, N., Scarcello, F.: On the complexity of some inductive logic programming problems. In: Džeroski, S., Lavrač, N. (eds.) ILP 1997. LNCS, vol. 1297, pp. 17–32. Springer, Heidelberg (1997)
5. Gutmann, B., Kersting, K.: TildeCRF: Conditional random fields for logical sequences. In: Fürnkranz, J., Scheffer, T., Spiliopoulou, M. (eds.) ECML 2006. LNCS (LNAI), vol. 4212, pp. 174–185. Springer, Heidelberg (2006)
6. Jawanpuria, P., Nath, J.S., Ramakrishnan, G.: Efficient rule ensemble learning using hierarchical kernels. In: ICML, pp. 161–168 (2011)
7. John, G.H., Kohavi, R., Pfleger, K.: Irrelevant features and the subset selection problem. In: Machine Learning: Proceedings of the Eleventh International, pp. 121–129. Morgan Kaufmann (1994)
8. Karwath, A., Kersting, K., Landwehr, N.: Boosting relational sequence alignments. In: Proceedings of the 2008 Eighth IEEE International Conference on Data Mining, ICDM 2008, pp. 857–862. IEEE Computer Society, Washington, DC (2008)
9. Kersting, K., Driessens, K.: Non-parametric policy gradients: a unified treatment of propositional and relational domains. In: Proceedings of the 25th International Conference on Machine Learning, ICML 2008, pp. 456–463. ACM, New York (2008)
10. Khot, T., Natarajan, S., Kersting, K., Shavlik, J.: Learning markov logic networks via functional gradient boosting. In: Proceedings of the IEEE 2011 11th International Conference on Data Mining, ICDM 2011, pp. 320–329. IEEE Computer Society, Washington, DC (2011)
11. Krishnapuram, B.I., Carin, L., Figueiredo, M.A.T., Hartemink, A.J.: Sparse multinomial logistic regression: Fast algorithms and generalization bounds. IEEE Trans. Pattern Anal. Mach. Intell. 27(6), 957–968 (2005)
12. Krogel, M.A., Rawles, S., Zelezny, F., Flach, P.A., Lavrac, N., Wrobel, S.: Comparative evaluation of approaches to propositionalization (2003)
13. Landwehr, N., Passerini, A., De Raedt, L., Frasconi, P.: kfoil: Learning simple relational kernels. In: AAAI, pp. 389–394. AAAI Press (2006)
14. Lavrac, N., Dzeroski, S.: Inductive Logic Programming: Techniques and Applications. Routledge, New York (1993)
15. Matwin, S., Sammut, C. (eds.): ILP 2002. LNCS (LNAI), vol. 2583. Springer, Heidelberg (2003)
16. Matwin, S., Sammut, C. (eds.): ILP 2002. LNCS (LNAI), vol. 2583. Springer, Heidelberg (2003)
17. McCreath, E., Sharma, A.: LIME: A system for learning relations. In: Richter, M.M., Smith, C.H., Wiehagen, R., Zeugmann, T. (eds.) ALT 1998. LNCS (LNAI), vol. 1501, pp. 336–374. Springer, Heidelberg (1998)
18. Muggleton, S.: Inverse entailment and progol. New Generation Computing 13, 245–286 (1995)

19. Natarajan, S., Khot, T., Kersting, K., Gutmann, B., Shavlik, J.: Gradient-based boosting for statistical relational learning: The relational dependency network case. Mach. Learn. 86(1), 25–56 (2012)
20. Richardson, M., Domingos, P.: Markov logic networks. Mach. Learn. 62(1-2), 107–136 (2006)
21. Tadepalli, P., Kristian, K., Natarajan, S., Joshi, S., Shavlik, J.: Imitation learning in relational domains: a functional-gradient boosting approach. In: Proceedings of the Twenty-Second International Joint Conference on Artificial Intelligence, IJCAI 2011, vol. 2, pp. 1414–1420. AAAI Press (2011)
22. Specia, L., Srinivasan, A., Joshi, S., Ramakrishnan, G., das Graças Volpe Nunes, M.: An investigation into feature construction to assist word sense disambiguation. Machine Learning 76(1), 109–136 (2009)
23. Srinivasan, A.: The aleph manual (1999)
24. Srinivasan, A., Muggleton, S., Sternberg, M.J.E., King, R.D.: Theories for mutagenicity: A study in first-order and feature-based induction. Artif. Intell. 85(1-2), 277–299 (1996)
25. Srinivasan, A., Ramakrishnan, G.: Parameter screening and optimisation for ILP using designed experiments. Journal of Machine Learning Research 12, 627–662 (2011)
26. Taskar, B., Abbeel, P., Wong, M.-F., Koller, D.: Relational markov networks. In: Getoor, L., Taskar, B. (eds.) Introduction to Statistical Relational Learning. MIT Press (2007)

Learning Dishonesty

Chiaki Sakama

Department of Computer and Communication Sciences
Wakayama University, Sakaedani, Wakayama 640-8510, Japan
sakama@sys.wakayama-u.ac.jp

Abstract. Children behave dishonestly as a way of managing problems in daily life. Then our primary interest of this paper is how children learn dishonesty and how one could model human acquisition of dishonesty using machine learning techniques. We first observe the structural similarities between dishonest reasoning and induction, and then characterize mental processes of dishonest reasoning using logic programming. We argue how one develops behavioral rules for dishonest acts and refines them to more advanced rules.

1 Introduction

"Lori, did you draw on your wall?" her mother asked, obviously upset.
"No," Lori answered, completely straight-faced.
"Well, who did it?"
"It wasn't me, Mommy," she replied, still the innocent angel.
"Was it a little ghost?" her mother asked sarcastically.
"Yeah, yeah," Lori said. "It was a ghost."

– Paul Ekman, "Why kids lie" [5]

Children learn dishonesty in their early ages. According to studies in psychology [2,5], children lie by four years or earlier, mainly in order to avoid punishment. As they grow up, children use lies not only for protecting the self but for benefiting others. Victoria Talwar, who is an expert of children's lying behavior, says that "Lying is related to intelligence . . . lying demands both advanced cognitive development and social skills that honesty simply doesn't require" [2]. Thus, learning dishonesty, including lying, is the process of socialization for children. Very young children start lying to their parents. In the dialog between a little girl and her mother at the beginning of the section, Lori, a three-and-a-half-year-old girl, scribbled on her bedroom wall with her crayons. She knows who did it, but denies her act in reply to the question of her mother.

The story is illustrative of several aspects of children's dishonest behavior. First, consider the reason why the little girl chose to lie. If her mother cheerfully asks the little girl that "Oh, what's a wonderful picture! Lori, did you draw it?", then her reply might be different. However, Lori observed that her mother appears displeased with the drawing. She then lied to avoid punishment. By this and other typical cases, we can say that *children come to behave dishonestly to*

F. Riguzzi and F. Železný (Eds.): ILP 2012, LNAI 7842, pp. 225–240, 2013.

avoid a unwanted outcome or to have a wanted outcome. Second, the little girl believes that she scribbled on the wall while behaves in a way that contradicts her belief. To resolve the inner conflict, she has to eliminate the believed-true fact "I scribbled on the wall" and instead introduce the believed-false fact "I did not scribble on the wall". Third, the little girl first just denied her act "It wasn't me". This is a simple and direct lie. Lori had no idea to whom she could impute the responsibility for the drawing. After Mom's sarcastic question, Lori put responsibility on a ghost. Compared with the first lie, "It was a ghost" is an indirect lie which is devised to cover the truth. Such an indirect lie requires advanced skills of creating a story. The little girl could not make the story herself but more aged children can do themselves.

Learning dishonesty is a process of human development in which children become adults. The topic has been studied by a number of researchers in the field of developmental psychology of children. On the other hand, the issue has been almost completely ignored in artificial intelligence and machine learning. Since the goal of machine learning research includes constructing a formal model of human learning and realizing the model on computers, it is important and meaningful to investigate the mechanism of learning dishonesty by humans. Then our question is: *How could one model human acquisition of dishonesty using machine learning techniques in AI?* The purpose of this study is providing an abstract framework for learning dishonesty. In this paper, we consider three different categories of dishonest acts. A *lie* is a statement of a believed-false sentence [1]. *Bullshit* is a statement that is grounded neither in a belief that it is true nor, as a lie must be, in a belief that it is not true [6]. *Withholding information* is to fail to offer information that would help someone acquire true beliefs and/or correct false beliefs [4]. We first characterize dishonest reasoning in terms of induction. We then argue how one acquires behavioral rules for dishonest acts and develops them to more advanced rules. To the best of our knowledge, this is the first attempt to formulate the process of learning dishonesty by humans using machine learning techniques.

The rest of this paper is organized as follows. Section 2 argues a logical formulation of dishonest reasoning. Section 3 provides connection between dishonest reasoning and induction. Section 4 characterizes behavioral rules for dishonest agents. Section 5 discusses related issues and Section 6 summarizes the paper.

2   Dishonest Reasoning

Dishonest reasoning made by a little girl in the introduction is represented in propositional logic. Lori initially believes that she draws something on the wall:

$$draw\_on\_wall. \tag{1}$$

She finds that Mom appears displeased with the drawing:

$$draw\_on\_wall \supset displeased\_Mom. \tag{2}$$

She experimentally believes that if she does something which makes Mom displeased, then she is scolded by Mom:

$$displeased\_Mom \supset scolded. \tag{3}$$

She wants to avoid to be scolded, then Lori decides to tell a lie. The process of reasoning by the little girl is formally presented as follows. Lori believes (1)–(3) which are represented as the background knowledge K:

$$K = \{\, draw\_on\_wall,$$
$$draw\_on\_wall \supset displeased\_Mom,$$
$$displeased\_Mom \supset scolded \,\}.$$

Mom asks whether she made the drawing, and Lori derives the fact $draw\_on\_wall$ in K. However, K also derives $scolded$ which she wants to avoid. Then, Lori removes the truth (1) and instead introduces the falsehood:

$$\neg\, draw\_on\_wall, \tag{4}$$

which results in the knowledge base K':

$$K' = (K \setminus \{\, draw\_on\_wall \,\}) \cup \{\neg\, draw\_on\_wall\}.$$

As a result, $K' \models \neg\, draw\_on\_wall$, and Lori replies "It wasn't me" to Mom.

Dishonest reasoning illustrated above is generalized in the following way. Suppose that a reasoner (or an agent) has background knowledge K and an unwanted outcome G (called *negative outcome*). When $K \models G$, the negative outcome G is obtained by honest reasoning. In this case, the agent tries to block the derivation of G by introducing disbelieved sentences I and removing believed sentences J:

$$(K \setminus J) \cup I \not\models G. \tag{5}$$

By contrast, suppose that an agent has background knowledge K and a wanted outcome G (called *positive outcome*). When $K \not\models G$, the positive outcome G is not obtained by honest reasoning. In this case, the agent tries to derive G by introducing disbelieved sentences I and removing believed sentences J:

$$(K \setminus J) \cup I \models G. \tag{6}$$

Here, I and J are sets of formulas which fill the gap between the current belief K of an agent and an (un)wanted outcome G. Dishonest reasoning is considered a process of revising K (not) to derive G. At this point, we observe some structural similarities between the process of dishonest reasoning and the problem of *induction* in machine learning. In fact, viewing G as a positive (resp. negative) evidence, the problem of constructing a new knowledge base $(K \setminus J) \cup I$ in (6) (resp. (5)) is considered a process of introducing a new hypothesis I and discarding an old belief J. [1]

---

[1] In the normal ILP setting, a hypothesis is only added to K to explain G, while it is more general to consider removal of current belief from K as well to explain G. This is especially the case when K is a nonmonotonic theory [8].

Induction

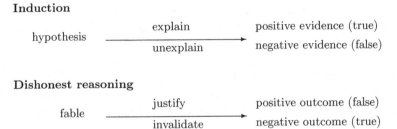

Fig. 1. induction vs. dishonest reasoning

An interesting contrast is that a positive evidence in induction is a true fact, and the goal is to construct a plausible hypotheses to explain it. On the other hand, a positive outcome in dishonest reasoning is a desired (and usually false) fact, and the goal is to invent a factitious story to justify it. Likewise, a negative evidence in induction is a false fact, and the goal is to construct a plausible hypotheses to unexplain it. On the other hand, a negative outcome in dishonest reasoning is an undesired (and usually true) fact, and the goal is to invent a factitious story to invalidate it. The correspondence between induction and dishonest reasoning is illustrated in Figure 1. Based on this intuition, we formulate dishonest reasoning as induction in the next section.

3 Dishonest Reasoning as Induction

An agent has a consistent propositional theory K as background knowledge. As usual, we identify a set of formulas with the conjunction of formulas included in the set. A consistent propositional formula is also called a *sentence*. A *positive outcome* is a sentence which an agent wants to obtain, while a *negative outcome* is a sentence which an agent wants to avoid.

Definition 3.1 (offensive/defensive dishonesty). Let K be background knowledge and G a positive outcome s.t. $K \not\models G$ (resp. a negative outcome s.t. $K \models G$). Suppose a pair (I, J) of sets of sentences satisfying the conditions:

1. $(K \setminus J) \cup I \models G$ (resp. $(K \setminus J) \cup I \not\models G$),
2. $(K \setminus J) \cup I$ is consistent.

Then,

(a) (I, J) (or simply I) is a *lie* for G if $I \neq \emptyset$ and $K \models \neg I$;
(b) (I, J) (or I) is *bullshit* (or BS) for G if $I \neq \emptyset$, $K \not\models \neg I$ and $K \not\models I$;
(c) (I, J) (or J) is *withholding information* (or WI) for G if $I = \emptyset$.

In each case, (I, J) is also called an *offensive dishonesty* for a positive outcome G (resp. a *defensive dishonesty* for a negative outcome G) wrt K.

In offensive dishonesty, a positive outcome G is not a consequence of a knowledge base K. To entail G, K is modified to a consistent theory $(K \setminus J) \cup I$ by removing a set J of believed sentences from K and introducing a set I of disbelieved sentences to K. In defensive dishonesty, on the other hand, a negative outcome G is a consequence of K. To invalidate G, K is modified to a consistent theory $(K \setminus J) \cup I$ that does not entail G.[2] In each case, (I, J) is called differently depending on its condition. I is called a lie if a nonempty set I is introduced to K and I is false in K. I is called bullshit if a nonempty set I is introduced to K and K has no belief wrt the truth nor falsity of I. J is called withholding information if I is empty (and hence, J is non-empty). Note that the entailment of a positive outcome (or non-entailment of a negative outcome) G from $(K \setminus J) \cup I$ does not mean the success of deceiving, however. Lying, BS, or WI is a dishonest act of an agent to attempt to mislead another agent. The success of these acts depends on whether the opponent believes information I (or disbelieves J) brought by the proponent. We do not argue the effect of a dishonest act of an agent that is produced in another agent. By the definition, we have the following properties.

Proposition 3.1. *Lies, BS and WI are mutually exclusive.*

Proposition 3.2. *If G is a positive outcome, then $G \not\equiv false$. If G is a negative outcome, then $G \not\equiv true$.*

Proof. If a positive outcome is $G \equiv false$ or a negative outcome is $G \equiv true$, the conditions 1 and 2 of Definition 3.1 are contradictory with each other.   □

Example 3.1. (a) Suppose the introductory example where Lori has the background knowledge:

$$K = \{\, draw\_on\_wall,$$
$$draw\_on\_wall \supset displeased\_Mom,$$
$$displeased\_Mom \supset scolded \,\}.$$

To avoid the negative outcome $G = scolded$, Lori introduces the falsehood $I = \{\neg draw\_on\_wall\}$ to K and eliminates the fact $J = \{draw\_on\_wall\}$. As a result, $(K \setminus J) \cup I$ does not entail G. In this case, I is a defensive lie. Lori also creates another story $I' = \{\, ghost\_did, \quad ghost\_did \supset \neg draw\_on\_wall \,\}$. As $K \models \neg (ghost\_did \wedge (ghost\_did \supset \neg draw\_on\_wall))$, I' is also a defensive lie.

(b) Suppose another story that a little boy Tom is playing a game on the iPad. Mom asks him if there is any email from Dad. Tom considers that his game will be interrupted if an email has been received, but he wants to keep playing anyway. The belief state of Tom is represented by the background knowledge:

$$K = \{\, mail \supset \neg playing, \quad \neg mail \supset playing \,\}.$$

---

[2] When K is a propositional theory, the introduction of I monotonically increases theorems and does not eliminate G. On the other hand, when K is a nonmonotonic theory, the introduction of I can eliminate G. Although we consider (monotonic) propositional theory here, we provide a general setting for defensive dishonesty.

To have the positive outcome $G = playing$, Tom introduces the unknown fact $I = \{\neg mail\}$ to K. As a result, $K \cup I$ entails G. In this case, I is offensive BS. (Tom replies that there is no email without checking the mailbox.)

In Example 3.1(a), withholding the fact $J = \{draw\_on\_wall\}$ from K also has the effect of making G underived $K \setminus J \not\models G$. This is defensive WI. This corresponds to the situation where Lori says nothing in response to mother's question. Some researchers consider that there is not much difference between saying something false and concealing the truth [5]. However, these two acts are generally considered different dishonest acts [4], so we also distinguish them.

We have already observed structural similarities between dishonest reasoning and induction in Section 2. We reformulate computation of offensive/defensive dishonesty as induction problems as follows.

Offensive Dishonesty as Induction

Given : background knowledge K as a consistent propositional theory, and a positive outcome G as a positive evidence such that $K \not\models G$,
Find : a pair of sentences (I, J) such that $(K \setminus J) \cup I \models G$ where $(K \setminus J) \cup I$ is consistent.

Defensive Dishonesty as Induction

Given : background knowledge K as a consistent propositional theory, and a negative outcome G as a negative evidence such that $K \models G$,
Find : a pair of sentences (I, J) such that $(K \setminus J) \cup I \not\models G$ where $(K \setminus J) \cup I$ is consistent.

In each problem, we call (I, J) a *dishonest solution* for G under K.

With this setting, we can use induction algorithms of propositional theories for computing offensive/defensive dishonesty.

Proposition 3.3. *Given background knowledge K and a positive (resp. negative) outcome G, suppose that an induction algorithm finds a dishonest solution (I, J) for G under K. Then, if $I \neq \emptyset$ and $K \models \neg I$, then (I, J) is a lie. Else if $I \neq \emptyset$, $K \not\models \neg I$ and $K \not\models I$, then (I, J) is BS. Else if $I = \emptyset$, then (I, J) is WI.*

Proof. The result directly follows by Definition 3.1. □

Proposition 3.4. *Given background knowledge K and a positive (resp. negative) outcome G, suppose that an induction algorithm finds a dishonest solution (I, J) for G under K. Let (I', J') be any pair such that $I \subseteq I'$ and $J \subseteq J'$. Then,*

1. *If (I, J) is a lie, $(K \setminus J') \cup I' \models G$ for a positive outcome G (resp. $(K \setminus J') \cup I' \not\models G$ for a negative outcome G), and $(K \setminus J') \cup I'$ is consistent, then (I', J') is also a lie.*
2. *If (I, J) is BS, $(K \setminus J') \cup I' \models G$ for a positive outcome G (resp. $(K \setminus J') \cup I' \not\models G$ for a negative outcome G), and $(K \setminus J') \cup I'$ is consistent, then (I', J') is either a lie or BS.*

3. *If $(\emptyset, J)$ is a WI, $K \not\models I'$, $(K \setminus J') \cup I' \models G$ for a positive outcome G (resp. $(K \setminus J') \cup I' \not\models G$ for a negative outcome G), and $(K \setminus J') \cup I'$ is consistent, then (I', J') is either a lie, BS or WI.*

Proof. (1) If (I, J) is a lie, $K \models \neg I$ and $I \subseteq I'$ imply $K \models \neg I'$. Then, (I', J') is also a lie. (2) If (I, J) is BS, $K \not\models I$ and $I \subseteq I'$ imply $K \not\models I'$. Then, (I', J') is a lie if $K \models \neg I'$, otherwise, it is BS. (3) If $(\emptyset, J)$ is WI, there are three cases. When $I' = \emptyset$, (I', J') is WI. Else when $I' \neq \emptyset$, (I', J') is a lie if $K \models \neg I'$; otherwise, if $K \not\models \neg I'$, $K \not\models I'$ implies that (I', J') is BS.   □

By Proposition 3.4, if a dishonest solution is computed, it could be extended for constructing another dishonest solution.

Example 3.2. In Example 3.1(a), a negative outcome G is proved in K. To compute (I, J) satisfying $(K \setminus J) \cup I \not\models G$, K is refined to K':

$$K' = \{\, p \supset draw\_on\_wall,$$
$$draw\_on\_wall \supset displeased\_Mom,$$
$$displeased\_Mom \supset scolded \,\}$$

where p is a newly introduced propositional sentence. Replacing $draw\_on\_wall$ by the sentence $p \supset draw\_on\_wall$ implies removing the fact $J = \{draw\_on\_wall\}$ from K, which is defensive WI $(\emptyset, \{draw\_on\_wall\})$. In this case, both $(\{\neg draw\_on\_wall\}, \{draw\_on\_wall\})$ and $(\{ghost\_did,\ ghost\_did \supset \neg draw\_on\_wall\}, \{draw\_on\_wall\})$ are defensive lies.

In Example 3.1(b), a positive outcome G is not proved in K. To compute (I, J) satisfying $(K \setminus J) \cup I \models G$, by inverting the entailment relation as $(K \setminus J) \cup \{\neg G\} \models \neg I$, we can obtain $I = \{\neg mail\}$, which is offensive BS.

The refinement technique presented above is used in [11,14], while the inverse entailment is used in [10].

4   Learning Dishonesty as Behavioral Rules

4.1   Behavioral Rules

Children who dishonestly get away with a problem, are likely to behave dishonestly again to cope with similar problems. In the introductory example, Lori lied to her Mom to avoid punishment. If Mom overlooks her act and does not scold the little girl, then Lori would lie again to avoid punishment in another situation. Thus, children begin to act dishonestly in particular situations, then learn dishonest acts as behavioral rules to manage difficulties.

In this section, we characterize the process of learning behavioral rules of dishonesty using logic programming. Let us review the mental process of dishonest reasoning by the little girl in Example 3.1(a). First, Lori believes that the negative outcome *scolded* is deduced in her background knowledge K. She believes that the negative outcome is caused by her act $draw\_on\_wall$. She then negates the fact and asserts $\neg draw\_on\_wall$. The act of defensive lying by the child is represented by the following meta-rule:

$$D\text{-}Lie(\neg draw\_on\_wall, scolded) \leftarrow neg(scolded),$$
$$prove(K, draw\_on\_wall \supset scolded), prove(K, draw\_on\_wall). \quad (7)$$

Here, $neg(G)$ means a negative outcome G, and $prove(K, F)$ holds iff $K \models F$. The rule (7) says if the negative outcome *scolded* is deduced by $draw\_on\_wall$ in K, then defensively lie on the sentence $\neg draw\_on\_wall$ to avoid *scolded*.

Another day, Lori watches TV in the dining room. Mom noticed a puddle on the floor under her feet and asked if she wets her pants. She denies the fact to avoid punishment. The belief state of Lori is represented by the background knowledge:

$$K' = \{ \ wet\_pants,$$
$$wet\_pants \supset displeased\_Mom,$$
$$displeased\_Mom \supset scolded \ \}.$$

To avoid the negative outcome $G = scolded$, Lori introduces the falsehood $I = \{\neg wet\_pants\}$ to K and eliminates the fact $J = \{wet\_pants\}$. Dishonest reasoning in this situation is represented by the meta-rule:

$$D\text{-}Lie(\neg wet\_pants, scolded) \leftarrow neg(scolded),$$
$$prove(K', wet\_pants \supset scolded), prove(K', wet\_pants). \quad (8)$$

Rules (7) and (8) represent two different situations for Lori not to be scolded. Using these rules, Lori can *induce* the new behavioral rule:

$$D\text{-}Lie(\neg F, scolded) \leftarrow neg(scolded), prove(K, F \supset scolded), prove(K, F) \quad (9)$$

where F is a variable representing any formula, or a more general rule:

$$D\text{-}Lie(\neg F, G) \leftarrow neg(G), prove(K, F \supset G), prove(K, F). \quad (10)$$

The rule (10) says if a negative outcome G is proved in background knowledge K using a believed-true sentence F, then lie on $\neg F$.[3] After obtaining such a behavioral rule, she can use the rule (10) to avoid negative outcomes in other circumstances. Suppose that a friend of Lori asks her to lend a toy. Lori does not want to lend it however. Then, she puts $neg(lend\_toy)$ and seeks a condition F which would deduce $lend\_toy$. If she finds the belief $have\_toy \wedge (have\_toy \supset lend\_toy)$ in her background knowledge, then she could lie on $\neg have\_toy$ ("I do not have it anymore"). Thus, once a behavioral rule is obtained by induction, it can be applied to individual situations by deduction.

A similar rule is obtained for offensive lying as

$$O\text{-}Lie(F, G) \leftarrow pos(G), not \ prove(K, G), prove(K, F \supset G), prove(K, \neg F)$$
$$(11)$$

---

[3] In (10), $prove(K, G)$ holds by $prove(K, F \supset G)$ and $prove(K, F)$.

where $pos(G)$ means a positive outcome G, and not is negation as failure to prove. The rule (11) says if a positive outcome G is not proved in background knowledge K but is proved using a believed-false sentence F, then lie on F. Rules are also constructed for offensive BS and defensive WI as follows:

$$O\text{-}BS(F,G) \leftarrow pos(G),\ not\ prove(K,G),\ prove(K, F \supset G),$$
$$not\ prove(K, \neg F). \tag{12}$$
$$D\text{-}WI(F,G) \leftarrow neg(G),\ prove(K, F \supset G),\ prove(K, F). \tag{13}$$

Some facts are observed on these rules.

- Putting $F \equiv G$ in (10)–(13), we can obtain simplified rules:

$$D\text{-}Lie(\neg G, G) \leftarrow neg(G),\ prove(K, G).$$
$$O\text{-}Lie(G, G) \leftarrow pos(G),\ not\ prove(K, G),\ prove(K, \neg G).$$
$$O\text{-}BS(G, G) \leftarrow pos(G),\ not\ prove(K, G),\ not\ prove(K, \neg G).$$
$$D\text{-}WI(G, G) \leftarrow neg(G),\ prove(K, G).$$

For instance, $D\text{-}Lie(\neg G, G)$ represents a defensive lie which just denies a negative outcome, and $O\text{-}Lie(G, G)$ represents an offensive lie which just asserts a positive outcome. These rules represent direct and unskillful lies. For instance, a child who has a homework but does not want to do it, lies "No homework".

- Comparing (11) and (12), $O\text{-}Lie(F, G)$ has the condition $prove(K, \neg F)$, while $O\text{-}BS(F, G)$ has the condition $not\ prove(K, \neg F)$. This means that a lier believes the falsehood of F, while a bullshitter has no belief on F.[4] A reasoner lies if $\neg F$ is proved in K, otherwise, he/she bullshits.

- Comparing (10) and (13), when a negative outcome G is proved using $F \supset G$ and F in K, a liar states $\neg F$ while a withholder just conceals F. That is, a reasoner can select one of the two dishonest acts under the same condition.

Note that we do not provide rules for defensive BS and offensive WI which do not happen in classical propositional theory. They would happen when background knowledge contains *nonmonotonic* rules. In defensive BS, $D\text{-}BS(F,G)$, a negative outcome G is proved in K in the absence of some formula F whose truth value is unknown. Then, a bullshitter states F to block the derivation of G. In offensive WI, $O\text{-}WI(F,G)$, on the other hand, a positive outcome G is not proved in K by the presence of a formula F. Then, a withholder conceals F to prove G. These two rules are useful when K is given as a nonmonotonic theory.

The meta-rules presented in this section instruct when to behave dishonestly. The process of taking each dishonest act is illustrated in Figure 2. In the figure, $A \rightarrow B$ means that if A holds then check B. We next consider rules for specifying how to behave dishonestly.

---

[4] In (12), $not\ prove(K, F)$ holds by $not\ prove(K, G)$ and $prove(K, F \supset G)$.

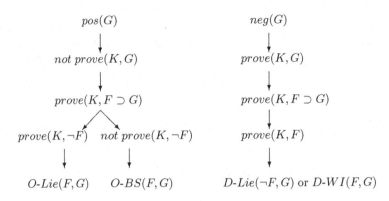

Fig. 2. Behavioral rules for dishonest reasoning

4.2 Selecting the Best Dishonest Act

In Example 3.1(a), to avoid the negative outcome $G = scolded$, Lori has two options for defensive lying: $I = \{\neg draw\_on\_wall\}$ and $I' = \{ghost\_did,\ ghost\_did \supset \neg draw\_on\_wall\}$. Then, which lie is better than the other? Comparing two lies, we can observe two facts. First, I' is stronger than I in the sense that I' implies I. Second, I' is more presumptive than I in the sense that I' requires an extra assumption $ghost\_did$. A liar normally wants to keep his/her lie as small as possible. This is because, lies make the belief state of a hearer deviate from the objective reality (or, at least from the reality as believed by a speaker) and a stronger lie would increase such deviation. This is undesirable for a speaker because it increases the chance of the lie being detected. Moreover, a smaller lie is less sinful than a bigger one. In [12], it is represented as the maxim: "*Never tell an unnecessarily strong lie*". Then, the rule of defensive lying (10) would be refined to the following rule:

$$D\text{-}Lie(\neg F, G) \leftarrow neg(G),\ prove(K,\ F \supset G),\ prove(K, F),$$
$$prove(K,\ \neg F \supset \neg H),\ not\ D\text{-}Lie(\neg H, G). \qquad (14)$$

The rule (14) says if a negative outcome G is proved in background knowledge K using F and there is no defensive lie $\neg H$ for G which is weaker than $\neg F$, then defensively lie on $\neg F$. Children would acquire such a rule by empirically learning that a stronger (or a bigger) lie is more detectable than a weaker (or a smaller) one.

Similarly, preference rules for $O\text{-}Lie$, $O\text{-}BS$, and $D\text{-}WI$ are constructed as follows.

$$O\text{-}Lie(F, G) \leftarrow pos(G),\ not\ prove(K, G),\ prove(K,\ F \supset G),$$
$$prove(K, \neg F),\ prove(K,\ F \supset H),\ not\ O\text{-}Lie(H, G). \qquad (15)$$
$$O\text{-}BS(F, G) \leftarrow pos(G),\ not\ prove(K, G),\ prove(K,\ F \supset G),$$
$$not\ prove(K, \neg F),\ prove(K,\ F \supset H),\ not\ O\text{-}BS(H, G). \qquad (16)$$

$$D\text{-}WI(F,G) \leftarrow neg(G), prove(K, F \supset G), prove(K,F),$$
$$prove(K, F \supset H), not\ D\text{-}WI(H,G). \qquad (17)$$

As mentioned in the previous section, one can choose $D\text{-}Lie$ (10) or $D\text{-}WI$ (13) to avoid a negative outcome. Then, a question is which dishonest act should one to choose? Suppose that Lori believes that both a lie and WI are effective to avoid punishment. In this case, WI is considered preferable to lies because WI introduces no disinformation. Hence, $WI = (\emptyset, \{draw\_on\_wall\})$ is preferred to lies $(\{\neg draw\_on\_wall\}, \{draw\_on\_wall\})$ or $(\{ghost\_did,\ ghost\_did \supset \neg draw\_on\_wall\}, \{draw\_on\_wall\})$. Such preference is represented as the maxim: "*Never lie if you can have your way by withholding information*". The rule (10) is then modified as

$$D\text{-}Lie(\neg F, G) \leftarrow neg(G), prove(K, F \supset G),$$
$$prove(K,F), not\ D\text{-}WI(F,G). \qquad (18)$$

The rule (18) represents that one selects the action $D\text{-}Lie(\neg F, G)$ if a negative outcome G is proved in background knowledge K and the action $D\text{-}WI(F,G)$ is not taken. Children would acquire such a preference by learning that providing false information is immoral and would lead to punishment when detected. Withholding information is also immoral in some cases, but is considered less sinful than lying. Such preferences are also considered between BS and lying. That is, BS is preferred to lying as the former provides unknown information, while the latter provides false information. In other words, "*Never lie if you can bullshit your way out of it*" [12].[5] The rule (11) is then modified as

$$O\text{-}Lie(F,G) \leftarrow pos(G), not\ prove(K,G),$$
$$prove(K, F \supset G), not\ O\text{-}BS(F,G). \qquad (19)$$

The rule (19) represents that one selects the action $O\text{-}Lie(F,G)$ if a positive outcome G is not proved in background knowledge K and the action $O\text{-}BS$ is not taken. That is, if *not prove*$(K, \neg F)$ holds then take the action of $O\text{-}BS(F,G)$, otherwise, take the action of $O\text{-}Lie(F,G)$.

The rules (14)–(17) represent qualitative guidelines for selecting effective dishonest acts, while the rules (18) and (19) represent quantitative guidelines for selecting best dishonest acts.

4.3   Reasoning about the Belief State of a Hearer

As they grow up, children become more skilled and careful in acting dishonestly. Paul Ekman says: "A successful liar considers the perspective of the target being lied to. Taking the role of the other person, considering what will seem credible or suspicious to that person, allows the liar to consider the impact of his own behavior on the target and to fine-tune and adjust his behavior accordingly. ...

---

[5] A similar imperative is mentioned in [6].

Preschoolers aren't very good at this because at such early ages children don't realize that there is more than one perspective—theirs—on an event. They think everyone thinks the way they do. As they move toward adolescence, kids become much more able to put themselves in someone else's shoes" [5].

The behavioral rules in Section 4.1 describe when to behave dishonestly based on the speaker's belief base, while they do not take into account of the belief of the hearer. In Ekman's viewpoint, those rules characterize a very early-stage of dishonest acts by children in which they identify their own belief with the hearer's one. We then consider how to develop those behavioral rules in a way which distinguishes a hearer's belief base from the speaker's one and reasons about the belief state of a hearer.

Example 4.1. Suppose a school child, John, got a bad score at an exam. Mom asks whether he did well on the exam. John then considers: if Mom knows that he got a bad score, then Mom will make him study hard. To this end, Mom may restrict him from watching TV. However, John does not want this restriction. The belief state of John is represented by

$$K_O = \{ \ bad\_score \ \},$$
$$K_S = \{ \ bad\_score \ \supset \ study\_hard,$$
$$study\_hard \ \supset \ restrict\_TV \ \}$$

where K_O represents objective facts, while K_S represents subjective beliefs with respect to the belief state of Mom. If John informs Mom of the objective fact *bad_score*, then $K_S \cup \{bad\_score\}$ derives *restrict_TV*. To avoid this, he would not inform Mom of the truth.

Dishonest reasoning by this school boy is more advanced than the one by the little girl Lori. John distinguishes his belief on the objective fact (K_O) and the subjective belief on the belief state of Mom (K_S). In his objective belief K_O, John believes that he got a bad score. He then surmises the belief state of Mom in his subjective belief K_S and conjectures that Mom will restrict his watching TV if she knows the score. As a result, John would make a decision to act dishonestly. Definition 3.1 of offensive/defensive dishonesty is modified for such an advanced reasoner as follows. Let K_O be a knowledge base representing an agent's belief on the objective fact, and let K_S be a knowledge base representing an agent's subjective belief on the belief state of another agent. Suppose a positive outcome G s.t. $K_O \cup K_S \not\models G$ (resp. a negative outcome s.t. $K_O \cup K_S \models G$). Then, a pair (I, J) of sets of sentences satisfying the conditions:

1. $(K_O \setminus J) \cup K_S \cup I \models G$ (resp. $(K_O \setminus J) \cup K_S \cup I \not\models G$),
2. $(K_O \setminus J) \cup K_S \cup I$ is consistent

is called an *offensive dishonesty* for a positive outcome G (resp. a *defensive dishonesty* for a negative outcome G) wrt $K_O \cup K_S$. With this modified definition, the results of Section 3 hold as they are by identifying K with $K_O \cup K_S$.

The behavioral rules of such an advanced reasoner take the mental state of a hearer into consideration and will become more sophisticated than those presented in Section 4.1. Suppose that there are two agents, a speaker a and a hearer b. To distinguish objective facts and subjective beliefs in the speaker's background knowledge K_a, we represent objective facts by sentences and subjective beliefs on the belief state of the hearer by $believe(b, F)$ which means "b believes F". With this setting, the behavioral rule for defensive lying is represented by the following rule:

$$D\text{-}Lie(\neg F, G) \leftarrow neg(G), prove(K_a, believe(b, F \supset G)), prove(K_a, F). \quad (20)$$

The rule (20) says if a believes that the believed-true sentence F leads b to believe a negative outcome G, then a defensively lies to b on $\neg F$. Comparing (10) and (20), $prove(K, F \supset G)$ in (10) is replaced by $prove(K_a, believe(b, F \supset G))$ in (20). This reflects the change that a speaker a considers the belief state of a hearer b and reasons by the hearer's viewpoint. Similarly, the behavioral rule of offensive lying is represented as the rule:

$$O\text{-}Lie(F, G) \leftarrow pos(G), not\, prove(K_a, believe(b, G)),$$
$$prove(K_a, believe(b, F \supset G)), prove(K_a, \neg F). \quad (21)$$

The rule (21) says if a disbelieves that b believes a positive outcome G, and a believes that the believed-false sentence F leads b to believe G, then a offensively lies to b on F. Comparing (11) and (21), $not\, prove(K, G)$ in (11) is replaced by $not\, prove(K_a, believe(b, G))$ in (21), and $prove(K, F \supset G)$ in (11) is replaced by $prove(K_a, believe(b, F \supset G))$ in (21). Behavioral rules are also constructed for offensive BS and defensive WI as follows:[6]

$$O\text{-}BS(F, G) \leftarrow pos(G), not\, prove(K_a, believe(b, G)),$$
$$prove(K_a, believe(b, F \supset G)), not\, prove(K_a, F), not\, prove(K_a, \neg F). (22)$$
$$D\text{-}WI(F, G) \leftarrow neg(G), prove(K_a, believe(b, F \supset G)), prove(K_a, F). \quad (23)$$

As mentioned by Ekman, very young children do not distinguish between their own belief and others' belief. Then, they identify $prove(K_a, believe(b, F))$ with $prove(K_a, F)$. With this identification, the rules (21)–(23) reduce to those rules (11)–(13) of Section 4.1. Children come to realize that different people have different belief bases and another person does not always think the way they do. To achieve their own goals, children need to reason about belief of the hearer as done in this section and will acquire advanced rules for dishonest acts.

5   Discussion

5.1   Scientific Hypotheses and Dishonesty

In Section 2 we argued structural similarities between dishonest reasoning and induction. Those similarities are not a coincidence because hypotheses are

---

[6] In (22) "$not\, prove(K_a, F)$" represents that the agent a has no belief on the truth of F. This is in contrast to (12) in which this fact is implicitly represented in the rule (cf. footnote 4).

sentences which have no logical reason to believe. Scientific discoveries have been achieved by inventing hypotheses. However, history shows that a number of hypotheses turned out false or incorrect—Aristotle's theory of elements, Ptolemaic model of the universe, and Dalton's atomic theory, to name a few. Are those philosophers or scientists are liars? We could answer negatively to this question because it is unlikely that they believed the falsity of their own hypotheses. If a person states a believed-true sentence, which is in fact false, the person is not lying. Saint Augustine, who was a Berber philosopher and theologian, says: "a person is to be judged as lying or not lying according to the intention of his own mind, not according to the truth or falsity of the matter itself" [1, p.55].[7]

If they are not liars, then are they bullshitters? To answer this question, consider how scientists build hypotheses. Carl G. Hempel says: "Perhaps there are no objective criteria of confirmation; perhaps the decision as to whether a given hypothesis is acceptable in the light of a given body of evidence is no more subject to rational, objective rules than is the process of inventing a scientific hypothesis or theory; perhaps, in the last analysis, it is a "sense of evidence", or feeling of plausibility in view of the relevant data, which ultimately decides whether a hypothesis is scientifically acceptable. ... it is often deceptive, and can certainly serve neither as a necessary nor as a sufficient condition for the soundness of the given assertion" [7]. The truth of any hypothesis is unknown; if it turns true/false, it is not a hypothesis anymore. In this respect, one could argue that "They are bullshit, in the precise sense that we cannot prove them to be true, as they are things we have to *assume* so we *can* prove things to be (provisionally) true" [3]. Nevertheless, we consider that scientific hypotheses are distinguished from bullshit. Generally speaking, scientists are kind of people who are interested in the truth of the world. Then, they build hypotheses to account for observed data in the world. By contrast, a bullshitter "does not care whether the things he says describe reality correctly" [6]. Harry G. Frankfurt says: "It is just this lack of connection to a concern with truth—this indifference to how things really are—that I regard as of the essence of bullshit" [6]. As such, there is no aspect of truth-seeking in bullshit. This is the difference between scientific hypotheses and bullshit of dishonest reasoners.

5.2   Related Work

Sakama *et al.* [12] have introduced a formal model of dishonesty. The study formulate lies, bullshit and deception using a propositional multimodal logic, and compare their formal properties. In his continuous study, Sakama [13] characterizes dishonest reasoning using *extended abduction* [8] and provides a method of computing dishonest reasoning in terms of abductive logic programming. In [13] the notion of *logic programs with disinformation* (LPD) is introduced, which is

---

[7] It is interesting to note that young children consider lies differently. "Up until about eight years of age, children consider any false statement a lie, regardless of whether the person who said it knew it was false. Intention is not the issue—only whether information is false or true" [5].

defined as a pair $\langle K, \mathcal{D} \rangle$ where K is a program representing believed-true information and $\mathcal{D}$ is a set of disinformation representing believed-false or disbelieved-true facts. Then, lies, BS and WI are characterized by introducing facts from $\mathcal{D}$ and removing facts from K. The logical framework of [13] is similar to Definition 3.1 of this paper, but there is an important difference. An LPD assumes the pre-specified set $\mathcal{D}$ of ground literals and selects appropriate disinformation from this set for dishonest reasoning. This is also the case in abductive logic programming where an abductive logic program is given as a pair $\langle K, \mathcal{A} \rangle$ where $\mathcal{A}$ is a set of *abducibles*. The paper [13] restricts possible disinformation to the ground facts in $\mathcal{D}$, which enables to compute dishonest acts of agents using abductive logic programming. The logical framework of this paper relaxes the condition and does not prepare possible disinformation in advance. We use induction instead of abduction which makes it possible to fabricate a story as a set of sentences like $I' = \{ ghost\_did, \; ghost\_did \supset \neg draw\_on\_wall \}$ of Example 3.1. We show possible computation of dishonest reasoning using existing induction algorithms (cf. Example 3.2), while appropriate inductive bias is needed for computing effective dishonest solutions. The paper [13] does not study how an agent develops behavioral rules of dishonesty nor how one refines those rules to more advanced ones.

To the best of our knowledge, there is no study which attempts to formulate the process of human acquisition of dishonesty. On the other hand, a recent study [9] simulates evolution and natural selection in robot learning. In this study, a group of robots have the task of finding a food source in a field. Once a robot found the food, it stays nearby and emits blue light, which informs other robots of the location and results in overcrowding around the food. After a few generations, robots become more secretive and learn to conceal food information for their own survival. The study shows the possibility of designing artificial agents which would acquire dishonest attitudes in their environment.

6   Conclusion

This paper focuses on the problem of children's learning to be dishonest. We related the process of fabricating a story in dishonest reasoning to the process of building a plausible hypothesis in induction. With this correspondence, there is a possibility of using induction algorithms for implementing learning dishonesty. We have also provided behavioral rules of dishonest reasoners which characterize when and how they would act dishonestly. Although those rules might represent the very early stage of dishonest acts by young children, those rules could be refined and sophisticated as argued in this paper. We also recognize that successful dishonest acts require various contingency plans including consideration of a plot to cover the truth and make a story credible, evaluation of the positive/negative effects of a dishonest act, preparation of responses to suspicious questions, etc. Such sophisticated planning matures with age, and children develop the ability as they grow older. The present paper focuses on the very early stage of learning dishonest acts and does not discuss the process of acquiring the ability of sophisticated planning. The issue is related to planning by dishonest agents and left for future research. This paper targets an issue that has not been thoroughly addressed from an AI point of view and the present study is at an early stage of the research. Further study is needed for development of the proposed model and its evaluation in reality.

References

1. Augustine: Lying. In: Treatises on Various Subjects, Fathers of the Church, vol. 56, pp. 45–110 (1952)
2. Bronson, P.: Learn to lie. New York Magazine (February 2008)
3. Brown, R.G.: Why science (natural philosophy) is bullshit. In: Axioms (2012), http://www.phy.duke.edu/~rgb/Philosophy/axioms/axioms/node44.html
4. Carson, T.L.: Lying and deception: theory and practice. Oxford University Press (2010)
5. Ekman, P.: Why kids lie: how parents can encourage truthfulness. Scribner (1989)
6. Frankfurt, H.G.: On Bullshit. Princeton University Press (2005)
7. Hempel, C.G.: Studies in the logic of confirmation. In: Brody, B.A. (ed.) Readings in the Philosophy of Science. McGraw-Hill (1970)
8. Inoue, K., Sakama, C.: Abductive framework for nonmonotonic theory change. In: Proc. 14th International Joint Conference on Artificial Intelligence (IJCAI), pp. 204–210 (1995)
9. Mitri, S., Floreano, D., Keller, L.: The evolution of information suppression in communicating robots with conflicting interests. Proc. National Academy of Sciences 106(37), 15786–15790 (2009)
10. Muggleton, S.: Inverse entailment and Progol. New Generation Computing 13, 245–286 (1995)
11. Quinlan, R.: Learning logical definitions from relations. Machine Learning 5, 239–266 (1990)
12. Sakama, C., Caminada, M., Herzig, A.: A logical account of lying. In: Janhunen, T., Niemelä, I. (eds.) JELIA 2010. LNCS, vol. 6341, pp. 286–299. Springer, Heidelberg (2010)
13. Sakama, C.: Dishonest reasoning by abduction. In: Proc. 22nd International Joint Conference on Artificial Intelligence (IJCAI), pp. 1063–1068 (2011)
14. Shapiro, E.Y.: Algorithmic Program Debugging. MIT Press, Cambridge (1983)

Heuristic Inverse Subsumption
in Full-Clausal Theories

Yoshitaka Yamamoto[1], Katsumi Inoue[2], and Koji Iwanuma[1]

[1] University of Yamanashi
4-3-11 Takeda, Kofu-shi, Yamanashi 400-8510, Japan
[2] National Institute of Informatics
2-1-2 Hitotsubashi, Chiyoda-ku, Tokyo 101-8430, Japan

Abstract. Inverse entailment (IE) is a fundamental approach for hypothesis finding in explanatory induction. Some IE systems can find any hypothesis in full-clausal theories, but need several non-deterministic operators that treat the inverse relation of entailment. In contrast, inverse subsumption (IS) is an alternative approach for finding hypotheses with the inverse relation of subsumption. Recently, it has been shown that IE can be logically reduced into a new form of IS, provided that it ensures the completeness of finding hypotheses in full-clausal theories. On the other hand, it is still open to clarify how the complete IS works well in the practical point of view. For the analysis, we have implemented it with heuristic lattice search techniques used in the state-of-the-art ILP systems. This paper first sketches our IS system, and then shows its experimental result suggesting that the complete IS can practically find better hypotheses with high predictive accuracies.

Keywords: explanatory induction, inverse entailment, inverse subsumption, heuristic lattice search.

1 Introduction

Given a background theory B and examples E, the task of *explanatory induction* [2,3,4,21] is to seek for a consistent hypothesis H such that $B \wedge H \models E$. Based on its equivalent relation $B \wedge \neg E \models \neg H$, H is derivable from the input theories B and E. This approach is called *inverse entailment* (IE) [4,7].

IE consists of two procedures: construction of a *bridge theory* F and generalization of F into H. Note that F is an intermediate theory satisfying the condition $B \wedge \neg E \models F$ and $F \models \neg H$. Once a bridge theory F is constructed, H is generated via the theory F based on the *generalization relation* $\neg F \dashv H$.

Every IE-based method [4,6,7,10,13,20] can be divided into two kinds in accordance with the generalization relation used in it. On the one hand, systems like CF-induction [4] use the inverse relation of entailment, called *anti-entailment*. They can find any hypothesis H such that $\neg F \models H$, but need several non-deterministic operators like *inverse resolution* [8]. On the other hand, systems like Progol [7,20], HAIL [10], Imparo [6] and Residure procedure [13] generate

F. Riguzzi and F. Železný (Eds.): ILP 2012, LNAI 7842, pp. 241–256, 2013.

H with the inverse relation of subsumption $\neg F \preceq H$, called *anti-subsumption*. In the following, we call this restricted approach *inverse subsumption* (IS) [18] to distinguish it from the general IE. IS methods focus the search space on the subsumption lattice bounded by $\neg F$ due to the computational efficiency, but may fail to generate relevant hypotheses, unlike complete IE methods.

For this trade-off between IE and IS, it has been shown that anti-entailment $\neg F \models H$ can be logically reduced to anti-subsumption $F^* \preceq H$ [18]. Note that F^* is an alternative theory logically equivalent to the original $\neg F$. This logical reduction means that IS can become a complete explanatory procedure in full-clausal theories only by replacing $\neg F$ with F^*. Indeed, CF-induction, which was originally an IE procedure, has been recently reconstructed into a complete IS one, and then evaluated by statistically characterizing the obtained hypotheses [19]. However, it is still unclear how the complete IS works well in practice, because the previous works [18,19] do not mention how to systematically search the subsumption lattice in full-clausal theories.

In this paper, we provide a heuristic IS algorithm for empirical evaluations, which is achieved with the notion of *bottom generalization* (BG) used in Progol-like ILP systems [7,14,20]. BG includes three key techniques to restrict the search space and efficiently search the subsumption lattice. The first is a language bias, called a *mode declaration*, which syntactically describes the target hypotheses. Secondly, BG uses a *heuristic function* that evaluates each hypothesis in accordance with its coverage of examples and description length. Finally, BG performs heuristic search, called A^*-*like* algorithm, which searches the subsumption lattice with the heuristic function in the best-first search manner.

These BG techniques are originally used for finding hypotheses in Horn theories. In turn, they are extended so as to be applicable for finding hypotheses in full-clausal theories. We have implemented them embedded into the complete IS method. In this paper, we first sketch our IS system in brief, and then show its experimental result suggesting that the complete IS method can generate better hypotheses which are beyond reach for the previous IS methods.

The rest of this paper is organized as follows. Section 2 introduces the background in this paper and reviews the previously proposed IE and IS methods. Section 3 describes the heuristic search algorithm used in the complete IS method. Section 4 shows experimental results and Section 5 discusses about the issue on scalability of the proposed system. Section 6 concludes.

2 Background

2.1 Notion and Terminologies

Here, we review the notion and terminology in ILP [21]. A *clause* is a finite disjunction of literals which is often identified with the set of its disjuncts. A clause $\{A_1, \ldots, A_n, \neg B_1, \ldots, \neg B_m\}$, where each A_i, B_j is an atom, is also written as $B_1 \wedge \cdots \wedge B_m \supset A_1 \vee \cdots \vee A_n$. A *Horn* clause is a clause which contains at most one positive literal; otherwise it is a *non-Horn* clause. It is known that a clause is *tautology* if it has two complementary literals. The *empty* clause, denoted by

$\square$, is the clause that contains no literal. A *clausal theory* is a finite set of clauses. A clausal theory is *full* if it contains at least one non-Horn clause. A *conjunctive normal form* (CNF) formula is a conjunction of clauses, and *disjunctive normal form* (DNF) is a disjunction of conjunctions of literals. Note that a CNF formula is written as the set of clauses in it for the simplicity.

Let C and D be two clauses. C *subsumes* D, denoted by $C \succeq D$, if there is a substitution θ such that $C\theta \subseteq D$. C *properly* subsumes D if $C \succeq D$ but $D \not\succeq C$. For a clausal theory S, μS denotes the set of clauses in S not properly subsumed by any clause in S. Let S and T be clausal theories. S *(theory-)subsumes* T, denoted by $S \succeq T$, if for every $D \in T$, there is a clause $C \in S$ such that $C \succeq D$.

When S is a clausal theory, the complement of S, denoted by $\overline{S}$, is defined as a clausal theory obtained by translating $\neg S$ into CNF using a standard translation procedure [21]. (In brief, $\overline{S}$ is obtained by converting $\neg S$ into prenex conjunctive normal form with standard equivalence-preserving operations and skolemizing it.) The complement $\overline{S}$ may contain redundant clauses like tautologies or subsumed ones. Especially, we call the clausal theory consisting of the non-tautological clauses in $\mu\overline{S}$ the *minimal complement* of S, denoted by $M(S)$.

We give the definition of *hypotheses* in the logical setting of ILP as follows:

Definition 1 (Hypotheses). Let B and E be clausal theories, representing a background theory and (positive) examples, respectively. Then H is a *hypothesis wrt B and E* if and only if H is a clausal theory such that $B \wedge H \models E$ and $B \wedge H$ is consistent. We simply call it a "hypothesis" if no confusion arises.

Example 1. Let B and E be as follows:

$$B = \{buy(john, diaper) \vee buy(john, beer)\}, \ E = \{shopping(john, at\_night)\}.$$

Then, the following theory H is a hypothesis wrt B and E, since $B \wedge H \models E$ and $B \wedge H$ is consistent.

$$H = \{buy(X, diaper) \supset buy(X, beer), \ buy(Y, beer) \supset shopping(Y, at\_night)\}.$$

Note that B contains one non-Horn clause, and thus requires full-clausal representation formalisms to treat this example.

Inverse entailment (IE) is one of the most fundamental approaches to finding hypotheses in Definition 1, which is used in many modern ILP methods [4,6,7,10,13,20]. In general, they find a hypothesis H in two steps. First, it constructs an intermediate theory F, called a *bridge theory*, and then generalizes F into a hypothesis H such that $\neg F =| H$, where $=|$ is called *generalization relation*. The bridge theory is formally defined as follows:

Definition 2 (Bridge theories). Let B and E be a background theory and examples. A ground clausal theory F is a *bridge theory* wrt B and E if $B \wedge \neg E \models F$ holds. If no confusion arises, we simply call it a "bridge theory".

Every IE-based method constructs F and generalizes $\neg F$ into H in its own way, which is summarized in Table 1.

Whereas Progol, HAILs and Imparo directly compute the *negation* of their own bridge theory, Residue procedure and CF-induction first construct a bridge theory and then compute its negation. We call the former (*resp.* the latter) the *direct* (*resp. indirect*) approach. The indirect approach requires the (non-monotone) *dualization* to translate the negation of the bridge theory (DNF) into CNF. The computational cost becomes expensive as the input size (i.e., the number of clauses in the bridge theory) increases. On the other hand, the direct approach can become incomplete for finding hypotheses[1].

In terms of the generalization relation, it is classified with two types: the one with anti-subsumption and the other with anti-entailment. The latter IE methods like CF-induction preserve the completeness of finding hypotheses, though it has to deal with many highly non-deterministic generalization operators based on anti-entailment, like inverse resolution. On the other hand, the former IS methods can focus the search space on the subsumption lattice that is more tractable than the latter based on entailment, though it can fail to find a relevant hypothesis worth considering due to its incompleteness.

Example 2. Recall Example 1. Now, we show how the target hypothesis is obtained using CF-induction that is the complete IE procedure. Let the bridge theory F be $B \wedge \neg E$ itself as follows:

$$F = \{ \ buy(john, diaper) \vee buy(john, beer), \ \neg shopping(john, at\_night) \ \}.$$

Note that F satisfies the condition of bridge theories of CF-induction, since F consists of instances from consequences of $B \wedge \neg E$. Next, we compute $M(F)$ by translating the DNF formula[2] of $\neg F$ into CNF as follows:

$$M(F) = \{ \ buy(john, diaper) \supset shopping(john, at\_night),$$
$$buy(john, beer) \supset shopping(john, at\_night) \ \}.$$

We notice that the target hypothesis H entails $M(F)$, but does not subsume it. This fact shows the incompleteness of IS methods that use anti-subsumption for generalizing $M(F)$. On the other hand, generalization with anti-entailment can derive H, whereas it requires inverse resolution which enables to find any parent clause of an input clause.

After constructing some F, IE methods generate H with anti-entailment $M(F) \models H$. However, they need to treat highly non-deterministic generalization operators that cause a huge search space of IE.

---

[1] To be exact, it is still an open problem whether or not Imparo ensures the completeness, though it has been shown that the other direct ones are incomplete [18].

[2] The DNF formula is easily obtained by De Morgan's laws interchanging the logical connections.

Table 1. Previously proposed IE-based methods

	Method	Formalism	Language	Bridge theory	Generalization relation	Valid ness
IS	Progol [7,20]	Horn	Mode declaration	Bottom clause (*)	Anti- subsumption	sound
	HAIL [10]	Horn	Mode declaration	Kernel set (*)	Anti- subsumption	sound
	FC-HAIL [11]	Full	Mode declaration	Kernel set (*)	Anti- subsumption	sound
	Imparo [6]	Horn	Mode declaration	Connection Theory (*)	Anti- subsumption	sound
	Residue Procedure [13]	Full	Nothing	Instances from $B \wedge \neg E$	Anti- subsumption	sound
IE	CF- induction [4,17]	Full	Production field	Instances from consequences of $B \wedge \neg E$	Anti- entailment	sound and complete

The above (*) label denotes the *negation* of the bridge theory for the direct approach.

2.2 Inverse Subsumption with Minimal Complements

For this problem, it has been shown that IE can be logically reduced into a complete IS form. This reduction is based on the notion of *induction fields*.

Definition 3 (Induction fields). An *induction field*, denoted by $\mathcal{I}_{\mathcal{H}} = \langle \mathbf{L} \rangle$, where $\mathbf{L}$ is a finite set of literals to appear in ground hypotheses. A clause C *belongs to* $\mathcal{I}_{\mathcal{H}}$ if there is a ground instance C_g such that every literal in C_g is included in $\mathbf{L}$. A clausal theory *belongs to* $\mathcal{I}_{\mathcal{H}}$ if every clause in it belongs to $\mathcal{I}_{\mathcal{H}}$. Given an induction field $\mathcal{I}_{\mathcal{H}} = \langle \mathbf{L} \rangle$, $Taut(\mathcal{I}_{\mathcal{H}})$ is defined as the set of tautologies $\{\neg A \vee A \mid A \in \mathbf{L} \text{ and } \neg A \in \mathbf{L}\}$.

Definition 4 (Hypotheses wrt $\mathcal{I}_{\mathcal{H}}$ and F). Let H be a hypothesis. H is a *hypothesis* wrt $\mathcal{I}_{\mathcal{H}}$ and F if there is a ground hypothesis H_g such that H_g consists of instances from H, $F \models \neg H_g$ and H_g belongs to $\mathcal{I}_{\mathcal{H}}$.

Theorem 1. [18] Let H be any hypothesis wrt $\mathcal{I}_{\mathcal{H}}$ and F. Then, it holds that

$$H \succeq M(F \cup Taut(\mathcal{I}_{\mathcal{H}})).$$

Theorem 1 shows that inverse subsumption (IS) can derive any hypothesis H from $M(F \cup Taut(\mathcal{I}_{\mathcal{H}}))$. This is the minimal complement of the theory obtained only by adding tautologies associated with $\mathcal{I}_{\mathcal{H}}$ to the original bridge theory F. In the following, we call $M(F \cup Taut(\mathcal{I}_{\mathcal{H}}))$ the *bottom theory* wrt F and $\mathcal{I}_{\mathcal{H}}$.

Example 3. Recall Example 2. Let an induction field $\mathcal{I}_{\mathcal{H}}$ be as follows:

$$\mathcal{I}_{\mathcal{H}} = \langle \{ \ \neg buy(john, diaper), \ buy(john, beer),$$
$$\neg buy(john, beer), \ shopping(john, at\_night) \ \} \rangle.$$

Suppose that the bridge theory F wrt B and E is set to $B \wedge \neg E$. Then, the target hypothesis H is a hypothesis wrt $\mathcal{I}_{\mathcal{H}}$ and F. Note here that $Taut(\mathcal{I}_{\mathcal{H}})$ contains one tautology: $buy(john, beer) \vee \neg buy(john, beer)$. The bottom theory wrt F and $\mathcal{I}_{\mathcal{H}}$ is as follows:

$$\{ \boxed{\neg buy(john, diaper) \vee buy(john, beer)} \vee shopping(john, at\_night),$$
$$\boxed{\neg buy(john, beer) \vee shopping(john, at\_night)} \}.$$

We notice that H subsumes this bottom theory (i.e., the dotted surrounding part), whereas it does not subsume the original $M(F)$.

As shown in Example 3, the complete IS method can generate better hypotheses which are beyond reach for the previous IS methods. On the other hand, it has not been clarified yet how this new IS actually works well in the practical point of view. For empirical analysis, we need to construct an efficient algorithm to search the subsumption lattice bounded by the bottom theory.

3   Heuristic Inverse Subsumption

We propose a lattice search algorithm for the complete IS method, which is based on the notion of *bottom generalization* (BG), BG is the key technique for efficiently finding a hypothesis in several state-of-the-art ILP systems like Progol and Aleph [7,14,20]. BG uses a language bias, called a *mode declaration*, syntactically describing the target hypotheses in order to reduce the search space. Each hypothesis is evaluated with a *heuristic function* determined by the description length of the hypothesis and its coverage of examples. BG then searches the subsumption lattice for one relevant hypothesis in the best-first search manner.

However, it is not straightforward to directly introduce BG to the complete IS method due to some limitations on its search strategy. Now, we extend the original BG so as to be suitable for the complete IS method.

3.1   Full-clausal Mode Language

We first recall the definition of mode declarations in Progol-like ILP systems.

Definition 5 (Mode declarations [7,20]). A mode declaration has either the form $modeh(n, atom)$ or $modeb(n, atom)$ where $n \geq 1$, or '$*$' and $atom$ is a ground atom. Terms in the atom are either normal or place-marker. A normal term is either a constant or a function symbol followed by a bracketed tuple of terms. A place-marker is either +type, -type or #type, where type is a constant. If m is a mode declaration, then $a(m)$ denotes an atom of m with the place-markers of +type and -type (*resp.* #type) replaced by distinct variables (*resp.* some constants).

Although the mode declarations are originally used in Horn representation formalisms, they have been lifted to describe full-clausal theories in [11] as follows:

Definition 6 (Full-clausal mode language $\mathcal{L}(M)$). Let M be a set of mode declarations. A clause $\{A_1, \ldots, A_n, \neg B_1, \ldots, \neg B_m\}$ *belongs to* a full-clausal mode language $\mathcal{L}(M)$ if each A_i (*resp.* B_j) is an atom $a(m)$ for some modeh (*resp.* modeb) declaration m in M. A clausal theory *belongs to* $\mathcal{L}(M)$ if every clause in it belongs to $\mathcal{L}(M)$.

We then introduce the following notion for describing the connectivity between variables that is commonly used in BG methods.

Definition 7 (Complete clauses). Let $\mathcal{L}(M)$ be a full-clausal mode language and C be a clause $\{A_1, \ldots, A_n, \neg B_1, \ldots, \neg B_m\}$ belonging to $\mathcal{L}(M)$. A variable v in C is *connected* wrt $\mathcal{L}(M)$ if v satisfies the following conditions:

- If v corresponds to the place-marker +type in some B_j, then v has the same variable either of +type in some A_i or of -type in some B_k ($k \neq j$).
- If v corresponds to the place-marker -type in some A_i, then v has the same variable of -type in some B_j.

C is *complete* wrt $\mathcal{L}(M)$ if every v is connected wrt $\mathcal{L}(M)$; otherwise *incomplete*. We denote by $incvar(C)$ the number of variables that are not connected in C.

Example 4. Let M be a set of mode declarations as follows:

$$M = \{modeh(1, buy(+man, \#item)), \; modeh(1, shopping(+man, \#date)),$$
$$modeb(1, buy(+man, \#item))\}.$$

Note that the value of first argument of each mode declaration m means the maximum number of $a(m)$ to be included in a hypothesis. Hence, for each m in the above, $a(m)$ appears at most once in the hypothesis.

We recall the hypothesis H in Example 1. Then, both clauses of H are complete wrt $\mathcal{L}(M)$. In contrast, the clause $C = buy(X, diaper) \supset buy(Y, beer)$ belongs to $\mathcal{L}(M)$ but incomplete. Note that $incvar(C) = 1$.

3.2 Evaluation Function

The basis of BG is a heuristic approach based on the best-first search strategy. It requires some relevant function for evaluating each target hypothesis. Our complete IE method seeks the *best hypothesis* H such that

(1) H subsumes the bottom theory $M(F \cup Taut(\mathcal{I}_\mathcal{H}))$,
(2) H is consistent with B,
(3) each clause of H is complete wrt $\mathcal{L}(M)$ and
(4) H has the minimal size in the hypotheses satisfying the above conditions.

Let the best hypothesis H_n be $\{C_1, \ldots, C_n\}$. For the minimality of H_n, each C_i has some *critical clause* in the bottom theory such that is subsumed only by C_i. Accordingly, C_i ($1 \leq i \leq n$) satisfies the following conditions:

(1) C_i has at least one critical clause in the bottom theory,
(2) the current hypothesis set $H_i = \{C_1, \ldots, C_i\}$ is consistent with B,

(3) C_i is complete wrt $\mathcal{L}(M)$ and

(4) C_i has the minimal length in the clauses satisfying the above conditions.

Based on these 4 criteria, we define the following function for evaluating each hypothesis clause belonging to a given mode language.

Definition 8 (Evaluation function).
Let B, F, $\mathcal{I}_\mathcal{H}$ and M be a background theory, a bridge theory, an induction field and a set of mode declarations. Let S be the current hypothesis and C be a new candidate clause belonging to $\mathcal{L}(M)$. We denote by $Rem(F, \mathcal{I}_\mathcal{H}, S)$ the clausal theory obtained by removing all the clauses from the bottom theory wrt F and $\mathcal{I}_\mathcal{H}$ subsumed by S. We then define the following three functions $cover(C, S)$, $length(C)$ and $inconst(C, S)$:

- $cover(C, S)$: the number of clauses in $Rem(F, \mathcal{I}_\mathcal{H}, S)$ subsumed by C.
- $length(C)$: the number of literals in C.
- $inconst(C, S)$: if C is inconsistent with $B \cup S$, then return 1; otherwise 0.

Now, the evaluation function $f(C, S)$ is defined as follows:

$$f(C, S) = p_1 * cover(C, S) - (p_2 * length(C) + p_3 * incvar(C) + p_4 * inconst(C, S)),$$

where each p_i ($1 \leq i \leq 4$) is a parameter with a non-negative value.

The first term in f is a reward one, whereas the others are penalty terms. Two of the penalty terms on inconsistency and incompleteness are also the conditions that every hypothesis clause should satisfy. When C does not satisfy all of them on the hypothesis search, those penalty terms play a role as the search bias influenced by two criteria: inconsistency and incompleteness. Note also that C is evaluated together with the current hypothesis S. In terms of $cover(C, S)$, we focus only on such clauses in the bottom theory that are newly subsumed by C, rather than S. Those newly subsumed ones are regarded as the critical clauses of C. In turn, $inconst(C, S)$ is determined if the current hypothesis S becomes inconsistent with B by adding the new clause C to S.

3.3 Heuristic IS Search Algorithm

Now, we describe our IS algorithm based on the best-first search strategy. It consists of the three functions as follows:

- $refine(h, \bot, M, \mathcal{I}_\mathcal{H})$ returns the clauses obtained by refining h wrt $\bot$, M and $\overline{\mathcal{I}_\mathcal{H}}$. Each refinement clause C subsumes $\bot$, belongs to $\mathcal{I}_\mathcal{H}$, and is written as the form of $h \cup \{l\}$, where l is a literal whose atom is $a(m)$ for some $m \in M$. If m is a modeh (*resp. modeb*), then l corresponds to $a(m)$ (*resp.* $\neg a(m)$).
- $\overline{terminate(h, H, B)}$ returns true if $B \cup H \cup \{h\}$ is consistent and h is complete; otherwise false.
- $\overline{best(S)}$ returns the clause h such that $f(h, H)$ has the maximal value in S.

Input: B, F, $\mathcal{I}_\mathcal{H}$ and M
Output: A consistent hypothesis H wrt $\mathcal{I}_\mathcal{H}$ and F belonging to $\mathcal{L}(M)$
    Step 1. $\perp := C$;   /* C is any clause in $M(F \cup Taut(\mathcal{I}_\mathcal{H}))$. */
    Step 2. $S := \{\Box\}$;   /* S is the set of candidate clauses. */
    Step 3. $h := \Box$;   /* h is the best candidate clause in S. */
    Step 4. $H := \{\Box\}$;   /* H is the current hypothesis. */
    Step 5. while $terminate(h, H, B)$ is false, do;
            /* Add the refinement clauses of h wrt $\perp$, M and $\mathcal{I}_\mathcal{H}$ to S. */
            $S := S \cup refine(h, \perp, M, \mathcal{I}_\mathcal{H})$;
            $h := best(S)$;   /* Select the best clause h among S. */
    Step 6. $H := H \cup \{h\}$;   /* Add h to H. */
    Step 7. Remove the clauses from $M(F \cup Taut(\mathcal{I}_\mathcal{H}))$ subsumed by h;
    Step 8. If $M(F \cup Taut(\mathcal{I}_\mathcal{H}))$ is empty, return H; otherwise go to Step 1;

Fig. 1. Heuristic IS search algorithm

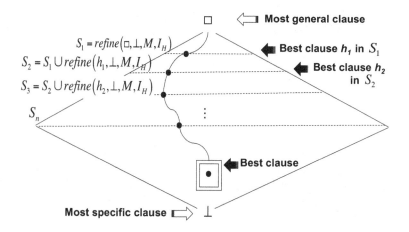

Fig. 2. Best-first search in the subsumption lattice

The above algorithm searches the subsumption lattice bounded by $\perp$ that is some clause in the bottom theory, and seek for each hypothesis clause one by one in the best-first manner. Fig. 2 shows this lattice. Eventually, we obtain the best hypothesis with respect to the evaluation function.

Example 5. We demonstrate how this algorithm works using Example 1-4. Recall B, F, $\mathcal{I}_\mathcal{H}$ and M. Let $\perp$ be the following clause in $M(F \cup Taut(\mathcal{I}_\mathcal{H}))$:

$$\perp = \neg buy(john, diaper) \vee buy(john, beer) \vee shopping(john, at\_night).$$

The algorithm first computes $S_1 = refine(\Box, \perp, M, \mathcal{I}_\mathcal{H})$ as follows:

$$\{C_1 = buy(X, beer),\ C_2 = shopping(X, at\_night),\ C_3 = \neg buy(X, diaper)\}.$$

By evaluating each clause, we have $f(C_1, \{\}) = 1 - (1 + 0 + 0) = 0$, $f(C_2, \{\}) = 2 - (1 + 0 + 0) = 1$, and $f(C_3, \{\}) = 1 - (1 + 1 + 0) = -1$. Note that each parameter

p_i is set to 1 in default. Then, C_2 is selected. Since C_2 is consistent with B and complete wrt $\mathcal{L}(M)$, the algorithm outputs $\{C_2\}$ as the best hypothesis.

Suppose that two new facts $\neg shopping(ken, at\_night)$ and $\neg buy(ken, beer)$ are added to B. Then, C_1 and C_2 become inconsistent with the new bridge theory B. If we set the coefficient parameter p_4 of the evaluation function to 5, the evaluation values of C_1 and C_2 are changed to -5 and -4, respectively. By selecting C_3 as the current best clause, we continue to refine C_3 until the best clause $h_1 = buy(X, diaper) \supset buy(X, beer)$ is found, as shown in Fig. 3.

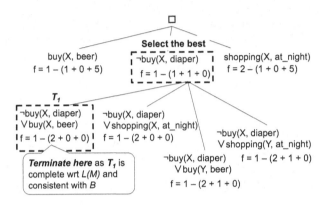

Fig. 3. Search for the best clause h_1

After removing the clause from the bottom theory that is subsumed by h_1, we select another $\bot = \neg buy(john, beer) \vee shopping(john, at\_night)$, and restart to search the subsumption lattice bounded by $\bot$ from the empty set until the best clause $h_2 = buy(X, beer) \supset shopping(X, at\_night)$ is reached, as shown in Fig. 4. Note that the target hypothesis H in Example 1 corresponds to the set $\{h_1, h_2\}$.

3.4   Implementation

We have implemented this heuristic IS algorithm into CF-induction[3] [4,19], which is a sound and complete IE method in full-clausal theories. Fig. 5 describes the system design of the latest CF-induction (ver. 0.45). CF-induction computes so-called characteristic clauses [5] that are consequences of $B \wedge \neg E$ belonging to some language bias reflected by a given induction field $\mathcal{I}_{\mathcal{H}}$.

In practice, those characteristic clauses are obtained by using SOLAR [9], which is a sophisticated deductive engine for consequence finding. CF-induction then constructs a bridge theory by selecting some instances from the characteristic clauses. In the current implementation, the bridge theory F consists of all the ground instances $Carc$ from the characteristic clauses obtained by SO-LAR. After adding the tautologies $taut$ to F, we use an efficient algorithm [15]

---

[3] Available from: http://www.iwlab.org/our-lab/our-staff/yy/software

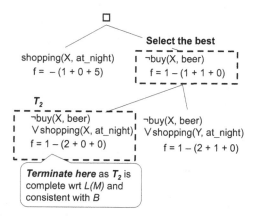

Fig. 4. Search for the best clause h_2

for non-monotone dualization (NMD) in order to compute the bottom theory $M(F \cup taut)$. Note that NMD is also used to compute the CNF formula $M(E)$ that is logically equivalent to $\neg E$. Finally, CF-induction performs the heuristic IS search algorithm with $M(F \cup taut)$, $\mathcal{L}(M)$ and $\mathcal{I}_{\mathcal{H}}$.

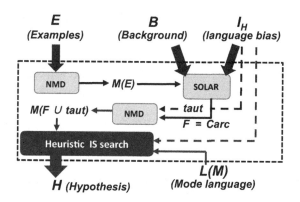

Fig. 5. System design of CF-induction (ver. 0.45)

4   Empirical Evaluation

Using CF-induction, we empirically investigate how the complete IS works well in practical examples. First, we have preliminarily confirmed that our system can solve all the 17 examples in [4,16,18]. We emphasize that some of them include full-clausal representation formalisms, and are difficult to solve by the previously proposed ILP systems.

Hereafter, we show the experimental result obtained by using two materials[4]. The first one is to learn the concept of "classes of animals", and the second one is to learn the concept of "addition of numbers".

4.1   Material 1: Finding Classification Rules

To compare our system with one of the state-of-art ILP systems: Progol5.0 [7,20], we use one concept learning problem that finds classification rules on animals. This problem contains 16 positive examples and 60 facts in the background theory, which are represented in Horn theories.

Using this material, we have evaluated the hypotheses obtained by our IS system in the viewpoint of their predictive accuracies. We apply a leave-one-out test strategy, that is, randomly leave out one example and use the other 15 examples as training data. The predictive accuracy is evaluated by varying the size of randomly chosen training data (9 points ranged from 20% to 100% of the 15 examples). For each training data size, we randomly generate 50 training sets and then compute their success ratio as the predictive accuracy.

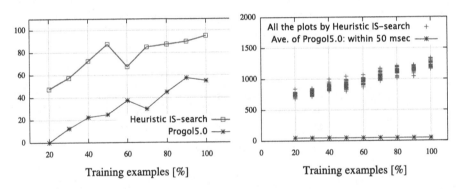

Fig. 6. Predictive accuracy [%]   **Fig. 7.** Executing time [msec]

Fig. 6 describes the predictive accuracies of hypotheses obtained by the heuristic IS-search system and Progol5.0. As far as this result shows, our system produces better hypotheses with higher predictive accuracies. On the other hand, Progol5.0 could generate every hypothesis within 50 (msec), though our system took around 1,000 (msec) as shown in Fig. 7.

4.2   Material 2: Finding Recursive Rules

Next, we empirically show that our complete IS method is useful than incomplete IS ones using another problem for finding recursive rules. The problem contains 10 positive examples E on the sum of two numbers, like $plus(1, 1, 2)$ meaning that

---

[4] Available from: http://www.iwlab.org/our-lab/our-staff/yy/SampleData

$1+1 = 2$. The background theory B has 16 facts that means the successor relation between numbers and the relation that $0 + X = X$ for each number X. Given B and E, we can consider the recursive rule that $X + Y = Z$ if $X + (Y-1) = (Z-1)$ as one correct hypothesis.

Though the problem size is small, the previously proposed incomplete IS methods cannot generate this recursive rule only with B in principle. Indeed, when we solve this problem by Progol5.0[5], it outputs the following hypothesis:

$$suc(D, C) \supset plus(A, B, C),$$

where $suc(D, C)$ means $C = D+1$, within 5 msec. This hypothesis may be caused by over-generalization. In contrast, if we add two negative examples $\neg plus(1, 2, 2)$ and $\neg plus(2, 3, 4)$ to B, the output has been changed to the positive examples themselves. This hypothesis may be caused by over-fitting. Since the bottom clause in BG does not subsume the target hypothesis, it is difficult for Progol to solve this problem in principle.

We have evaluated the hypotheses obtained by our IS system in the viewpoint of their predictive accuracies. Like the previous experiment, we apply a leave-one-out test strategy that randomly leaves out one example and uses the other 9 examples as training data. The predictive accuracy is evaluated by varying the size of randomly chosen training data (9 points ranged from 20% to 100% of the 9 examples). For each training data size, we randomly generate 10 training sets and then compute their success ratio as the predictive accuracy.

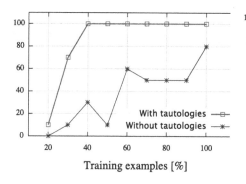

Fig. 8. Predictive accuracy [%]

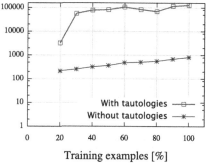

Fig. 9. Generalization time [msec]

For the comparison, we prepare one incomplete variation of our IS method, which do not use the tautologies in the bridge theory. Fig. 8 describes the performance of two cases: the one (thick line pointed with squares) generating H from $M(F \cup Taut(\mathcal{I}_\mathcal{H}))$ and the other (thin line pointed with stars) from $M(F)$. In other words, the former (*resp.* latter) case is for the complete (*resp.* incomplete) IS method. We then notice that the complete IS succeeds in generating better

---

[5] We use it in the default setting and also with splitting and posonly options.

hypotheses with high (almost 100%) predictive accuracies. In fact, it generated
the target hypothesis in more than 40% training data size. On the other hand,
its computational cost was much more expensive than the incomplete IS without
adding tautologies, as shown in Fig. 9. Note that $Taut(\mathcal{I}_\mathcal{H})$ contains 10 tautologies. $M(F \cup Taut(\mathcal{I}_\mathcal{H}))$ can be blow-up in size as increasing the size of those
tautologies, which make the performance inefficient.

5   Discussion

In the previous section, we have empirically shown that the complete IS system
can find better hypotheses than incomplete IS ones. On the other hand, our
system takes much more time even for small examples. Although it can provide
better solutions, it must be inconvenient for users to jump from 50 msec (for
Progol) like the previous examples. In belief, our current system can be too
late to be used, whereas the previously proposed ones may often be too fast to
provide good solutions.

To address this problem, we intend to embed the notion of so-called *Limited
Resource Strategy* (LRS) into the system [12]. Softwares are usually limited in
their memory usage or execution time. Given the information on this computational resource in advance, LRS provides the best search strategy where the
resource is used in the most effective way. LRS is a well-established approach in
modern theorem-provers like Vampire and SOLAR, and is expected for improving the scalability of our system too. Fig. 10 shows the correlation between the
number of literals included in each bottom theory and its generalization time
by our IS method over all the instances in the experiment. We notice that the
executing time increases as the bottom theory is blow-up in size. For example,
if the time resource is limited to 1,000 msec, we should construct some bottom
theory whose size is at least less than 100 according to Fig. 10. In this way, we
intend to use LRS for making the current system more scalable.

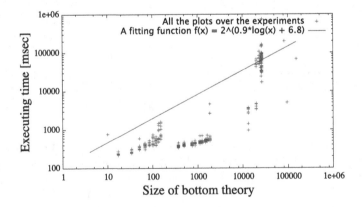

Fig. 10. Correlated distribution on the search time and size of bottom theory

It is also considerable for the scalability to limit the hypotheses to be searched. In this paper, we define the evaluation function based on the common one in BG methods. On the other hand, there is another criterion that can be used for evaluating a hypothesis H. For example, H should be eliminated if it contains such a variable that appears only once for over-generalization. Though we can regulate some connectivity between variables by Definition. 7, it does not necessarily take out those "independent" variables from the target hypotheses. In Example 4, the variable X of $buy(X, at\_night)$ is connected wrt $\mathcal{L}(\mathcal{M})$ but independent. Indeed, $buy(X, at\_night)$ is regarded as a over-generalized hypothesis. The current CF-induction has implemented this evaluation criterion with an additional option ("-onInd"), and confirmed its availability using examples.

6   Conclusion and Future Work

There are two generalization approaches for hypothesis finding: inverse entailment (IE) and inverse subsumption (IS). Recently, it has been shown that IE can be logically reduced into a new form of IS, provided that it ensures the completeness of IE. This paper aims at investigating how this complete IS method works well in the practical point of view. For the analysis, we have provided and implemented a heuristic IS algorithm based on the techniques of bottom generalization used in the state-of-the-art ILP systems. We have shown two experimental results suggesting that the complete IS method actually finds better hypotheses than incomplete IS methods like Progol and a variation of our IS method that does not contain tautologies to the original bridge theory F.

It is an important future work to improve the scalability of the current IS system. In case that the size of the bridge theory with tautologies increases, the task for computing the bottom theory (i.e., non-monotone dualization) becomes exponentially expensive. The computational cost of the heuristic IS search depends strongly on the size of the bottom theory. Then, it will be necessary to limit the number of clauses to be included in the bridge theory with some relevant criteria like LRS as we discussed before.

On behalf of the computational efficiency, it will be fruitful to consider if $M(F \cup Taut(\mathcal{I}_{\mathcal{H}}))$ can be treated in some compressed representation formalization like BDD [1]. A further experimental comparison is another important future work. In this paper, our system is compared only with Progol5.0 (mainly in the default setting) using two materials. We then intend to empirically analysis our system by comparison of other state-of-the-art ILP systems using more practical data sets.

Acknowledgements. This research is supported by 2010-2012 JSPS Grant-in-Aid for Young Scientists (B) (No. 22700141). The authors would like to thank the anonymous reviewers for giving us useful and constructive comments. We would like to thank Pr. H. Nabeshima for his precious comments on this research.

References

1. Bryant, R.E.: Graph-based algorithms for Boolean function manipulation. IEEE Trans. Comput. 35(8), 677–691 (1986)
2. De Raedt, L.: Logical setting for concept-learning. Artificial Intelligence 95, 187–201 (1997)
3. Flach, P.A.: Rationality postulates for induction. In: Proceedings of the 6th International Conference on Theoretical Aspects of Rationality and Knowledge, pp. 267–281 (1996)
4. Inoue, K.: Induction as consequence finding. Machine Learning 55(2), 109–135 (2004)
5. Inoue, K.: Linear resolution for consequence finding. Artificial Intelligence 56(2-3), 301–353 (1992)
6. Kimber, T., Broda, K., Russo, A.: Induction on failure: learning connected Horn theories. In: Erdem, E., Lin, F., Schaub, T. (eds.) LPNMR 2009. LNCS, vol. 5753, pp. 169–181. Springer, Heidelberg (2009)
7. Muggleton, S.H.: Inverse entailment and Progol. New Generation Computing 13, 245–286 (1995)
8. Muggleton, S.H., Buntine, W.L.: Machine invention of first order predicates by inverting resolution. In: Proc. of the 5th Int. Conf. on ML, pp. 339–352 (1988)
9. Nabeshima, H., Iwanuma, K., Inoue, K., Ray, O.: SOLAR: An automated deduction system for consequence finding. AI Commun. 23(2-3), 183–203 (2010)
10. Ray, O., Broda, K., Russo, A.: Hybrid abductive inductive learning: A generalisation of progol. In: Horváth, T., Yamamoto, A. (eds.) ILP 2003. LNCS (LNAI), vol. 2835, pp. 311–328. Springer, Heidelberg (2003)
11. Ray, O., Inoue, K.: Mode-directed inverse entailment for full clausal theories. In: Blockeel, H., Ramon, J., Shavlik, J., Tadepalli, P. (eds.) ILP 2007. LNCS (LNAI), vol. 4894, pp. 225–238. Springer, Heidelberg (2008)
12. Riazanov, A., Voronkov, A.: Limited Resource Strategy in Resolution Theorem Proving. Symbolic Computations 36, 101–115 (2003)
13. Yamamoto, A.: Hypothesis finding based on upward refinement of residue hypotheses. Theoretical Computer Science 298, 5–19 (2003)
14. Srinivasan, A.: The Aleph Manual. University of Oxford, Oxford (2007)
15. Uno, T.: A practical fast algorithm for enumerating minimal set coverings. IPSJ SIG Notes (29), 9–16 (2002)
16. Yamamoto, Y., Inoue, K., Doncescu, A.: Integrating abduction and induction in biological inference using CF-induction. In: Lodhi, H., Muggleton, S. (eds.) Elements of Computational Systems Biology, ch. 9, pp. 213–234 (2009)
17. Yamamoto, Y., Ray, O., Inoue, K.: Towards a logical reconstruction of CF-induction. In: Satoh, K., Inokuchi, A., Nagao, K., Kawamura, T. (eds.) JSAI 2007. LNCS (LNAI), vol. 4914, pp. 330–343. Springer, Heidelberg (2008)
18. Yamamoto, Y., Inoue, K., Iwanuma, K.: Inverse subsumption for complete explanatory induction. Machine Learning 86, 115–139 (2012)
19. Yamamoto, Y., Inoue, K., Iwanuma, K.: Comparison of upward and downward generalizations in CF-induction. In: Muggleton, S.H., Tamaddoni-Nezhad, A., Lisi, F.A. (eds.) ILP 2011. LNCS (LNAI), vol. 7207, pp. 373–388. Springer, Heidelberg (2012)
20. Tamaddoni-Nezhad, A., Muggleton, S.H.: The lattice structure and refinement operators for the hypothesis space bounded by a bottom clause. Machine Learning 76, 37–72 (2009)
21. Nienhuys-Cheng, S.-H., de Wolf, R. (eds.): Foundations of Inductive Logic Programming. LNCS, vol. 1228. Springer, Heidelberg (1997)

Learning Unordered Tree Contraction Patterns in Polynomial Time

Yuta Yoshimura and Takayoshi Shoudai

Department of Informatics, Kyushu University, Fukuoka 819-0395, Japan
{yuuta.yoshimura,shoudai}@inf.kyushu-u.ac.jp

Abstract. In this paper, we present a concept of edge contraction-based tree-structured patterns as a graph pattern suited to represent tree-structured data. A *tree contraction pattern* (*TC-pattern*) is an unordered tree-structured pattern common to a given tree-structured data, which is obtained by merging every uncommon connected substructure into one vertex by edge contraction. In this paper, in order to establish an algorithmic foundation for the discovery of knowledge from tree-structured data, we show that TC-patterns are learnable in polynomial time.

1 Introduction

Many documents such as Web documents or XML files have tree structures. In order to extract meaningful and hidden knowledge from such documents, we need tree-structured patterns that can explain them. As unordered tree-structured patterns, an unordered tree pattern [1,4], a type of object [7], a tree-expression pattern [11], and an unordered term tree [6] have been proposed. An unordered tree pattern represents a rooted tree pattern that may contain structural variables as leaves. A linear unordered term tree (a term tree for short) is a rooted tree pattern that consists of internal structural variables that can be replaced with arbitrary trees [6,9,10]. We show examples of those tree-structured patterns in Fig. 1, which represent a characteristic common to the unordered trees in $S = \{T_1, \ldots, T_4\}$. In this paper, a tree means a rooted unordered labeled tree. In the upper-middle figure of Fig. 1, we give all maximal subtrees that appear in all trees in S. All unordered tree patterns and term trees are given in the lower-middle and right figures, respectively. We note that all unordered trees in S consist of a root that has one child and exactly three leaves. Unfortunately, previously proposed unordered tree-structured patterns are difficult to represent such a common characteristic as a single pattern in a natural way. Thus, in this paper, we introduce an edge contraction-based tree-structured pattern, called a *tree contraction pattern*, as a new graph pattern that is suitable for a graph-structural expression.

A *tree contraction pattern* (*TC-pattern*) is a triplet $t = (V_t, E_t, U_t)$ where (V_t, E_t) is a tree with a specified root r_t and U_t is a subset of V_t. A vertex in U_t is called a *contractible vertex*. A tree $T = (V_T, E_T)$ with root r_T matches a TC-pattern t with root r_t if there exists a partition $\mathcal{W} = \{W(u) \mid u \in V_t\}$ of

F. Riguzzi and F. Železný (Eds.): ILP 2012, LNAI 7842, pp. 257–272, 2013.
© Springer-Verlag Berlin Heidelberg 2013

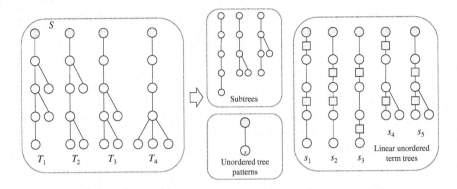

Fig. 1. A set of unordered trees $S = \{T_1, \ldots, T_4\}$ and tree-like structures common to all trees in S: All $T_1, \ldots, T_4$ match all term trees $s_1, \ldots, s_5$. A variable of a term tree is represented by a box with lines connecting to its elements.

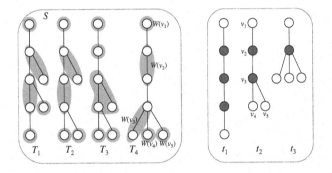

Fig. 2. TC-patterns t_1, t_2, t_3: A contractible vertex of a TC-pattern is represented by dark-colored circles

V_T such that (i) for $u \in V_t \setminus U_t$, $W(u)$ includes exactly one vertex, (ii) for any $u \in V_t$, any pair of vertices in $W(u)$ is connected, (iii) $W(r_t)$ includes r_T, and (iv) the tree obtained from T by merging all vertices in $W(u)$ into one vertex for each $u \in U_t$ is isomorphic to (V_t, E_t). In the right frame of Fig. 2, we present three TC-patterns. A vertex in U_t is represented by a dark-colored circle. Unordered tree T_4 matches TC-pattern t_2 because there exists a partition $\mathcal{W}$ of the vertex set of T_4, which is described by dark-colored areas, satisfying the above conditions (i)–(iv). Moreover, t_2 is a minimally generalized TC-pattern that is matched by all $T_1, \ldots, T_4$. It can be clearly observed that all trees $T_1, \ldots, T_4$ match all t_1, t_2, t_3. In general, TC-patterns express different patterns from previously proposed unordered tree-structured patterns. Actually, t_3 represents a tree-like structure common to all trees in S each of which contains a root with exactly one child and three leaves.

In this paper, we first prove that the TC-pattern matching problem is NP-complete unless the degree of every contractible vertex of t is bounded by a constant. Second we present an algorithm for computing the TC-pattern matching problem that is executed in $O(nN(n + \sqrt{N}))$ time, when the degree of a contractible vertex is bounded by a constant. Finally, we present a polynomial time algorithm for finding a minimally generalized TC-pattern explaining a given set of trees. Then, we conclude that the class of tree languages obtained from TC-patterns is polynomial-time inductively inferable from positive data.

2   Preliminaries

2.1   Linear Unordered Term Trees

We briefly state the definition and facts about linear unordered term trees for our motivation in this paper. Here we start with a concept of term graphs. For a set or a list S, the number of elements in S is denoted by $|S|$.

Let Σ and Λ be finite alphabets, and let X be an alphabet. We assume that $(\Sigma \cup \Lambda) \cap X = \emptyset$. A *term graph* $g = (V, E, H)$ consists of a vertex set V, an edge set E and a multi-set H. Each element in H is a list of distinct vertices in V and is called a *variable*. All vertices, edges, and variables are labeled with symbols Σ, Λ, and X, respectively. The *dimension* of a term graph g, denoted by $dim(g)$, is the maximum number of $|h|$ over all variables h in g.

Let g be a term graph and σ a list of distinct vertices in g. We call the form $x := [g, \sigma]$ a *binding* for a variable label $x \in X$. Let $g_1, \ldots, g_n$ be term graphs. A *substitution* θ is a finite collection of bindings $\{x_1 := [g_1, \sigma_1], \ldots, x_n := [g_n, \sigma_n]\}$, where x_i's are mutually distinct variable labels in X and each g_i has no variable labeled with an element in $\{x_1, \ldots, x_n\}$. We obtain a new term graph f by applying a substitution $\theta = \{x_1 := [g_1, \sigma_1], \ldots, x_n := [g_n, \sigma_n]\}$ to a term graph $g = (V, E, H)$ in the following way. For each binding $x_i := [g_i, \sigma_i] \in \theta$ $(1 \leq i \leq n)$ in parallel, we attach g_i to g by removing all variables $h_1, \ldots, h_k$ labeled with x_i from H, and by identifying the m-th vertex of h_j $(1 \leq j \leq k)$ and the m-th vertex of σ_i $(1 \leq m \leq |t_j| = |\sigma_i|)$. The resulting term graph is denoted by $g\theta$.

A substitution $\theta = \{x_1 := [g_1, \sigma_1], \ldots, x_n := [g_n, \sigma_n]\}$ is called a *tree substitution* if all of the g_i are trees. A term graph g is called an *unordered term tree* if for any tree substitution θ which contains all variable labels in g, $g\theta$ is also an unordered tree. An unordered term tree g is called *linear* if each variable label in g occurs exactly once. For a linear unordered term tree t and an unordered tree T, we say that T *matches* t if there is a tree substitution θ such that T is isomorphic to $t\theta$ as unordered trees. **LUTT MATCHING** is defined as the problem, given a linear unordered term tree t and an unordered tree T, of deciding whether or not T matches t. In Fig. 1, for example, all trees $T_1, \ldots, T_4$ match all $s_1, \ldots, s_5$. That is, $s_1, \ldots, s_5$ show common structures between $T_1, \ldots, T_4$. We have the following theorem.

Theorem 1 ([6,9]). *Let t and T be a linear unordered term tree with n vertices and an unordered tree with N vertices, respectively.*

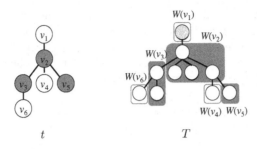

Fig. 3. A TC-pattern $t = (V(t), E(t), U(t))$ and an unordered tree $T = (V(T), E(T))$:
$V(t) = \{v_1, v_2, v_3, v_4, v_5, v_6\}$, $E(t) = \{\{v_1, v_2\}, \{v_2, v_3\}, \{v_2, v_4\}, \{v_2, v_5\}, \{v_3, v_6\}\}$,
$U(t) = \{v_2, v_3, v_5\}$, and v_1 is the root of t. $\mathcal{W} = \{W(v_1), \ldots, W(v_6)\}$ is a t-witness
structure of T.

1. *LUTT MATCHING is NP-complete if $dim(t) > 3$.*
2. *If $dim(t) \le K$ for some constant K and the degree of vertices in T is bounded by constants, then LUTT MATCHING is solvable in $O(nN^2)$ time.*
3. *If $dim(t) = 2$, then LUTT MATCHING is solvable in $O(n^2 N^{3.5})$ time.*

Next we introduce a new tree-structured pattern that has an expression power similar to a term tree and a more efficient pattern matching algorithm.

2.2   Tree Contraction Patterns

For a graph G, we denote by $V(G)$ the set of vertices of G and by $E(G)$ the set of edges. For $V' \subseteq V(G)$, we denote by $G[V']$ the subgraph of G induced by V'.

Definition 1. *Let G and H be connected graphs. An H-witness structure $\mathcal{W} = \{W(u) \mid u \in V(H)\}$ is a partition of $V(G)$ satisfying the following conditions.*

1. *For any $u \in V(H)$, $G[W(u)]$ is connected, and*
2. *for any $u, u' \in V(H)$ $(u \ne u')$, $\{u, u'\} \in E(H)$ if and only if there is an edge $\{v, v'\} \in E(G)$ such that $v \in W(u)$ and $v' \in W(u')$*

We call each set $W(u) \in \mathcal{W}$ the H-*witness set* of u. When G has an H-witness structure, G can be transformed to H by contracting each of the H-witness sets into one vertex by edge contractions. A *graph contraction pattern* h (*GC-pattern*, for short) is a triplet $h = (V, E, U)$ where (V, E) is a connected graph and U is a subset of V. Let Σ be a finite alphabets. A GC-pattern h has a vertex labeling $\varphi_h : V \to \Sigma$. For $V' \subseteq V$, we denote by $h[V']$ the graph contraction subpattern (*GC-subpattern*, for short) induced by V', i.e., $h[V'] = (V', E \cap \{\{v, w\} \mid v, w \in V'\}, U \cap V')$. We call an element of U a *contractible vertex*, and an element of $V(h) \setminus U$ an *uncontractible vertex*.

For a GC-pattern h, we denote by $V(h)$ the set of vertices of h, by $E(h)$ the set of edges of h, and by $U(h)$ the set of contractible vertices of h.

Definition 2. *Let h and g be GC-patterns with vertex labelings φ_h and φ_g, respectively. Let H and G be graphs $(V(h), E(h))$ and $(V(g), E(g))$, respectively. We write $h \preceq g$ if there is an H-witness structure $\mathcal{W} = \{W(u) \mid u \in V(h)\}$ of G such that for all $v \in V(h) \setminus U(h)$, $W(v)$ contains exactly one vertex $u \in V(g) \setminus U(g)$ with $\varphi_g(u) = \varphi_h(v)$. Such an H-witness structure of G is called an h-witness structure of g.*

We regard a GC-pattern h as a standard graph $(V(h), E(h))$ if $U(h) = \emptyset$. For a GC-pattern h and a connected graph G, we say that h *matches* G if $h \preceq G$.

Definition 3. *We say that a GC-pattern t is a TC-pattern if $(V(t), E(t))$ is an unordered tree with a specified root. A TC-pattern t with root r_t matches an unordered tree T with root r_T if there is a t-witness structure $\mathcal{W} = \{W(u) \mid u \in V(t)\}$ of T such that $r_T \in W(r_t)$.*

For example, we show in Fig. 3 a TC-pattern t and an unordered tree T that has a t-witness structures $\mathcal{W} = \{W(v_1), \ldots, W(v_6)\}$ of T. Each witness set in $\mathcal{W}$ is specified with a squared area. Thus, T matches t.

3   Tree Contraction Pattern Matching in Polynomial Time

We formally define the pattern matching problem for TC-patterns as follows:

TC-PATTERN MATCHING
Instance: A TC-pattern t and an unordered tree T.
Question: Does T match t?

3.1   The Hardness of TC-Patten Matching

We first show the following theorem:

Theorem 2. *TC-PATTERN MATCHING is NP-complete even if the maximum degree of a given unordered tree is equal to 3.*

Proof. Membership in NP is obvious. We transform EXACT COVER BY 3-SETS (X3C) [3, page 221] to this problem.

EXACT COVER BY 3-SETS (X3C)
Instance: Set X with $|X| = 3q$ for a natural number q and a collection $\mathcal{C}$ of 3-element subsets of X.
Question: Does $\mathcal{C}$ contain an exact cover for X, i.e., a subcollection $\mathcal{C}' \subseteq \mathcal{C}$ such that every element of X occurs in exactly one member of $\mathcal{C}'$.

We give a transformation from an instance of X3C to a TC-pattern and an unordered tree with vertex labels. Let $X = \{a_1, \ldots, a_n\}$ where $n = 3q$. Let

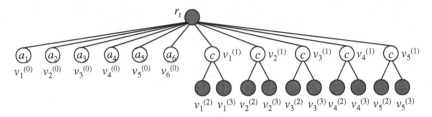

TC-pattern t

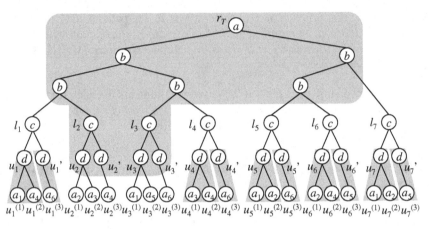

Unordered tree T

Fig. 4. A transformation from an instance $(X, \mathcal{C})$ of X3C to an instance (t, T) of TC-PATTERN MATCHING. $X = \{a_1, \ldots, a_6\}$, $\mathcal{C} = \{C_1, \ldots, C_7\}$ where $C_1 = \{a_1, a_4, a_6\}$, $C_2 = \{a_2, a_3, a_4\}$, $C_3 = \{a_1, a_5, a_6\}$, $C_4 = \{a_3, a_4, a_6\}$, $C_5 = \{a_1, a_2, a_6\}$, $C_6 = \{a_2, a_4, a_5\}$, $C_7 = \{a_1, a_2, a_5\}$.

$\mathcal{C} = \{C_1, \ldots, C_m\}$ where $C_i \subseteq X$ with $|C_i| = 3$ and let $C_i = \{c_i^{(1)}, c_i^{(2)}, c_i^{(3)}\}$ for $i = 1, \ldots, m$. Let t be a TC-pattern with root r_t, where

$V(t) = \{r_t\} \cup \bigcup_{i=1}^{n} \{v_i^{(0)}\} \cup \bigcup_{i=1}^{m-q} \{v_i^{(1)}, v_i^{(2)}, v_i^{(3)}\}$,

$E(t) = \bigcup_{i=1}^{n} \{\{r_t, v_i^{(0)}\}\} \cup \bigcup_{i=1}^{m-q} \{\{r_t, v_i^{(1)}\}, \{v_i^{(1)}, v_i^{(2)}\}, \{v_i^{(1)}, v_i^{(3)}\}\}$, and

$U(t) = \{r_t\} \cup \bigcup_{i=1}^{m-q} \{v_i^{(2)}, v_i^{(3)}\}$.

Let $\Sigma = \{a, b, c, d\} \cup \{a_1, \ldots, a_n\}$ where a, b, c, d are special vertex labels which do not appear in $\{a_1, \ldots, a_n\}$. The vertex labeling $\varphi_t : V(t) \backslash U(t) \to \Sigma$ is defined as $\varphi_t(v_i^{(0)}) = a_i$ $(1 \le i \le n)$ and $\varphi_t(v_i^{(1)}) = c$ $(1 \le i \le m - q)$.

Let T_0 be a binary tree with exactly m leaves. Let r_T be the root of T_0 and $\ell_1, \ldots, \ell_m$ the leaves of T_0. Let T be a tree, where

$V(T) = V(T_0) \cup \bigcup_{i=1}^{m} \{u_i, u_i', u_i^{(1)}, u_i^{(2)}, u_i^{(3)}\}$, and

$E(T) = E(T_0) \cup \bigcup_{i=1}^{m} \{\{\ell_i, u_i\}, \{\ell_i, u_i'\}, \{u_i, u_i^{(1)}\}, \{u_i, u_i^{(2)}\}, \{u_i', u_i^{(3)}\}\}$.

The vertex labeling $\varphi_T : V(T) \to \Sigma$ is defined as follows: $\varphi_T(r) = a$, $\varphi_T(u) = b$ for any $u \in V(T_0) \setminus \{r, \ell_1, \dots, \ell_m\}$, and $\varphi_T(\ell_i) = c$, $\varphi_T(u_i) = \varphi_T(u'_i) = d$, $\varphi_T(u_i^{(j)}) = c_i^{(j)}$ $(j = 1, 2, 3)$ for any i $(1 \leq i \leq m)$.

We assume that there is a subcollection $\mathcal{C}' \subseteq \mathcal{C}$ such that every element of X occurs in exactly one member of $\mathcal{C}'$. Let $\mathcal{C}' = \{C_{i_1}, \dots, C_{i_q}\}$. We construct a t-witness structure $\mathcal{W}$ of T in the following way: $W(v) = (V(T_0) \setminus \{\ell_1, \dots, \ell_m\}) \cup \bigcup_{j=1}^{q} \{\ell_{i_j}, u_{i_j}, u'_{i_j}\}$. For any i $(1 \leq i \leq n)$, $W(v_i^{(0)}) = \varphi_T^{-1}(a_i) \cap (\bigcup_{j=1}^{q} C_{i_j})$. For any $i \in \{1, \dots, m\} \setminus \{i_1, \dots, i_q\}$, $W(v_i^{(1)}) = \{\ell_i\}$, $W(v_i^{(2)}) = \{u_i, u_i^{(1)}, u_i^{(2)}\}$, and $W(v_i^{(3)}) = \{u'_i, u_i^{(3)}\}$. We easily see that $\mathcal{W} = \{W(v) \mid v \in V(t)\}$ satisfies conditions 1 and 2 of Def. 1.

Conversely we assume that there is an h-witness structure $\mathcal{W} = \{W(v) \mid v \in V(t)\}$. For all vertices $v_1^{(1)}, \dots, v_{m-q}^{(1)}$, there are $m - q$ vertices $\ell_{k_1}, \dots, \ell_{k_{m-q}}$ $(1 \leq k_i \leq m)$ such that $W(v_i^{(1)}) = \{\ell_{k_i}\}$ $(1 \leq i \leq m - q)$, since $\ell_1, \dots, \ell_m$ are only the vertices in T that have a vertex label c. For each ℓ_{k_i} $(1 \leq i \leq m - q)$, u_{k_i} and u'_{k_i}, which are two of three adjacent vertices of ℓ_{k_i}, and $u_{k_i}^{(1)}, u_{k_i}^{(2)}, u_{k_i}^{(3)}$ must be contained in $W(v_i^{(2)}) \cup W(v_i^{(3)})$. On the other hand, every ℓ_j $(j \in \{1, \dots, m\} \setminus \{k_1, \dots, k_{m-q}\})$ and its children must be contained in $W(v)$. Let $\mathcal{C}' = \{C_j \in \mathcal{C} \mid j \in \{1, \dots, m\} \setminus \{k_1, \dots, k_{m-q}\}\}$. Since $W(v_i^{(0)})$ $(1 \leq i \leq n)$ has one vertex that has a vertex label a_i, every element of $a_1, \dots, a_n$ occurs in exactly one member of $\mathcal{C}'$.

3.2   Set Cover by Collections

The following problem plays an important role in our TC-pattern matching algorithm.

SET COVER BY COLLECTIONS (SCC)
Instance: Set X with $|X| = d$ for a natural number d, and m collections $\mathcal{C}_1, \dots, \mathcal{C}_m$ of subsets of X with $m \geq 1$.
Question: Are there collections $\mathcal{C}_{j_1}, \dots, \mathcal{C}_{j_r}$ $(1 \leq j_1 < \dots < j_r \leq m)$ and sets $X_{j_\ell} \in \mathcal{C}_{j_\ell}$ for each ℓ $(1 \leq \ell \leq r)$ such that $\bigcup_{\ell=1}^{r} X_{j_\ell} = X$.

We easily see that SCC is NP-complete. Let $(X, \mathcal{C})$ be an instance for X3C with $|X| = 3q$. For all j $(1 \leq j \leq m)$, we let $m = q$ and $\mathcal{C}_j = \mathcal{C}$. Then, X3C for $(X, \mathcal{C})$ returns yes if and only if SCC for $(X, \mathcal{C}_1, \dots, \mathcal{C}_m)$ returns yes.

We have two restrictions in which SCC are computed in polynomial time.

Lemma 1. *Let $(X, \mathcal{C}_1, \dots, \mathcal{C}_m)$ with $|X| = d$ for natural numbers m and d be an instance for SCC.*

1. *If all $\mathcal{C}_j$ $(1 \leq j \leq m)$ contain singleton sets of X only, SCC is computed in $O(d \sum_{j=1}^{m} |\mathcal{C}_j| + dm\sqrt{d + m})$ time.*
2. *If d is bounded by a constant, SCC is computed in $O(\sum_{j=1}^{m} |\mathcal{C}_j| + m\sqrt{m})$ time.*

Proof. 1. Let G be a bipartite graph where $V(G) = X \cup \{C_1, \ldots, C_m\}$ and $E(G) = \{\{x, C_j\} \mid x \in X \text{ and } \{x\} \in C_j \ (1 \leq j \leq m)\}$. We use $O(d\sum_{j=1}^{m} |C_j|)$ time to construct G from an input. The answer of SCC for $(X, C_1, \ldots, C_m)$ is obtained by finding a maximum bipartite matching for G. The Hopcroft-Karp algorithm [5] executes this work in $O(E(G)\sqrt{V(G)}) = O(dm\sqrt{d+m})$ time. The size of the maximum matching is equal to d if and only if the answer of SCC is *yes*.

2. For each d' $(d' = 1, \ldots, d)$ and each partition $\{X_1, \ldots, X_{d'}\}$ of X, we construct a bipartite graph G which has a vertex set $V(G) = \{X_1, \ldots, X_{d'}\} \cup \{C_1, \ldots, C_m\}$ and an edge set $E(G) = \{\{X_i, C_j\} \mid \exists C \in C_j \text{ s.t. } X_i \subseteq C\}$. We use $O(\sum_{i=1}^{d'} \sum_{j=1}^{m} |X_i||C_j|) = O(d\sum_{j=1}^{m} |C_j|)$ time to construct such a bipartite graph G. Next we find a maximum bipartite matching for G. The answer of SCC is *yes* if and only if there is a partition $\{X_1, \ldots, X_{d'}\}$ of X such that the size of the maximum matching for its corresponding bipartite graph G is equal to d'. Since the total number of partitions of X with $|X| = d$ is the d-th Bell number B_d, the total time complexity is $O(B_d \cdot (d\sum_{j=1}^{m} |C_j| + dm\sqrt{d+m}))$ time. If d is bounded by a constant, it becomes $O(\sum_{j=1}^{m} |C_j| + m\sqrt{m})$ time.

Below we use $p_{scc}(X, C_1, \ldots, C_m)$ as a predicate that returns *true* if and only if the problem SCC for an instance $(X, C_1, \ldots, C_m)$ answers *yes*.

3.3   A Pattern Matching Algorithm for Restricted TC-Patterns

Let t be a TC-pattern or an unordered tree. We denote the *root* of t by r_t. For a vertex v of t, we denote the number of children of v of t by $d_t(v)$. For a vertex v of t, $t[v]$ denotes the subtree induced by v and all the descendants of v.

In this section, we propose a TC-pattern matching algorithm that determines whether an unordered tree T matches a TC-pattern t. We assume that every vertex v of t has a different identifier, denoted by $I_t(v)$. Let $c_{v,1}, \ldots, c_{v,d_t(v)}$ be the children of v and $X_t(v) = \{I_t(c_{v,1}), \ldots, I_t(c_{v,d_t(v)})\}$

We assign two collections $\mathcal{I}_T(u)$ and $\mathcal{J}_T(u)$ to every vertex u of T, where

$$\mathcal{I}_T(u) \subseteq \{\{I_t(v)\} \mid v \in V(t)\} \text{ and } \mathcal{J}_T(u) \subseteq \{P \subsetneq X_t(v) \mid v \in U(t)\}.$$

Let $\mathcal{C}_T(u) = \mathcal{I}_T(u) \cup \mathcal{J}_T(u)$.

Algorithm MATCHING;
Input: A TC-pattern t and an unordered tree T.
Step 1: (Initialization) For all $u \in V(T)$, if u is a leaf then set $\mathcal{I}_T(u) := \{\{I(v)\} \mid v \text{ is a leaf of } t\}$ and $\mathcal{J}_T(u) := \emptyset$, else $\mathcal{I}_T(u) := \emptyset$ and $\mathcal{J}_T(u) := \emptyset$. Set $h := h_T - 1$ where h_T is the height of T.
Step 2: (Iteration) For all $u \in V(T)$ of height h, do Steps 2.1–2.4:
    2.1 For all $v \in V(t) \setminus U(t)$, if $p_{scc}(X_t(v), \mathcal{I}_T(c_{u,1}), \ldots, \mathcal{I}_T(c_{u,d_T(u)})) = true$ and $|X_t(v)| = d_T(u)$, then $\mathcal{I}_T(u) := \mathcal{I}_T(u) \cup \{\{I_t(v)\}\}$. (We show a simple example of Step 2.1 in Fig. 5.)

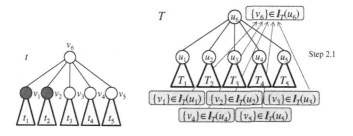

Fig. 5. An example of Step 2.1: We assume that $v_6 \in V(t) \backslash U(t)$ and subtrees $T_1, \ldots, T_5$ match TC-subpatterns $t_1, \ldots, t_5$, respectively.

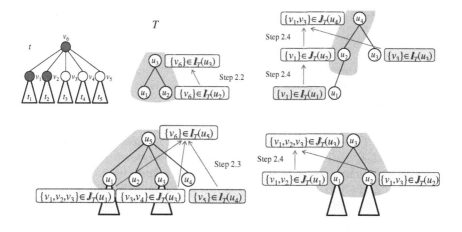

Fig. 6. Examples of Steps 2.2–2.4: We assume that $v_6 \in U(t)$. Each shaded area of T might be a part of a witness set of v_6.

2.2 For all $v \in U(t)$, if there is an index $j \in \{1, \ldots, d_T(u)\}$ such that $\{I_t(v)\} \in \mathcal{I}_T(c_{u,j})$, then $\mathcal{I}_T(u) := \mathcal{I}_T(u) \cup \{\{I_t(v)\}\}$.

2.3 For all $v \in U(t)$, if $p_{scc}(X_t(v), \mathcal{C}_T(c_{u,1}), \ldots, \mathcal{C}_T(c_{u,d_T(u)})) = true$, then $\mathcal{I}_T(u) := \mathcal{I}_T(u) \cup \{\{I_t(v)\}\}$.

2.4 For all $v \in U(t)$ and all subsets X' of $X_t(v)$ with $1 \le |X'| < |X_t(v)|$, if $p_{scc}(X', \mathcal{C}_T(c_{u,1}), \ldots, \mathcal{C}_T(c_{u,d_T(u)})) = true$, then $\mathcal{J}_T(u) := \mathcal{J}_T(u) \cup \{X'\}$. (We show examples of Steps 2.1–2.4 in Fig. 6.)

Step 3: (Decision) If $h > 0$ then $h := h-1$ and goto Step 2. If $\{I_t(r_t)\} \in \mathcal{I}_T(r_T)$, then T matches t, otherwise T does not match t.

We show in Fig. 7 an example process of Algorithm MATCHING for t and T. Next we prove the correctness and time complexity of Algorithm MATCHING.

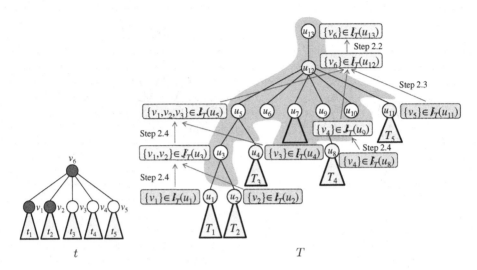

Fig. 7. An example process of Algorithm MATCHING: We assume that $v_6 \in V(t) \setminus U(t)$ and subtrees $T_1, \ldots, T_5$ match TC-subpatterns $t_1, \ldots, t_5$, respectively. The shaded area in the background of T specifies a witness set of v_6.

Lemma 2. *Let t be a TC-pattern and T an unordered tree. For any $u \in V(T)$, $X' = \{I_1, \ldots, I_d\} \in \mathcal{J}_T(u)$ if and only if there is a vertex $v \in U(t)$ such that $X' \subsetneq X_t(v)$ and there are d different proper descendants $u_1, \ldots, u_d$ of u such that for any i and j ($1 \leq i \neq j \leq d$), u_i is neither an ancestor nor a descendant of u_j, and $\{I_i\} \in \mathcal{I}_T(u_i)$.*

Proof. Since only Step 2.4 of Algorithm MATCHING can add $X' = \{I_1, \ldots, I_d\}$ to $\mathcal{J}_T(u)$, $X' \in \mathcal{J}_T(u)$ if and only if for a natural number m ($m \leq d$), there is a partition $\{S_1, \ldots, S_m\}$ of X' and children $c_{u,j_1}, \ldots, c_{u,j_m}$ ($1 \leq j_1 < \cdots < j_m \leq d_T(u)$) of u such that $S_i \in \mathcal{C}_T(c_{u,j_i})$ for all $i = 1, \ldots, m$. Therefore we can prove this lemma by induction on the height of u. ☐

Lemma 3. *Let t be a TC-pattern and T an unordered tree. For any $v \in V(t)$ and $u \in V(T)$, $\{I_t(v)\} \in \mathcal{I}_T(u)$ if and only if either of the following four conditions holds:*

1. *$v \in V(t) \setminus U(t)$ and u and v are leaves of T and t, respectively.*
2. *$v \in U(t)$ and v is a leaf of t.*
3. *$v \in V(t) \setminus U(t)$ and there is a bijection $\psi : \{1, \ldots, d_t(v)\} \rightarrow \{1, \ldots, d_T(u)\}$ such that $\{I_t(c_{v,i})\} \in \mathcal{I}_T(c_{u,\psi(i)})$ for all i ($1 \leq i \leq d_t(v)$).*
4. *$v \in U(t)$ and there are $d_t(v)$ different proper descendants $u_1, \ldots, u_{d_t(v)}$ of u such that for any i and j ($1 \leq i \neq j \leq d_t(v)$), u_i is neither an ancestor nor a descendant of u_j, and $\{I_t(c_{v,i})\} \in \mathcal{I}_T(u_i)$ ($1 \leq i \leq d_t(v)$).*

Proof. From Lemma 2 and by Steps 1 and 2.1–2.3 of Algorithm MATCHING, we can prove this lemma by induction on the height of u. ☐

Lemma 4. *Let t be a TC-pattern and T an unordered tree. For any pair of $v \in V(t)$ and $u \in V(T)$, $\{I_t(v)\} \in \mathcal{I}_T(u)$ if and only if $T[u]$ matches $t[v]$.*

Proof. We prove this by induction on the height of u. The following cases correspond to the 4 cases in Lemma 3. The first 2 cases are the basic steps of the induction and the last 2 cases are the inductive steps.

1. Suppose that $v \in V(t) \setminus U(t)$ and u and v are leaves of T and t, respectively. From Lemma 3, $\{I_t(v))\} \in \mathcal{I}_T(u)$. On the other hand, by letting $W(v) = \{u\}$, $\mathcal{W} = \{W(v)\}$ is a $t[v]$-witness structure of $T[u]$. Therefore this lemma holds.
2. Suppose that $v \in U(t)$ and v is a leaf of t. From Lemma 3, $\{I_t(v))\} \in \mathcal{I}_T(u)$. On the other hand, by letting $W(v) = V(T[u])$, $\mathcal{W} = \{W(v)\}$ is a $t[v]$-witness structure of $T[u]$. Therefore this lemma holds.
3. Suppose that $v \in V(t) \setminus U(t)$ and there is a bijection $\psi : \{1, \ldots, d_t(v)\} \to \{1, \ldots, d_T(u)\}$ such that $\{I_t(c_{v,i})\} \in \mathcal{I}_T(c_{u,\psi(i)})$ for all i $(1 \leq i \leq d_t(v))$. From Lemma 3, $\{I_t(v))\} \in \mathcal{I}_T(u)$. On the other hand, by the inductive assumption, there is a $t[c_{v,i}]$-witness structure of $T[c_{u,\psi(i)}]$, denoted by $\mathcal{W}_i$, for each i $(1 \leq i \leq d_t(v))$. Then we define $\mathcal{W} = \{W(v)\} \cup \bigcup_{i=1}^{d_t(v)} \mathcal{W}_i$ with $W(v) = \{u\}$. Since $\mathcal{W}$ is a $t[v]$-witness structure of $T[u]$, this lemma holds.
4. Suppose that $v \in U(t)$ and there are $d_t(v)$ different proper descendants $u_1, \ldots, u_{d_t(v)}$ of u such that for any i and j $(1 \leq i \neq j \leq d_t(v))$, u_i is neither an ancestor nor a descendant of u_j, and $\{I_t(c_{v,i})\} \in \mathcal{I}_T(u_i)$ $(1 \leq i \leq d_t(v))$. From Lemma 3, $\{I_t(v))\} \in \mathcal{I}_T(u)$. On the other hand, by the inductive assumption, there is a $t[c_{v,i}]$-witness structure of $T[u_i]$, denoted by $\mathcal{W}_i$, for each i $(1 \leq i \leq d_t(v))$. Then, we let $W(v) = V(T[u]) \setminus (\bigcup_{i=1}^{d_t(v)} V(T[u_i]))$ and $\mathcal{W} = \{W(v)\} \cup \bigcup_{i=1}^{d_t(v)} \mathcal{W}_i$. We observe that $\mathcal{W}$ is a $t[v]$-witness structure of $T[u]$. Then this lemma holds. □

From Lemma 4, we observe that $\{I_t(r_t)\} \in \mathcal{I}_T(r_T)$ if and only if T matches t. Thus, we have the following theorem.

Theorem 3. *We assume that the number of children of every contractible vertex in TC-patterns is bounded by a constant. Then, TC-PATTERN MATCHING for a given TC-pattern t and a given tree T is solved in $O(nN(n+\sqrt{N}))$ time, where $n = |V(t)|$ and $N = |V(T)|$.*

Proof. The correctness of Algorithm MATCHING follows from Lemma 4. We show the time complexity of Algorithm MATCHING. Step 1 uses $O(nN)$ time. We note that for any $u \in V(T)$, $|\mathcal{I}_T(u)| \leq |V(t)| = n$ and $|\mathcal{J}_T(u)| \leq 2^d |U(t)|$. Next we analyze the total time complexity of Step 2.

1. From (1) of Lemma 1, for each $v \in V(t) \setminus U(t)$ and $u \in V(T)$, Step 2.1 spends $O(|X_t(v)| \sum_{j=1}^{d_T(u)} |\mathcal{I}_T(c_{u,j})| + |X_t(v)| d_T(u) \sqrt{|X_t(v)| + d_T(u)}) = O(d_t(v) d_T(u)(n + \sqrt{d_t(v) + d_T(u)}))$ time. The total time complexity of Step 2.1 for all $v \in V(t) \setminus U(t)$ and $u \in V(T)$ is

$$\sum_{v \in V(t) \setminus U(t)} \sum_{u \in V(T)} O(d_t(v) d_T(u)(n + \sqrt{d_t(v) + d_T(u)})) = O(nN(n + \sqrt{N})).$$

2. For each $v \in U(t)$ and $u \in V(T)$, Step 2.2 can search $\mathcal{I}_T(c_{u,1}), \ldots,$ $\mathcal{I}_T(c_{u,d_T(u)})$ for $\{I_t(v)\}$ linearly. For all $v \in U(t)$, we execute them at the same time. Then the total time complexity of Step 2.2 is $\sum_{u \in V(T)} O(\sum_{j=1}^{d_T(u)} | \mathcal{I}_T(c_{u,j})|) = O(nN)$ time.

3. From (2) of Lemma 1, for each $v \in U(t)$ and $u \in V(T)$, Steps 2.3 and 2.4 spend $2^d \cdot O((\sum_{j=1}^{d_T(u)} |\mathcal{C}_T(c_{u,j})| + d_T(u)\sqrt{d_T(u)})) = O(d_T(u)(n + \sqrt{d_T(u)}))$ time. The total time complexity of Steps 2.3 and 2.4 for all $v \in U(t)$ and $u \in V(T)$ is

$$\sum_{v \in U(t)} \sum_{u \in V(T)} O(d_T(u)(n + \sqrt{d_T(u)}) = O(nN(n + \sqrt{N})).$$

From the above analysis, we conclude that the time complexity of Algorithm MATCHING is $O(nN(n + \sqrt{N}))$. □

4   A Polynomial-Time Algorithm for a Minimally Generalized TC-Pattern

We say that two TC-patterns t and t' are *isomorphic*, denoted by $t \equiv t'$, if there is a bijection φ from $V(t)$ to $V(t')$ such that (i) $v \in U(t)$ if and only if $\varphi(v) \in U(t')$, (ii) $\{u, v\} \in E(t)$ if and only if $\{\varphi(u), \varphi(v)\} \in E(t')$, (iii) the root of t is mapped to the root of t' by φ, and (iv) $v \in V(t) \setminus U(t)$ and $\varphi(v) \in V(t') \setminus U(t')$ have the same vertex label.

For a TC-pattern t, we define the TC-pattern language of t as $L(t) = \{T \mid T \text{ matches } t\}$. For a finite set of trees S, a *minimally generalized TC-pattern* explaining S is defined as a TC-pattern t such that $S \subseteq L(t)$ and there exists no TC-pattern t' that satisfies $S \subseteq L(t') \subsetneq L(t)$. For example, in Fig. 2, t_1, t_2, t_3 are all the minimally generalized TC-patterns explaining S. We present a polynomial-time algorithm for the following problem.

MINIMALLY GENERALIZED TC-PATTERN PROBLEM
Instance: A finite set S of trees.
Problem: Find a minimally generalized TC-pattern explaining S.

We present the two refinement operators, as shown in Fig. 8, that are used to extend a TC-pattern as much as possible while $S \subseteq L(t)$ holds. We start with a TC-pattern t consisting of only one contractible vertex, and try to replace every contractible vertex v in t with one of the structures on the right-hand sides of the arrows in Fig. 8 if it is possible. We prove the following lemmas.

Lemma 5. *Let t and t' be two TC-patterns with $t \preceq t'$ and $|V(t)| < |V(t')|$. Let $k = |V(t')| - |V(t)|$. Then a TC-pattern that is isomorphic to t' is obtained from t by applying k refinement operators in Fig. 8 to t.*

Proof. Let $\mathcal{W} = \{W(v) \mid v \in V(t)\}$ be a t-witness structure of t'. Let v be a contractible vertex in $U(t)$ with $|W(v)| \geq 2$. Let $c_1, \ldots, c_p$ be the children of v. Let v' be the vertex in $W(v)$ nearest to the root t'.

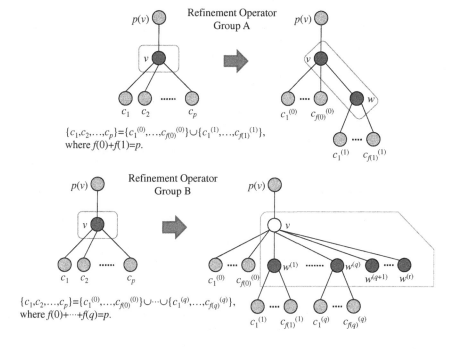

Fig. 8. Refinement operator groups A and B: A and B contain the $O(2^d)$ and $O(N_{\min}^d |\Sigma(S)|)$ refinement operators, respectively, where $N_{\min} = \min_{T \in S} |V(T)|$ and $\Sigma(S) = \{\sigma \in \Sigma \mid \sigma \text{ appears in all trees in } S\}$.

Suppose that $v' \in U(t')$. Let $\{v', w\}$ be an edge of t' that is contained in $W(v)$. Then we can divide the witness sets $W(c_1), \ldots, W(c_p)$ into two groups: one group consists of witness sets $W(c_1^{(1)}), \ldots, W(c_{f(1)}^{(1)})$ that contain proper descendants of w and another group consists of the other witness sets $W(c_1^{(0)}), \ldots, W(c_{f(0)}^{(0)})$, where $f(0) + f(1) = p$. Therefore, by applying a refinement operator in Group A in Fig. 8 to v, we obtain a TC-pattern t'' with $|V(t'')| = |V(t)| + 1$ and $t'' \preceq t'$.

Suppose that $v' \in V(t') \setminus U(t')$. Let $\{v', w^{(1)}\}, \ldots, \{v', w^{(r)}\}$ be all the edges of t' that are contained in $W(v)$. We assume that $w^{(1)}, \ldots, w^{(q)}$ ($q \leq r$) have children and $w^{(q+1)}, \ldots, w^{(r)}$ have no child in t'. Then we can divide the witness sets $W(c_1), \ldots, W(c_p)$ into $q+1$ groups: for j ($1 \leq j \leq q$), the j-th group consists of witness sets $W(c_1^{(j)}), \ldots, W(c_{f(j)}^{(j)})$ that contain proper descendants of $w^{(j)}$ and the remaining one group consists of the other witness sets $W(c_1^{(0)}), \ldots, W(c_{f(0)}^{(0)})$, where $f(0) + \cdots + f(q) = p$. Therefore, by applying a refinement operator in Group B in Fig. 8 to v, we obtain a TC-pattern t'' with $|V(t'')| = |V(t)| + r$ and $t'' \preceq t'$.

By repeating refinement operations like these, we finally obtain from t a TC-pattern that is isomorphic to t'. $\qquad \square$

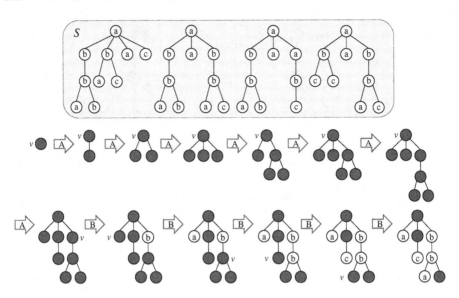

Fig. 9. A refinement process of our algorithm for finding a minimally generalized TC-pattern explaining S

We present a polynomial-time algorithm for computing MINIMALLY GENERALIZED TC-PATTERN PROBLEM. We start with only one contractible vertex which becomes a root and add a new contractible vertex by using refinement operators in Fig. 8.

Algorithm MINL;
Input: A nonempty finite set S of unordered trees.
Step 1: (Initialization) Let v be a new contractible vertex, and $t := (\{v\}, \emptyset, \{v\})$.
Step 2: (Iteration) Until a TC-pattern t cannot be updated anymore, do the next step: Choose a contractible vertex $v \in U(t)$ and a refinement operator σ in Fig. 8. Construct a new TC-pattern t' from t by applying σ to v. If all $T \in S$ matches t', then $t := t'$.

We describe in Fig. 9 an example of the refinement processes. We have the following lemma and theorem on Algorithm MINL.

Lemma 6. *Let Σ be a vertex label set with $|\Sigma| = \infty$. Let t be an output of Algorithm MINL for an input S. If there is a TC-pattern t' that satisfies $S \subseteq L(t') \subseteq L(t)$, then $t \equiv t'$.*

Proof. Since $|\Sigma| = \infty$, there is a vertex label λ that does not appear in S. Let $T_\lambda(t')$ be the unordered tree obtained by replacing all contractible vertices in t' with λ-labeled vertices, i.e., $V(T_\lambda(t')) = V(t')$, $E(T_\lambda(t')) = E(t')$, and the vertex label of $v \in V(T_\lambda(t'))$ is λ if $v \in U(t)$, otherwise the same vertex label of $v \in V(t')$. Since $L(t') \subseteq L(t)$, $T_\lambda(t') \in L(t)$ holds. Since there is no occurrence of λ in t, all λ-labeled vertices must appear in t-witness sets of some contractible vertices in t. Let $\mathcal{W} = \{W(v) \mid v \in V(t)\}$ be a t-witness structure of $T_\lambda(t')$. It is easy to see that $\mathcal{W}$ is also a t-witness structure of t' such that for each contractible vertex $v' \in U(t')$, there is a contractible vertex $v \in U(t)$ with $v' \in W(v)$. Therefore, from Lemma 5, Algorithm MINL has applied some of the refinement operators to t so that t becomes isomorphic to t'. Hence, we conclude that $t \equiv t'$. $\square$

Theorem 4. *We assume that there are infinitely many vertex labels in Σ, and that the number of children of every contractible vertex in a TC-pattern is bounded by a constant d. Then, MINIMALLY GENERALIZED TC-PATTERN PROBLEM for a given set of trees S is computed in $O(N_{\min}^{d+3} N_{\max}(N_{\min} + \sqrt{N_{\max}})|S|)$ time, where $N_{\min} = \min_{T \in S} |V(T)|$, $N_{\max} = \max_{T \in S} |V(T)|$.*

Proof. The correctness of Algorithm MINL is shown from Lemma 6. We show the time complexity of the algorithm. The number of vertices of a minimally generalized TC-pattern explaining S does not exceed $N_{\min}$. From Theorem 3, Algorithm MATCHING is executed in $O(nN(n + \sqrt{N}))$ time for an n-vertex TC-pattern and an N-vertex tree. We have the $O(N_{\min}^d |\Sigma(S)|)$ refinement operators, where $\Sigma(S) = \{\sigma \in \Sigma \mid \sigma$ appears in all trees in $S\}$. It is easy to see that $|\Sigma(S)| \leq N_{\min}$. Then, since we need to compute TC-PATTERN MATCHING $O(N_{\min}|S| \cdot N_{\min}^{d+1})$ times and every Algorithm MATCHING is executed in $O(N_{\min}N_{\max}(N_{\min} + \sqrt{N_{\max}}))$ time, the total time complexity of Algorithm MINL is $O(N_{\min}^{d+3} N_{\max}(N_{\min} + \sqrt{N_{\max}})|S|)$. $\square$

5   Conclusions

Let $\mathcal{RTCP}$ be the set of all TC-patterns whose contractible vertices have a constant number of children and $\mathcal{L_{RTCP}} = \{L(t) \mid t \in \mathcal{RTCP}\}$. Angluin [2] and Shinohara [8] showed that if a class of languages $\mathcal{C}$ has finite thickness, and the membership problem and the minimal language problem for $\mathcal{C}$ are solvable in polynomial time, then $\mathcal{C}$ is polynomial time inductively inferable from positive data. We consider the class $\mathcal{L_{RTCP}}$ as a target of inductive inference.

For any TC-pattern t and an unordered tree T that matches t, $|V(t)| \leq |V(T)|$. Hence, for any nonempty finite set $S \subseteq \mathcal{RTCP}$, the cardinality of the set $\{L \in \mathcal{L_{RTCP}} \mid S \subseteq L\}$ is finite. That is, the class $\mathcal{L_{RTCP}}$ has finite thickness. The membership problem for $\mathcal{L_{RTCP}}$ and the minimal language problem for $\mathcal{L_{RTCP}}$ are the same problems as TC-PATTERN MATCHING in Section 3 and MINIMALLY GENERALIZED TC-PATTERN PROBLEM in Section 4, respectively. From Theorems 3 and 4, these two problems are solvable in polynomial time. Therefore, we have the following theorem.

Theorem 5. *We assume that there are infinitely many vertex labels in Σ. The class $\mathcal{L_{RTCP}}$ is polynomial time inductively inferable from positive data.*

In this paper, we established an algorithmic foundation for discovery of knowledge from tree-structured data. Currently, we are developing general data mining techniques for a variety of real-world data that can be modeled by unordered trees.

Acknowledgments. This work was supported by Grant-in-Aid for Scientific Research (C) (Grant Numbers 23500182, 24500178) from Japan Society for the Promotion of Science (JSPS), and Grant-in-Aid for Scientific Research on Innovative Areas (Grant Number 24106010) from the Ministry of Education, Culture, Sports, Science and Technology (MEXT), Japan.

References

1. Amoth, T.R., Cull, P., Tadepalli, P.: On exact learning of unordered tree patterns. Machine Learning 44(3), 211–243 (2001)
2. Angluin, D.: Inductive inference of formal languages from positive data. Information and Control 45, 117–135 (1980)
3. Garey, M., Johnson, D.: Computers and Intractability: A Guide to the Theory of NP-Completeness. Freeman (1979)
4. Goldman, S.A., Kwek, S.S.: On learning unions of pattern languages and tree patterns. In: Watanabe, O., Yokomori, T. (eds.) ALT 1999. LNCS (LNAI), vol. 1720, pp. 347–363. Springer, Heidelberg (1999)
5. Hopcroft, J.E., Karp, R.M.: An $n^{5/2}$ algorithm for maximum matchings in bipartite graphs. SIAM Journal on Computing 2, 225–231 (1973)
6. Miyahara, T., Shoudai, T., Uchida, T., Takahashi, K., Ueda, H.: Polynomial time matching algorithms for tree-like structured patterns in knowledge discovery. In: Terano, T., Liu, H., Chen, A.L.P. (eds.) PAKDD 2000. LNCS, vol. 1805, pp. 5–16. Springer, Heidelberg (2000)
7. Nestorov, S., Abiteboul, S., Motwani, R.: Extracting schema from semistructured data. In: Proc. ACM SIGMOD Int. Conf. Management of Data, pp. 295–306 (1998)
8. Shinohara, T.: Polynomial time inductive inference of extended regular pattern languages. In: Goto, E., Furukawa, K., Nakajima, R., Nakata, I., Yonezawa, A. (eds.) RIMS 1982. LNCS, vol. 147, pp. 115–127. Springer, Heidelberg (1983)
9. Shoudai, T., Uchida, T., Miyahara, T.: Polynomial time algorithms for finding unordered tree patterns with internal variables. In: Freivalds, R. (ed.) FCT 2001. LNCS, vol. 2138, pp. 335–346. Springer, Heidelberg (2001)
10. Suzuki, Y., Shoudai, T., Matsumoto, S., Uchida, T., Miyahara, T.: Efficient learning of ordered and unordered tree patterns with contractible variables. In: Gavaldá, R., Jantke, K.P., Takimoto, E. (eds.) ALT 2003. LNCS (LNAI), vol. 2842, pp. 114–128. Springer, Heidelberg (2003)
11. Wang, K., Liu, H.: Discovering structural association of semistructured data. IEEE Trans. Knowledge and Data Engineering 12, 353–371 (2000)

Author Index